COHERENCE OF LIGHT

THE MODERN UNIVERSITY PHYSICS SERIES

Editor PROFESSOR G. K. T. CONN
Department of Physics, University of Exeter

This series is intended for readers whose main interest is in physics, or who need the methods of physics in the study of science and technology. Some of the books will provide a sound treatment of topics essential in any physics training, while other, more advanced, volumes will be suitable as preliminary reading for research in the field covered. New titles will be added from time to time.

LITTLEFIELD and THORLEY: *Atomic and Nuclear Physics* (2nd edn)
PEŘINA: *Coherence of Light*

COHERENCE OF LIGHT

JAN PEŘINA, Rer. Nat. Dr., Ph. D.

Laboratory of Optics
Palacký University, Olomouc, Czechoslovakia

Translation Editor: Dr. T. W. PREIST
University of Exeter

VAN NOSTRAND REINHOLD COMPANY
LONDON

NEW YORK CINCINNATI TORONTO MELBOURNE

VAN NOSTRAND REINHOLD COMPANY LTD
Windsor House, 46 Victoria Street, London S. W. 1

INTERNATIONAL OFFICES

New York Cincinnati Toronto Melbourne

Translated by the author
First published in 1972
by Van Nostrand Reinhold Company Ltd,
Windsor House, 46 Victoria Street, London S. W. 1
in co-edition with SNTL-Publishers of
Technical Literature, Prague

Library of Congress Catalog No. 77 – 141981

ISBN 0 442 06342 3

Printed in Czechoslovakia by SNTL, Prague

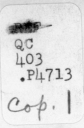

CONTENTS

19.95

Acknowledgement

The author is grateful to Professors F. T. Arecchi, V. Korenman, L. Mandel and to Dr. A. W. Smith for permission to reprint the following figures:

Fig. 10.1, reprinted from F. T. Arecchi, E. Gatti and A. Sona, *Phys. Lett.* ***20,*** 27 (1966).
Fig. 10.2, reprinted from B. L. Morgan and L. Mandel, *Phys. Rev. Lett.* ***16,*** 1012 (1966).
Fig. 17.1, reprinted from A. W. Smith and J. A. Armstrong, *Phys. Rev. Lett.* ***16,*** 1169 (1966).
Fig. 17.2, reprinted from R. F. Chang, V. Korenman, C. O. Alley and R. W. Detenbeck, *Phys. Rev.* ***178,*** 612 (1969).
Fig. 17.3, reprinted from F. T. Arecchi, A. Berné and P. Burlamacchi, *Phys. Rev. Lett.* ***16,*** 32 (1966).
Fig. 17.4, reprinted from F. T. Arecchi, V. Degiorgio and B. Querzola, *Phys. Rev. Lett.* ***19,*** 1168 (1967).

The author thanks Prof. B. Havelka and Drs. R. Horák and V. Peřinová for their critical reading of the manuscript and for their comments. He would like to thank his wife Dr. Vlasta Peřinová for encouragement and deep understanding during the writing of this book. Calculations performed in the Computer Centre of the Palacký University, particularly with the help of Dr. Z. Braunerová, are also acknowledged with thanks.

Functions Used in the Text

Auto-correlation function, $\Gamma_{jj}(\tau)$
Characteristic function, $C^{(n)}(y)$
  multimode, $\langle \exp ix\hat{n} \rangle_{\mathcal{N}}$
  arbitrary ordering, $C(y, \Omega)$
Characteristic functional, $C\{y(x)\}$
Coherence matrix, $\mathscr{J}$
  spectral $\mathscr{R}(v)$
Correlation function, $(m + n)$th order, $\Gamma^{(m,n)}_{\mu_1,\ldots,\mu_{m+n}}(\boldsymbol{x}_1, \ldots, \boldsymbol{x}_{m+n}; t_1, \ldots, t_{m+n})$
Correlation function, spectral $G^{(m,n)}(\boldsymbol{x}_1, \ldots, \boldsymbol{x}_{m+n}; \tau)$
Correlation tensor $\mathscr{K}_{jk}(x_1, x_2)$
Cross-spectral tensor $\tilde{\mathscr{K}}_{jk}(\boldsymbol{x}_1, \boldsymbol{x}_2; v)$
Cross-spectral density $G(\boldsymbol{x}_1, \boldsymbol{x}_2; v)$
Degree of coherence, $\gamma^{(n,n)}(x_1, \ldots, x_{2n})$
  second-order, $\gamma(\boldsymbol{x}_1, \boldsymbol{x}_2; \tau)$
Density matrix, Fock matrix elements, $\varrho(m, n)$
Detection operator $\hat{\mathbf{A}}(x)$
  eigenvalue, $\mathscr{A}(x)$
Device matrix, $\mathscr{L}(v)$
Diffraction function, $K(Q, P; v)$
Displacement operator, $\hat{D}(\alpha)$
Electromagnetic field operator, $\hat{A}_\mu(x)$
  eigenvalue-scalar component, $V(\boldsymbol{x}, t)$
  positive frequency part, $\hat{A}^{(+)}_\mu(x)$
  negative frequency part, $\hat{A}^{(-)}_\mu(x)$
Filter function, $\Omega(\beta^*, \beta)$
Glauber-Sudarshan quasi-probability $\Phi_{\mathcal{N}}(\alpha)$
Green's function, $\mathscr{G}_j(Q_j, P_j; v)$
Intensity matrix, $\hat{A}$
Intensity, mutual $\Gamma(\boldsymbol{x}_1, \boldsymbol{x}_2) \equiv \Gamma(\boldsymbol{x}_1, \boldsymbol{x}_2; 0)$
  spectral $\tilde{I}'(v)$
Light tensor, T_{ij}
Multimode quasi-probability, $\phi_{\mathcal{N}}(\{\alpha_\lambda\})$

for the integrated intensity W, $P_{\mathcal{N}}(W)$
for the integrated intensity W (s-ordered), $P(W, s)$
Mutual coherence function, $\Gamma(\boldsymbol{x}_1, \boldsymbol{x}_2; \tau)$
Mutual spectral density, $G(\boldsymbol{x}_1, \boldsymbol{x}_2; \nu)$
Photon-counting distribution, $p(n, T)$
Photon-number distribution, $p(n)$
Power spectrum, $G(\boldsymbol{x}, \nu) \equiv G(\boldsymbol{x}, \boldsymbol{x}; \nu)$
Probability density, $P_n(V_1, V_2, \ldots, V_n)$
Pupil function, $\tilde{K}(\boldsymbol{\mu}, \nu)$
Quasi-distribution $\Phi(\alpha, \Omega)$
s-ordered, $\Phi(\alpha, s)$
Quasi-probability,
antinormal, $\Phi_{\mathcal{A}}(\alpha)$
normal, $\Phi_{\mathcal{N}}(\alpha)$
symmetric, $\Phi_{\mathrm{Weyl}}(\alpha)$
Response function, $\mathscr{S}(x_j, x_j')$
S-matrix, $(S_{mn}) \equiv \hat{S} \equiv \hat{S}(+\infty)$
Spectral field, $\mathscr{F}(\nu)$
Spectral density, $G(\nu)$
normalized, $g(\nu)$
Time development operator $\hat{S}(t)$
Transfer function coherent $\tilde{K}(\mu, \nu)$
Transmission cross-coefficient $\mathscr{L}^{(1,1)}(\boldsymbol{\eta}_1, \boldsymbol{\eta}_2)$
Transmission matrix, $(K_{mn}) \equiv \hat{K}$
Vector potential operator $\hat{\boldsymbol{A}}(x)$
positive frequency part $\hat{\boldsymbol{A}}^{(+)}(x)$
negative frequency part $\hat{\boldsymbol{A}}^{(-)}(x)$
Visibility $\mathscr{V} \equiv |\gamma(\boldsymbol{x}_1, \boldsymbol{x}_2; \tau)|$

Chapter 1

INTRODUCTION

Interference and diffraction phenomena of electromagnetic waves are usually described with the aid of *ideally coherent* or *ideally incoherent* light beams. In the first case the amplitudes of beams are superimposed and the superposition of such beams gives an observable interference pattern on the screen. In the second case the intensities are superimposed and the interference pattern is not observable. In fact both these cases are a mathematical idealization since real beams partially influence one another, i.e. they are correlated. Thus the actual case is an intermediate one, involving partially coherent light beams. A superposition of such beams gives an interference pattern whose visibility is less than the visibility of the interference pattern formed by coherent beams. The inadequacy of the description of light by ideally coherent or ideally incoherent beams was first proved in about 1869 by Verdet [33] who showed that light from two pinholes illuminated by the Sun creates an observable interference pattern on the screen if the separation of the pinholes is less than about 1/20 mm. The Sun must be considered as an incoherent source composed of many elementary radiators (atoms) which do not influence one another in practice. Since coherence is the property of coherent radiators (which are mutually synchronized), it is obvious that states intermediate to the states of coherence and incoherence, i.e. states of *partial coherence*, must be considered.

Nevertheless, little attention was devoted to partial coherence until recently. A few years ago the concept of partial coherence became important in many branches of physics: in the theory of the electromagnetic field in all spectral regions, but especially in optics (image formation, interferometry), in radioastronomy and at present in the theory of masers and lasers.

Every optical field found in nature has certain statistical features because such an electromagnetic field is generated by many uncorrelated elementary radiators (atoms) and so it represents a *statistical dynamic system.* Thus the *theory of coherence,* from a general point of view, is concerned with the *statistical description* of the electromagnetic field, just as the *statistical properties* of optical fields manifest themselves as the *coherence properties* of these fields.

The concept of optical coherence was first introduced in connection with the description of interference and diffraction phenomena. However new possibilities of detecting optical fields as well as the preparation of new types of sources (masers and lasers) led to the need for a systematic and complete description and classification of optical coherence phenomena including coherence effects of all orders. This problem can be solved either from a *classical* (wave) or a *quantum* (particle) point of view. In the first case the theory of coherence is based upon the Maxwell wave theory and the theory of stochastic functions, whereas in the second case it is necessary to formulate the theory of coherence in terms of quantum electrodynamics. From this it is obvious that the theory of coherence may be regarded as a component of a more general theory – information theory – which investigates the ability of the field, considered as a statistical dynamic system, to transfer information. Consequently the theory of coherence is closely connected with the noise theory (in classical terms) and with the theory of quantum fluctuations of the field (in quantum terms).

The earliest investigations of partial coherence and polarization, limited to the second order effects (in our classification), were carried out by Verdet [33], von Laue [245], Berek [72–75], Michelson [312], Van Cittert [457, 458], Zernike [487], Wiener [465–467], Hopkins [200–202], Wolf [470–473], Blanc-Lapierre and Dumontet [83] and Pancharatnam [331, 332]. Their results were further continued in papers by Wolf [476], Mandel [266, 267], Mandel and Wolf [282], Parrent [334, 335], Beran and Parrent [70, 71], Roman and Wolf [407, 408], Roman [404, 405], Parrent and Roman [337], Barakat [53], Gamo [165] and Gabor [159].

A new period in the development of the theory of coherence began after the experiments of Hanbury Brown and Twiss [91–96] in which the fourth-order correlation effects were measured. In principle experiments of this kind make it possible to consider correlation effects of arbitrary order. This is particularly important for the complete statistical description of non-chaotic (e.g. laser) light, chaotic light being completely described by the second-order moment-correlation function. Such a general description of the statistical properties of optical fields, in which coherence effects of all orders are included, was proposed in the classical terms of the theory of stochastic processes by Wolf [479].

As information about the statistics of an optical field is obtained mainly using photodetectors (a special type of a more general class of quadratic detectors) it was necessary to find a relation between the statistics of photons in the field and the statistics of photoelectrons emitted by a photodetector exposed in this field. This was done by Mandel [263, 264, 269], who derived classically the so-called photodetection equation relating the probability distribution of the time-integrated intensity to the distribution of emitted photoelectrons.

In 1963 Glauber [172–175] developed the quantum theory of coherence based on quantum electrodynamics. Quantum correlation functions introduced by him represent expectation values of normally ordered products of field operators and they are closely related to the quantities measured by means of photoelectric detectors. The "diagonal" representation of the density matrix obtained by introducing the so-called coherent states (obtained by Glauber [172–174] and Sudarshan [435, 436]) enables one to study the relation between the quantum and classical descriptions of the statistical properties of optical fields. In this representation the classical and quantum correlation functions are formally equivalent if a generalized phase-space distribution is introduced. In the quantum theory of coherence the problem of correspondence between functions of operators (q-numbers) and functions of classical quantities (c-numbers) plays an important role. This correspondence was recently investigated by Agarwal and Wolf [37, 40], Lax [250] and also by Cahill and Glauber [105, 106]. In particular it was shown that antinormally ordered products of field operators are closely related to quantities measured by the so-called quantum counters [278], which operate by stimulated emission rather than by absorption. A number of further problems in the quantum formulation of coherence properties of optical fields and of the laser theory were solved by various authors: Glauber, Mandel, Wolf, Mehta, Klauder, Sudarshan, Haken and Lax, among others.

We also note here that the mathematical technique used in connection with partial coherence is also useful in the analysis of partial polarization. Both these phenomena may be unified in the correlation theory using the correlation tensors.

An interesting feature of the theory of coherence is that it operates with measurable quantities only. Classical Maxwell theory of the electromagnetic field assumes that the electric and magnetic fields are measurable functions in space and time. In fact electromagnetic vibrations, such as light, are described by rapidly oscillating quantities and no real detector can follow such rapid changes. Apart from this, the field represents a statistical dynamic system, as has already been mentioned. Therefore it is necessary to introduce an averaging process for physical quantities and only such averaged quantities can be measured. Thus the laws of the electromagnetic field are formulated in the theory of coherence in terms of *measurable quantities* only and, because one must observe every field through a kind of detector, it seems that any theory of every field – including fields in interaction – will have to be formulated in such a realistic form.

Finally let us note that the main goal of this book is to study the coherence properties of free electromagnetic fields and their detection. The role of coherence in the process of interaction of the electromagnetic field with matter as well as the description of the statistical properties of optical fields in the interaction lie outside the framework of this book. These questions have been treated for example in the book by Louisell [19], in papers by Mollow and Glauber [322, 323], in papers of Haken's school [191, 192], and in papers by Lax [247 – 250], Lax and Louisell [251], Lax and Yuen [252], Willis [469], Picard and Willis [382], Paul [339], Fleck [145 – 147], Scully and Lamb [415 – 417], and Risken [398, 534], etc. Optical pumping has been discussed for example by Cohen-Tannoudji and Kastler [116] and by Series [420]. The main subject of our investigations will be light but most of our results will be valid for the electromagnetic field regardless of the spectral region.

There exist several sources on the subject of the theory of coherence. The classical theory has been treated in the following books: "Principles of Optics" by M. Born and E. Wolf [8], Chapter 10; "Diffraction. Structure des images" by A. Maréchal and M. Françon [20], Chapter 7; "Theory of Partial Coherence" by M. Beran and G. B. Parrent [3]; "Cohérence en optique" by M. Françon and S. Slansky [11]; and "Introduction to Statistical Optics" by E. L. O'Neill [24]. A survey of the quantum theory of coherence can be found in lectures by R. J. Glauber given at the Summer Schools at Les Houches (France) [176], at Noordwijk aan Zee (The Netherlands) [180] and at Carberry Tower (Scotland) [181]. Both the classical and quantum theories of coherence and their connection have been treated by L. Mandel and E. Wolf [285], by J. Peřina [355, 362] and by J. R. Klauder and E. C. G. Sudarshan in the book "Fundamentals of Quantum Optics" [17]. A brief review of the theory of coherence can also be found in "Optical Coherence Theory" by G. J. Troup [32].

Chapter 2

DEFINITIONS AND MATHEMATICAL PRELIMINARIES

2.1 Complex value representation of real polychromatic fields

It is well known that the electromagnetic field is a real physical field in contrast to the electron-positron field which is naturally complex. However in classical coherence theory it is advantageous to represent the real electromagnetic field by a complex quantity because of its mathematical simplicity and also because it serves to emphasize that the coherence theory deals with phenomena which are sensitive to the "envelope" or to the "average intensity" of the field. In spite of the fact that such a complex representation is rather artificial in the classical theory it has a deep physical meaning in the quantum theory providing insight into the detection process. This complex representation represents a bridge between the classical and quantum formulations of coherence phenomena and the complex representation of the real polychromatic field, introduced in the following, represents a natural generalization of the well-known complex representation of monochromatic fields used in classical optics.

Let us consider the scalar quantity $V^{(r)}(t)$ which may represent for example a Cartesian component of the electric vector $\boldsymbol{E}$ of the electromagnetic field. If $V^{(r)}(t)$ possesses a Fourier transform we can write

$$V^{(r)}(t) = \int_{-\infty}^{+\infty} \tilde{V}^{(r)}(\nu) \exp(-i2\pi\nu t)\, d\nu =$$

$$= 2\int_{0}^{\infty} |\tilde{V}^{(r)}(\nu)| \cos(2\pi\nu t - \arg \tilde{V}^{(r)}(\nu))\, d\nu \,, \qquad (2.1)$$

where

$$\tilde{V}^{(r)}(\nu) = \int_{-\infty}^{+\infty} V^{(r)}(t) \exp i2\pi\nu t\, dt \qquad (2.2)$$

and the condition $\tilde{V}^{(r)*}(\nu) = \tilde{V}^{(r)}(-\nu)$ has been used. This condition is a consequence of the reality of $V^{(r)}(t)$ (the asterisk denotes complex conjugation). Hence, the negative frequency components ($\nu < 0$) do not carry any physical information which is not already contained in the positive frequency ones ($\nu > 0$) and consequently they can be omitted. Thus we may employ the complex function

$$V(t) = \int_0^{\infty} \tilde{V}^{(r)}(\nu) \exp(-i2\pi\nu t)\, d\nu\,, \tag{2.3}$$

where

$$\begin{aligned} \tilde{V}^{(r)}(\nu) &= \int_{-\infty}^{+\infty} V(t) \exp i2\pi\nu t\, dt\,, \quad \nu \geqq 0\,, \\ &= 0 \qquad\qquad\qquad\qquad\quad , \quad \nu < 0\,. \end{aligned} \tag{2.4}$$

The function $V(t)$, called the *complex analytic signal*, was introduced by Gabor [157] (see also [8], Sec. 10.2 and [3], Chap. 2, and also [280]). The function $V(t)$ can be written in the form

$$V(t) = \tfrac{1}{2}[V^{(r)}(t) + i\, V^{(i)}(t)]\,, \tag{2.5}$$

where

$$V^{(i)}(t) = -2 \int_0^{\infty} \left|\tilde{V}^{(r)}(\nu)\right| \sin(2\pi\nu t - \arg \tilde{V}^{(r)}(\nu))\, d\nu\,. \tag{2.6}$$

As the total energy $\int_0^{\infty} |\tilde{V}^{(r)}(\nu)|^2\, d\nu$ (which using the Parceval's equality, equals $\int_{-\infty}^{+\infty} |V(t)|^2\, dt$) is assumed to be finite the function (2.3) is (using the Schwarz inequality) analytic in the lower half of the complex z-plane ($z = t + i\vartheta$) and $V(t)$ is the boundary value of such an analytic function for $\vartheta \to -0$. Hence, making use of the Cauchy theorem on analytic functions, we obtain the *dispersion relations* which are also called *Hilbert transforms* [31]

$$V^{(i)}(t) = \frac{1}{\pi} P \int_{-\infty}^{+\infty} \frac{V^{(r)}(t')}{t' - t}\, dt'\,, \quad V^{(r)}(t) = -\frac{1}{\pi} P \int_{-\infty}^{+\infty} \frac{V^{(i)}(t')}{t' - t}\, dt'\,, \tag{2.7}$$

where P denotes the Cauchy principal value of the integral at $t = t'$. (A more detailed derivation of dispersion relations is contained in Sec. 4.4.) Equations (2.7) show that functions $V^{(r)}(t)$ and $V^{(i)}(t)$ are connected to each other and we see again that $V(t)$ given by (2.5) cannot contain more physical information than $V^{(r)}(t)$.

Defining the *generalized functions*

$$\delta_\pm(t) = \int_0^\infty \exp(\pm i2\pi\nu t)\, d\nu = \lim_{\varepsilon\to+0}\left[-\frac{1}{2\pi i(\pm t + i\varepsilon)}\right] = = \frac{1}{2}\left[\delta(t) \pm \frac{i}{\pi}\frac{P}{t}\right],$$

($\delta(t)$ is the Dirac function), we obtain by substituting (2.2) into (2.3)

$$V(t) = \int_{-\infty}^{+\infty} V^{(r)}(t')\,\delta_-(t-t')\,dt' = \lim_{\varepsilon\to+0}\frac{1}{2\pi i}\int_{-\infty}^{+\infty}\frac{V^{(r)}(t')}{t-t'-i\varepsilon}\,dt' = = \frac{1}{2}\left[V^{(r)}(t) - \frac{i}{\pi}P\int_{-\infty}^{+\infty}\frac{V^{(r)}(t')}{t-t'}\,dt'\right] \tag{2.8a}$$

and

$$V^*(t) = \int_{-\infty}^{+\infty} V^{(r)}(t')\,\delta_+(t-t')\,dt' = \lim_{\varepsilon\to+0}\frac{1}{2\pi i}\int_{-\infty}^{+\infty}\frac{V^{(r)}(t')}{t'-t-i\varepsilon}\,dt' = = \frac{1}{2}\left[V^{(r)}(t) + \frac{i}{\pi}P\int_{-\infty}^{+\infty}\frac{V^{(r)}(t')}{t-t'}\,dt'\right]. \tag{2.8b}$$

Thus if the function $V^{(r)}(t)$ fulfils the wave equation in vacuo,

$$\nabla^2 V^{(r)}(t') - \frac{1}{c^2}\frac{\partial^2 V^{(r)}(t')}{\partial t'^2} = 0\,, \tag{2.9}$$

where ∇^2 is the Laplace operator, we obtain the same equation for the complex function $V(t)$ multiplying (2.9) by $\delta_-(t-t')$ and integrating over t':

$$\nabla^2 V(t) - \frac{1}{c^2}\int_{-\infty}^{+\infty}\frac{\partial^2 V^{(r)}(t')}{\partial t'^2}\,\delta_-(t-t')\,dt' = = \nabla^2 V(t) - \frac{1}{c^2}\int_{-\infty}^{+\infty} V^{(r)}(t')\frac{\partial^2}{\partial t'^2}\delta_-(t-t')\,dt' = = \nabla^2 V(t) - \frac{1}{c^2}\frac{\partial^2}{\partial t^2}V(t) = 0\,, \tag{2.10}$$

where we have twice integrated by parts; c in (2.9) and (2.10) denotes the velocity of light *in vacuo*.

2.2 Correlation functions and properties

Let the (complex) electromagnetic field with polarization μ at a point $\boldsymbol{x}$ at time t be $V_\mu(\boldsymbol{x}, t)$. We define the $(m + n)$th-order *correlation function* as an ensemble average of the product of these field functions considered at different space-time points and for different polarizations as follows [479, 481–483]:

$$\Gamma^{(m,n)}_{\mu_1 \dots \mu_{m+n}}(\boldsymbol{x}_1 \dots, \boldsymbol{x}_{m+n}; t_1, \dots, t_{m+n}) = \langle \prod_{j=1}^{m} V^*_{\mu_j}(\boldsymbol{x}_j, t_j) \prod_{k=m+1}^{m+n} V_{\mu_k}(\boldsymbol{x}_k, t_k) \rangle , \tag{2.11}$$

where the brackets $\langle \dots \rangle$ denote an overall average. Quantum correlation functions corresponding to (2.11) suitable for the description of the coherence properties of fields determined using photodetectors can be written in the form

$$\Gamma^{(m,n)}_{\mathcal{N},\mu_1 \dots \mu_{m+n}}(x_1, \dots, x_{m+n}) = \mathrm{Tr}\,\{\hat{\varrho} \prod_{j=1}^{m} \hat{A}^{(-)}_{\mu_j}(x_j) \prod_{k=m+1}^{m+n} \hat{A}^{(+)}_{\mu_k}(x_k)\} , \tag{2.12}$$

where $x \equiv (\boldsymbol{x}, t)$, $\hat{\varrho}$ is the density matrix, Tr denotes the trace and $\hat{A}^{(+)}_\mu$ and $\hat{A}^{(-)}_\mu$ are the annihilation and creation operators of a photon respectively [173–175]. The suffix $\mathcal{N}$ indicates that the product of operators $\hat{A}^{(+)}$ and $\hat{A}^{(-)}$ is in a *normal form*, i.e. all annihilation operators $\hat{A}^{(+)}$ stand to the right of all creation operators $\hat{A}^{(-)}$. The correspondence between the correlation functions (2.11) and (2.12) is given by substitutions $\langle \dots \rangle \rightleftarrows \mathrm{Tr}\,\{\hat{\varrho}\, \dots\}$, $V^* \rightleftarrows \hat{A}^{(-)}$ and $V \rightleftarrows \hat{A}^{(+)}$. Of course, this correspondence is not unique since the operators $\hat{A}^{(+)}$ and $\hat{A}^{(-)}$ do not commute. A rigorous definition of the correlation functions (2.11) will be given in Chapter 8. The correlation functions (2.12) and quantum correlation functions involving other orderings of the field operators $\hat{A}^{(+)}$ and $\hat{A}^{(-)}$ will be defined in Chapters 12, 14 and 16, together with their correspondence to the classical correlation functions (2.11) and to various types of physical measurements.

An important class of optical fields appearing in nature is the class of *stationary* and *ergodic* fields. For a stationary field the correlation functions (as well as the probability densities describing the statistics of the field (Sec. 8.1)) are independent of translations of the time origin. For an ergodic field the *ensemble* average can be replaced by the *time* average, i.e.

$$\langle \dots \rangle = \lim_{T \to \infty} \frac{1}{2T} \int_{-T}^{+T} \dots \, dt . \tag{2.13}$$

For this class of fields the correlation function (2.11) can be rewritten in the form

$$\Gamma^{(m,n)}(\boldsymbol{x}_1, \ldots, \boldsymbol{x}_{m+n}; t_1, \ldots, t_{m+n}) \equiv \Gamma^{(m,n)}(\boldsymbol{x}_1, \ldots, \boldsymbol{x}_{m+n}; \tau_2, \ldots, \tau_{m+n}) =$$

$$= \lim_{T\to\infty} \frac{1}{2T} \int_{-T}^{+T} \prod_{j=1}^{m} V^*(\boldsymbol{x}_j, t + \tau_j) \prod_{k=m+1}^{m+n} V(\boldsymbol{x}_k, t + \tau_k)\, dt\,; \qquad (2.14)$$

τ_1 can be put to zero and then $\tau_j = t_j - t_1$ for $j = 1, 2, \ldots, m + n$. In (2.14) we have omitted the polarization indices; we can assume that light at a space-time point $x_j \equiv (\boldsymbol{x}_j, t_j)$ is of polarization μ_j without explicitly writing the polarization indices $(x_j \equiv (\boldsymbol{x}_j, t_j, \mu_j))$.†

The correlation function of the second order ($m = n = 1$) for stationary and ergodic fields is called the *mutual coherence function* and it can be written as

$$\Gamma(\boldsymbol{x}_1, \boldsymbol{x}_2; \tau) \equiv \Gamma^{(1,1)}(\boldsymbol{x}_1, \boldsymbol{x}_2; \tau) = \lim_{T\to\infty} \frac{1}{2T} \int_{-T}^{+T} V^*(\boldsymbol{x}_1, t)\, V(\boldsymbol{x}_2, t + \tau)\, dt\,. \qquad (2.15)$$

The correlation functions for stationary fields depend on time differences $\tau_j = t_j - t_1$ and the mutual coherence function depends on the time difference $\tau = t_2 - t_1$ only.

The mutual coherence function, which can be concisely denoted as $\Gamma_{12}(\tau)$, will play an important role in the description of the second-order coherence effects connected with classical interference and diffraction phenomena. The higher-order correlation functions will describe higher-order coherence phenomena observed with the help of a number of quadratic detectors (photoelectric detectors) whose photocurrents are correlated. The earlier investigations of the coherence properties of light used only the function $\Gamma_{12}(\tau)$ [8, 3, 11].

Let us note here that the difference between the classical and quantum correlation functions (2.11) and (2.12) arises from the averaging, i.e. from the "quality" of the brackets. Consequently results concerning the space-time polarization behaviour of the correlation functions such as

(i) the spectral decomposition of correlation functions,
(ii) the interference law,
(iii) wave equations,
(iv) the propagation laws of the correlation functions,

† For another definition of the time average using the so-called truncated functions ($V_T^{(r)}(t) = V^{(r)}(t)$ for $|t| < T$ and $V_T^{(r)}(t) = 0$ for $|t| > T$) we refer the reader to a discussion in [8] and [3].

(v) the concept of cross-spectral purity of light,
(vi) the conservation laws for correlation functions,
(vii) the matrix formulation of the theory and
(viii) the formulation of the polarization properties of light, etc.,

will be independent of the "quality" of the brackets and therefore they are valid for classical as well as quantum correlation functions. This independence will be demonstrated using Glauber-Sudarshan diagonal representation of the density matrix which allows the quantum average of operators (q-numbers) to be transformed into a "classical" average of classical fields (c-numbers) in a phase space.

Spectral and analytic properties of the correlation functions

Assuming that the field $V(x, t)$ possesses a Fourier transform

$$V(x, t) = \int_0^\infty \tilde{V}(x, \nu) \exp(-i2\pi\nu t)\, d\nu \tag{2.16}$$

(we write $\tilde{V}$ instead of $\tilde{V}^{(r)}$) we can derive the following spectral decomposition of the correlation function. Substituting (2.16) into (2.11),

$$\Gamma^{(m,n)}(\boldsymbol{x}_1, \ldots, \boldsymbol{x}_{m+n}; \boldsymbol{t}) = \int_0^\infty G^{(m,n)}(\boldsymbol{x}_1, \ldots, \boldsymbol{x}_{m+n}; \boldsymbol{\nu}) \exp i2\pi(\boldsymbol{\nu}, \boldsymbol{t})\, d\boldsymbol{\nu}\,, \tag{2.17}$$

where

$$G^{(m,n)}(\boldsymbol{x}_1, \ldots, \boldsymbol{x}_{m+n}; \boldsymbol{\nu}) = \langle \prod_{j=1}^{m} \tilde{V}^*(\boldsymbol{x}_j, \nu_j) \prod_{k=m+1}^{m+n} \tilde{V}(\boldsymbol{x}_k, \nu_k)\rangle =$$

$$= \int_{-\infty}^{+\infty} \Gamma^{(m,n)}(\boldsymbol{x}_1, \ldots, \boldsymbol{x}_{m+n}; \boldsymbol{t}) \exp[-i2\pi(\boldsymbol{\nu}, \boldsymbol{t})]\, d\boldsymbol{t}\,, \quad (\nu_j \geqq 0 \text{ for all } j) \tag{2.18}$$

is the *spectral correlation function*. Here $\boldsymbol{t} \equiv (t_1, \ldots, t_{m+n})$,

$$\boldsymbol{\nu} \equiv (\nu_1, \ldots, \nu_m, -\nu_{m+1}, \ldots, -\nu_{m+n})\,, \quad d\boldsymbol{t} \equiv \prod_{j=1}^{m+n} dt_j\,, \quad d\boldsymbol{\nu} \equiv \prod_{j=1}^{m+n} d\nu_j$$

and

$$(\boldsymbol{\nu}, \boldsymbol{t}) \equiv \sum_{j=1}^{m+n} \varepsilon_j \nu_j t_j\,, \quad \varepsilon_j = \begin{cases} +1\,, & (j = 1, \ldots, m)\,, \\ -1\,, & (j = m+1, \ldots, m+n)\,. \end{cases}$$

For a stationary and ergodic field we obtain the corresponding relations using the definition (2.14) of the correlation function and the spectral decomposition (2.16)

$$\Gamma^{(m,n)}(\mathbf{x}_1, \ldots, \mathbf{x}_{m+n}; \boldsymbol{\tau}) =$$

$$= \int_0^\infty G^{(m,n)}(\mathbf{x}_1, \ldots, \mathbf{x}_{m+n}; \boldsymbol{\nu})\, \delta(\sum_{j=1}^{m+n} \varepsilon_j \nu_j) \exp i2\pi(\boldsymbol{\nu}, \boldsymbol{\tau})\, d\boldsymbol{\nu}\,, \qquad (2.19)$$

where the spectral correlation function is given by

$$G^{(m,n)}(\mathbf{x}_1, \ldots, \mathbf{x}_{m+n}; \boldsymbol{\nu}) = \lim_{T\to\infty} \frac{1}{2T} \langle \prod_{j=1}^{m} \tilde{V}^*(\mathbf{x}_j, \nu_j) \prod_{k=m+1}^{m+n} \tilde{V}(\mathbf{x}_k, \nu_k)\rangle_e \qquad (2.20)$$

and

$$\boldsymbol{\tau} \equiv (\tau_2, \tau_3, \ldots, \tau_{m+n})\,, \quad (\boldsymbol{\nu}, \boldsymbol{\tau}) = \sum_{j=2}^{m+n} \varepsilon_j \nu_j \tau_j\,.$$

The brackets $\langle\ldots\rangle_e$ in (2.20) denote an overall average ensuring the existence of the above limit for stationary fields (cf. a discussion in [8] and [3]). Performing the integration in (2.19) over the variable ν_1 and introducing the quantity

$$G_H^{(m,n)}(\mathbf{x}_1, \ldots, \mathbf{x}_{m+n}; \boldsymbol{\nu}) \equiv$$

$$\equiv G^{(m,n)}(\mathbf{x}_1, \ldots, \mathbf{x}_{m+n}; -\sum_{j=2}^{m+n} \varepsilon_j \nu_j, \nu_2, \ldots, \nu_m, -\nu_{m+1}, \ldots, -\nu_{m+n})\,,$$

which is the spectral correlation function considered on the hypersurface

$$\sum_{j=1}^{m+n} \varepsilon_j \nu_j = 0 \qquad (2.21)$$

of the space $\nu_1 \otimes \nu_2 \otimes \ldots \otimes \nu_{m+n}$, we can invert (2.19) giving

$$G_H^{(m,n)}(\mathbf{x}_1, \ldots, \mathbf{x}_{m+n}; \boldsymbol{\nu}) = \int_{-\infty}^{+\infty} \Gamma^{(m,n)}(\mathbf{x}_1, \ldots, \mathbf{x}_{m+n}; \boldsymbol{\tau}) \exp[-i2\pi(\boldsymbol{\nu}, \boldsymbol{\tau})]\, d\boldsymbol{\tau}\,,$$

$$(\nu_j \geqq 0 \text{ for all } j)\,. \qquad (2.22)$$

Equations (2.17) and (2.19) represent the *generalized Wiener-Khintchine theorem.*

From (2.17) and (2.19) it can be seen that

$$G^{(m,n)}(\boldsymbol{\nu}) = G_H^{(m,n)}(\boldsymbol{\nu})\, \delta(\sum_{j=1}^{m+n} \varepsilon_j \nu_j)\,, \qquad (2.23)$$

where we have suppressed the space dependence. Substituting this expression into (2.17) we obtain (2.19) with $\tau_j = t_j - t_1$. The δ-function dependence implies that $(m + n)$ frequency components $\tilde{V}(\nu_1), \ldots, \tilde{V}(\nu_{m+n})$ of a stationary field may be correlated if and only if the frequencies $\nu_1, \ldots, \nu_{m+n}$ are coupled by the relation (2.21).

Of course, the same frequency condition (2.21) is a necessary condition for the non-vanishing of the correlation function of a stationary field. This follows from the definitions (2.11) and (2.12) of the correlation functions using the ensemble and quantum average, respectively, taking into account that such correlation functions are independent of the translation of the time origin, i.e.

$$\Gamma^{(m,n)}(t_1 + \tau, t_2 + \tau, \ldots, t_{m+n} + \tau) = \Gamma^{(m,n)}(t_1, t_2, \ldots, t_{m+n}),$$

which follows using the stationary condition and the spectral decomposition (2.16). The connection between this definition of the stationary property and the quantum-mechanical definition $[\hat{\varrho}, \hat{H}] = 0$, where $\hat{\varrho}$ is the density matrix, $\hat{H}$ is the Hamiltonian of the field and $[\hat{\varrho}, \hat{H}] = \hat{\varrho}\hat{H} - \hat{H}\hat{\varrho}$ is the commutator, will be discussed in Chapter 15.

We note that (2.22) may be useful for getting information about mode coupling in laser light from the measured correlation functions. Some further details about the spectral decompositions of the correlation functions and their applications in optical imagery and spectroscopy, etc., can be found in [363], [303], [272] and [482].

For the description of second-order coherence and polarization phenomena the mutual coherence function (2.15) will be appropriate. In this case we obtain from (2.19)

$$\Gamma(\mathbf{x}_1, \mathbf{x}_2; \tau) = \int_0^\infty G(\mathbf{x}_1, \mathbf{x}_2; \nu) \exp(-i2\pi\nu\tau)\, d\nu, \tag{2.24}$$

where

$$\begin{aligned} G(\mathbf{x}_1, \mathbf{x}_2; \nu) &\equiv G^{(1,1)}(\mathbf{x}_1, \mathbf{x}_2; \nu) \equiv G^{(1,1)}(\mathbf{x}_1, \mathbf{x}_2; \nu, -\nu) = \\ &= \lim_{T\to\infty} \frac{1}{2T} \langle \tilde{V}^*(\mathbf{x}_1, \nu)\, \tilde{V}(\mathbf{x}_2, \nu)\rangle_e = \int_{-\infty}^{+\infty} \Gamma(\mathbf{x}_1, \mathbf{x}_2; \tau) \exp i2\pi\nu\tau\, d\tau, \quad \nu \geqq 0, \\ &= 0 \qquad\qquad\qquad\qquad\qquad\qquad\qquad\qquad\qquad\qquad\qquad\qquad\qquad\qquad\quad, \quad \nu < 0, \end{aligned} \tag{2.25}$$

which follows from (2.20) and (2.22). The function (2.25) is called the *cross-spectral density* or the *mutual spectral density*. Equation (2.24) is known in the theory of stationary stochastic processes as the *Wiener-Khintchine theorem*.

One can see from (2.24) that the mutual coherence function $\Gamma_{12}(\tau) \equiv \Gamma(\boldsymbol{x}_1, \boldsymbol{x}_2; \tau)$ is analytic in the lower half-plane of the complex τ-plane since the analytic signal (2.3) is. Therefore the dispersion relations hold for the mutual coherence function:

$$\Gamma_{12}^{(i)}(\tau) = \frac{1}{\pi} P \int_{-\infty}^{+\infty} \frac{\Gamma_{12}^{(r)}(\tau')}{\tau' - \tau} d\tau', \quad \Gamma_{12}^{(r)}(\tau) = -\frac{1}{\pi} P \int_{-\infty}^{+\infty} \frac{\Gamma_{12}^{(i)}(\tau')}{\tau' - \tau} d\tau'. \tag{2.26}$$

The functions $\Gamma_{12}^{(r)}$ and $\Gamma_{12}^{(i)}$ may be determined in the following way. Using (2.5) we have

$$\Gamma_{12}(\tau) = \langle V_1^*(t) V_2(t+\tau)\rangle = \tfrac{1}{4}\{\langle V_1^{(r)}(t) V_2^{(r)}(t+\tau)\rangle + \langle V_1^{(i)}(t) V_2^{(i)}(t+\tau)\rangle\}$$
$$- \frac{i}{4}\{\langle V_1^{(i)}(t) V_2^{(r)}(t+\tau)\rangle - \langle V_1^{(r)}(t) V_2^{(i)}(t+\tau)\rangle\}. \tag{2.27}$$

However, from the direct and inverted dispersion relations we obtain

$$\langle V_1^{(r)}(t) V_2^{(r)}(t+\tau'')\rangle = -\frac{1}{\pi^2} P \int_{-\infty}^{+\infty} \frac{d\tau'}{\tau' - \tau''} P \int_{-\infty}^{+\infty} \frac{\langle V_1^{(r)}(t) V_2^{(r)}(t+\tau)\rangle}{\tau - \tau'} d\tau$$
$$= -\frac{1}{\pi} P \int_{-\infty}^{+\infty} \frac{d\tau'}{\tau' - \tau''} \langle V_1^{(r)}(t) V_2^{(i)}(t+\tau')\rangle$$
$$= -\frac{1}{\pi} P \int_{-\infty}^{+\infty} \frac{d\tau'}{\tau' - \tau''} \langle V_1^{(r)}(t-\tau') V_2^{(i)}(t)\rangle = \langle V_1^{(i)}(t) V_2^{(i)}(t+\tau'')\rangle, \tag{2.28a}$$

where we have used the invariance property of the mutual coherence function with respect to translations of the time origin. We can obtain in the same way

$$\langle V_1^{(r)}(t) V_2^{(i)}(t+\tau)\rangle = -\langle V_1^{(i)}(t) V_2^{(r)}(t+\tau)\rangle, \tag{2.28b}$$

so that (2.27) gives for the mutual coherence function

$$\Gamma_{12}(\tau) = \tfrac{1}{2}\langle V_1^{(r)}(t) V_2^{(r)}(t+\tau)\rangle + \frac{i}{2}\langle V_1^{(r)}(t) V_2^{(i)}(t+\tau)\rangle. \tag{2.29}$$

Hence

$$\Gamma_{12}^{(r)}(\tau) = \tfrac{1}{2}\langle V_1^{(r)}(t) V_2^{(r)}(t+\tau)\rangle = \tfrac{1}{2}\langle V_1^{(i)}(t) V_2^{(i)}(t+\tau)\rangle,$$
$$\Gamma_{12}^{(i)}(\tau) = \tfrac{1}{2}\langle V_1^{(r)}(t) V_2^{(i)}(t+\tau)\rangle = -\tfrac{1}{2}\langle V_1^{(i)}(t) V_2^{(r)}(t+\tau)\rangle, \tag{2.30}$$

and we note that the same result follows from the condition

$$\langle V_1(t) V_2(t+\tau)\rangle = 0,$$

which is a consequence of the non-negativeness of frequencies [407, 269]. In particular for $\boldsymbol{x}_1 \equiv \boldsymbol{x}_2$ and $\tau = 0$

$$\langle [V^{(r)}(\boldsymbol{x}, t)]^2 \rangle = \langle [V^{(i)}(\boldsymbol{x}, t)]^2 \rangle = \tfrac{1}{2} \langle |V(\boldsymbol{x}, t)|^2 \rangle , \tag{2.31a}$$

$$\langle V^{(r)}(\boldsymbol{x}, t)\, V^{(i)}(\boldsymbol{x}, t) \rangle = 0 . \tag{2.31b}$$

From the definitions of $\Gamma_{12}(\tau)$ and $G_{12}(\nu)$ the following identities follow immediately:

$$\Gamma^*_{12}(\tau) = \Gamma_{21}(-\tau) , \tag{2.32a}$$

$$G^*_{12}(\nu) = G_{21}(\nu) . \tag{2.32b}$$

The first condition is called the *cross-symmetry condition* and the second expresses the hermiticity of G as a matrix.

Analogous results for the third and fourth-order correlation functions have been obtained in [67].

One can see from (2.17) that the correlation function $\Gamma^{(m,n)}(t_1, \ldots, t_{m+n})$ is an analytic function in the upper half of the complex t-plane in variables $t_1, \ldots, t_m$ and in the lower half of the complex t-plane in variables $t_{m+1}, \ldots$ $\ldots, t_{m+n}$ and consequently the dispersion relations hold:

$$\operatorname{Im} \Gamma^{(m,n)}(t_1, \ldots, t_{m+n}) = -\varepsilon_k \frac{1}{\pi} P \int_{-\infty}^{+\infty} \frac{\operatorname{Re} \Gamma^{(m,n)}(t_1, \ldots, t'_k, \ldots, t_{m+n})}{t'_k - t_k} \, dt'_k ,$$

$$\operatorname{Re} \Gamma^{(m,n)}(t_1, \ldots, t_{m+n}) = \varepsilon_k \frac{1}{\pi} P \int_{-\infty}^{+\infty} \frac{\operatorname{Im} \Gamma^{(m,n)}(t_1, \ldots, t'_k, \ldots, t_{m+n})}{t'_k - t_k} \, dt'_k , \tag{2.33}$$

where $\varepsilon_k = 1$ for $k = 1, \ldots, m$ and $\varepsilon_k = -1$ for $k = m + 1, \ldots, m + n$. In (2.33) Im and Re denote the imaginary and real parts respectively. Analogous relations are valid for stationary fields in the complex τ-plane (cf. (2.19)).

The wave equations *in vacuo*

As a consequence of the wave equation (2.10) for the complex amplitude of the field, the following system of wave equations holds *in vacuo* for the correlation functions and the spectral correlation functions:

$$\Box_j \, \Gamma^{(m,n)}(\boldsymbol{x}_1, \ldots, \boldsymbol{x}_{m+n}; t) = 0 , \tag{2.34}$$

$$j = 1, \ldots, m + n ,$$

$$\left[\nabla_j^2 + \left(\frac{2\pi\nu_j}{c} \right)^2 \right] G^{(m,n)}(\boldsymbol{x}_1, \ldots, \boldsymbol{x}_{m+n}; \nu) = 0 , \tag{2.35}$$

where the spectral decomposition (2.17) has been used and $\Box_j$ and ∇_j^2 are D'Alembert and Laplace operators respectively, for the j-th coordinates.

In the case of stationary fields the following system of wave equations holds for the correlation functions:

$$\left[\nabla_1^2 - \frac{1}{c^2}\left(\frac{\partial}{\partial \tau_2} + \frac{\partial}{\partial \tau_3} + \ldots + \frac{\partial}{\partial \tau_{m+n}}\right)^2\right] \Gamma^{(m,n)}(\boldsymbol{x}_1, \ldots, \boldsymbol{x}_{m+n}; \tau) = 0\,,$$

$$\left[\nabla_j^2 - \frac{1}{c^2}\frac{\partial^2}{\partial \tau_j^2}\right] \Gamma^{(m,n)}(\boldsymbol{x}_1, \ldots, \boldsymbol{x}_{m+n}; \tau) = 0\,, \quad j = 2, 3, \ldots, m + n\,. \tag{2.36}$$

These correspond to the following system of equations for the spectral quantities

$$\left[\nabla_j^2 + \left(\frac{2\pi\nu_j}{c}\right)^2\right] G_H^{(m,n)}(\boldsymbol{x}_1, \ldots, \boldsymbol{x}_{m+n}; \nu) = 0\,, \quad j = 1, 2, \ldots, m + n\,. \tag{2.37}$$

Here we have used (2.19) and the condition (2.21) holds for frequencies.

The wave equations for the mutual coherence function $\Gamma_{12}(\tau)$ were first derived by Wolf in 1955 [472] (see also [8] and [3]).

Degree of coherence

Before defining the quantity which can be called the degree of coherence we summarize some simple inequalities for the correlation functions.

From the definition (2.11) and also from (2.12) it follows immediately that

$$[\Gamma^{(m,n)}(x_1, \ldots, x_{m+n})]^* = \Gamma^{(n,m)}(x_{m+n}, \ldots, x_1) \tag{2.38}$$

and condition (2.32a) represents a special case of this identity ($m = n = 1$, $t_2 - t_1 = \tau$ for stationary fields).

A further property is that an exchange of arguments $x_1, \ldots, x_m$ and also of arguments $x_{m+1}, \ldots, x_{m+n}$ does not change the correlation function $\Gamma^{(m,n)}$.

We can easily see that

$$\Gamma^{(n,n)}(x_1, \ldots, x_n, x_n, \ldots, x_1) \geqq 0\,. \tag{2.39}$$

Using Hölder's inequality we can show that

$$|\Gamma^{(n,n)}(x_1, \ldots, x_{2n})|^{2n} \leqq \prod_{j=1}^{2n} \Gamma^{(n,n)}(x_j, \ldots, x_j)\,. \tag{2.40}$$

Further it follows that

$$\Gamma^{(n,n)}(x_1, \ldots, x_n, x_n, \ldots, x_1)\, \Gamma^{(n,n)}(x_{n+1}, \ldots, x_{2n}, x_{2n}, \ldots, x_{n+1}) \geqq$$
$$\geqq \left|\Gamma^{(n,n)}(x_1, \ldots, x_{2n})\right|^2 . \tag{2.41}$$

These as well as a number of further properties of the correlation functions will be derived in Chapter 12 [173, 176].

Now we can define the following quantities

$$\gamma^{(n,n)}(x_1, \ldots, x_{2n}) = \frac{\Gamma^{(n,n)}(x_1, \ldots, x_{2n})}{\left\{\prod_{j=1}^{2n} \Gamma^{(n,n)}(x_j, \ldots, x_j)\right\}^{1/2n}}, \tag{2.42a}$$

$$^{(G)}\gamma^{(n,n)}(x_1, \ldots, x_{2n}) = \frac{\Gamma^{(n,n)}(x_1, \ldots, x_{2n})}{\left\{\prod_{j=1}^{2n} \Gamma^{(1,1)}(x_j, x_j)\right\}^{1/2}}, \tag{2.42b}$$

$$^{(S)}\gamma^{(n,n)}(x_1, \ldots, x_{2n}) =$$
$$= \frac{\Gamma^{(n,n)}(x_1, \ldots, x_{2n})}{\left\{\Gamma^{(n,n)}(x_1, \ldots, x_n, x_n, \ldots, x_1)\, \Gamma^{(n,n)}(x_{n+1}, \ldots, x_{2n}, x_{2n}, \ldots, x_{n+1})\right\}^{1/2}} . \tag{2.42c}$$

Each of these we call the *degree of coherence*. The quantity (2.42a) was introduced in [372] and [298]. The quantity (2.42b) was introduced by Glauber [173, 175, 176] and the quantity (2.42c) was introduced by Sudarshan (see [17]). One can see from (2.40) and (2.41) that

$$\left|\gamma^{(n,n)}(x_1, \ldots, x_{2n})\right| \leqq 1 \tag{2.43a}$$

and

$$\left|^{(S)}\gamma^{(n,n)}(x_1, \ldots, x_{2n})\right| \leqq 1 \tag{2.43b}$$

respectively but such an inequality does not hold for $^{(G)}\gamma^{(n,n)}$. The degree of coherence for stationary fields, which depends on $\tau_j = t_j - t_1, j = 2, \ldots \ldots, 2n$, can be defined in an identical way.

The correlation functions defined here represent the basic mathematical as well as physical quantities of this book and they will serve as a powerful tool for investigations of the coherence and polarization properties of the electromagnetic field.

2.3 Elementary ideas of temporal and spatial coherence

Temporal coherence

Suppose that a light beam from a point source σ (Fig. 2.1) is divided into two beams in a Michelson interferometer and that these two beams are united after a path delay $\Delta s = c\,\Delta t$ is introduced between them (c is

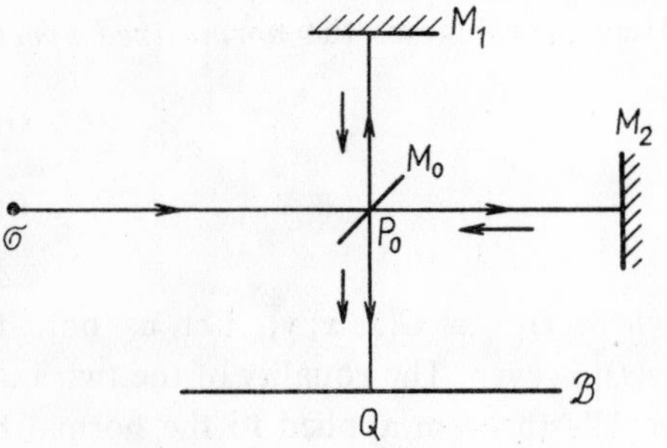

Fig.2.1. Illustration of temporal coherence by means of the Michelson interferometer; M_1 and M_2 are mirrors, M_0 is a half-silvered mirror, σ is a point source, $\mathscr{B}$ is the screen of observation.

velocity of light). If Δs is sufficiently small, interference fringes are formed in the plane $\mathscr{B}$. The appearance of the fringes is a manifestation of *temporal coherence* between the two beams, since the visibility of fringes depends on the time delay Δt introduced between them. In general, interference fringes will be observed if

$$\Delta t\,\Delta\nu \lesssim 1\,, \tag{2.44}$$

where $\Delta\nu$ is the effective bandwidth of the light. The time delay

$$\Delta t \approx \frac{1}{\Delta\nu} \tag{2.45}$$

is called the *coherence time* of the light and the corresponding path $c\,\Delta t$ is called the *coherence length*. A detailed discussion of the terms 'coherence time' and 'coherence length' based on wave packets can be found in [11].

Using the second-order degree of coherence $\gamma(\boldsymbol{x}, \boldsymbol{x}; \tau) \equiv \gamma(\tau)$ of stationary fields we can give a somewhat more precise definition of the coherence time [475, 8].

Considering the simple experiment illustrated in Fig. 2.1 again, where the time delay between two beams is τ, then the visibility of interference fringes observed in the plane $\mathscr{B}$ is proportional to the absolute value of the degree of coherence $\gamma(\boldsymbol{x}, \boldsymbol{x}; \tau)$ ($\boldsymbol{x}$ specifies the position of the point P_0 of the mirror M_0), as we shall see in Chapter 3. Therefore one can define the coherence time by means of the formula [475]

$$\tau_c^2 = (\Delta t)^2 = N^{-1} \int_{-\infty}^{+\infty} \tau^2 |\gamma(\tau)|^2 \, d\tau = 2N^{-1} \int_0^{\infty} \tau^2 |\gamma(\tau)|^2 \, d\tau \,, \tag{2.46}$$

where

$$N = \int_{-\infty}^{+\infty} |\gamma(\tau)|^2 \, d\tau = \int_0^{\infty} g^2(\nu) \, d\nu \,. \tag{2.47}$$

Here $g(\nu)$ denotes the *normalized spectral density* of the light

$$g(\nu) = \frac{G(\nu)}{\int_0^{\infty} G(\nu) \, d\nu} \,, \tag{2.48}$$

where $G(\nu) \equiv G(\boldsymbol{x}, \boldsymbol{x}; \nu)$. Let us note that $\bar{\tau} \equiv \int_{-\infty}^{+\infty} \tau |\gamma(\tau)|^2 \, d\tau = 0$ since $|\gamma(\tau)|$ is even. The equality of the two integrals in (2.47) follows from Parceval's theorem applied to the normalized spectral decomposition (2.24),

$$\gamma(\tau) = \int_0^{\infty} g(\nu) \exp(-i2\pi\nu\tau) \, d\nu \,, \tag{2.49}$$

where obviously $\int_0^{\infty} g(\nu) \, d\nu = 1$.

If the effective bandwidth of the light is defined by

$$(\Delta\nu)^2 = N^{-1} \int_0^{\infty} (\nu - \bar{\nu})^2 \, g^2(\nu) \, d\nu \tag{2.50}$$

with

$$\bar{\nu} = N^{-1} \int_0^{\infty} \nu \, g^2(\nu) \, d\nu \,, \tag{2.51}$$

then one may show that [8]

$$\Delta t \, \Delta\nu \geqq \frac{1}{4\pi} \,. \tag{2.52}$$

For quasi-monochromatic light with a Gaussian spectral profile the equality sign in (2.52) approximately holds.

Another definition of the coherence time can also be adopted, [264, 269], namely

$$\tau_c = \Delta t = \int_{-\infty}^{+\infty} |\gamma(\tau)|^2 \, d\tau \,. \tag{2.53}$$

If the bandwidth is defined as

$$\Delta\nu = \frac{1}{\int_0^\infty g^2(\nu)\, d\nu}, \tag{2.54}$$

then we have

$$\Delta t\, \Delta\nu = 1 \tag{2.55}$$

applying Parceval's equality (2.47). This definition is useful for a theory of the Hanbury Brown – Twiss experiment measuring the fourth-order correlation function [269]. For simple types of spectral profiles both these definitions are approximately equivalent [284, 295].

For experiments involving the division of light beams at two points (e.g. Young's interference experiment, Chapter 3) more general definitions can be given [475,8]. Still more general definitions of the coherence time which apply to non-stationary fields have been proposed in [110].

Spatial coherence

Let us consider another interference experiment of the Young type, illustrated by Fig. 2.2. Assuming quasi-monochromatic light (for which $\Delta\nu/\bar{\nu} \ll 1$) from an extended chaotic source σ in the form of a square of

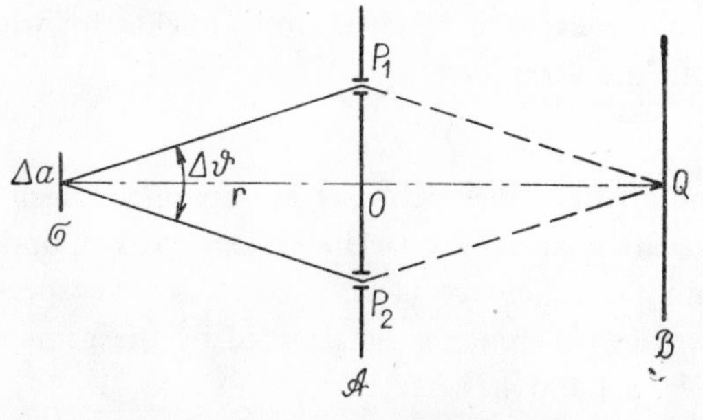

Fig. 2.2. Illustration of spatial coherence by means of Young's interference experiment; σ is an extended chaotic quasi-monochromatic source, $\mathscr{A}$ is a screen with two pinholes P_1 and P_2 and $\mathscr{B}$ is the screen of observation.

side Δa we can observe the interference fringes near the point Q of the screen $\mathscr{B}$ if the pinholes P_1 and P_2 are close enough to each other. The appearance of these fringes is a manifestation of *spatial coherence* between beams arriving at Q from P_1 and P_2, since the visibility of these fringes depends on the spatial separation of the pinholes. Interference fringes are generally observable if

$$\Delta\vartheta\,\Delta a < \bar{\lambda}\,, \tag{2.56}$$

where $\bar{\lambda} = c/\bar{\nu}$ is the effective wavelength of the light. In order to observe the fringes near Q the pinholes must be situated within an area around the point O of size

$$\Delta A \approx (r\,\Delta\vartheta)^2 \approx \frac{r^2\bar{\lambda}^2}{(\Delta a)^2} = \frac{c^2 r^2}{\bar{\nu}^2 S}\,, \tag{2.57}$$

where $S = (\Delta a)^2$ is the area of the source. The area ΔA is called the *coherence area* of the light in the plane $\mathscr{A}$, around the point O. The solid angle $\Delta\Omega$ which the area ΔA subtends at the source is given by

$$\Delta\Omega \approx \frac{\Delta A}{r^2} \approx \frac{c^2}{\bar{\nu}^2 S}\,. \tag{2.58}$$

Of course, the assumption of a chaotic source is not important and the concept of the area of coherence applies generally.

Both these types of coherence are connected in the function $\gamma(\boldsymbol{x}_1, \boldsymbol{x}_2; \tau)$; temporal coherence is described by the function $\gamma(\boldsymbol{x}_1, \boldsymbol{x}_1; \tau)$ and spatial coherence is described by $\gamma(\boldsymbol{x}_1, \boldsymbol{x}_2; 0)$. In general it is not possible to separate temporal and spatial coherence (to decompose the degree of coherence $\gamma_{12}(\tau)$ into the product of $\gamma_{11}(\tau)$ and $\gamma_{12}(0)$) due to the fact that temporal and spatial coherence are mutually connected, since $\Gamma_{12}(\tau)$ obeys *in vacuo* the wave equations, which relate the space and time variation of Γ. However, there exists a class of optical fields, for which $\gamma_{12}(\tau)$ may be factorized into the form

$$\gamma_{12}(\tau) = \gamma_{11}(\tau)\,\gamma_{12}(0)\,, \tag{2.59}$$

i.e. temporal and spatial coherence may be separated. Such fields are said to be *cross-spectrally pure* [266]. Under certain assumptions it can be shown that temporal coherence is determined by the spectral properties of light while spatial coherence is determined by the geometric properties of the source (Sec. 3.3 and 4.3).

Volume of coherence

The *volume of coherence* can be defined as a right-angled cylinder, whose base is the area of coherence and whose height is the coherence length, i.e.

$$\Delta V = c\,\Delta t\,\Delta A \approx \frac{c}{\Delta\nu}\frac{c^2}{\bar{\nu}^2}\frac{r^2}{S} = \frac{c\bar{\lambda}^2 r^2 c}{\Delta\lambda\,\bar{\nu}^2 S} = \frac{\bar{\lambda}^4 r^2}{\Delta\lambda\, S} = \left(\frac{\bar{\lambda}}{\Delta\lambda}\right)\left(\frac{r}{\Delta a}\right)^2 \bar{\lambda}^3\,, \tag{2.60}$$

where $\Delta\lambda = \Delta(c/\bar{\nu}) = (c/\bar{\nu}^2)\,\Delta\nu$. This is also the volume corresponding to one cell of phase space of photons [229].

A very important parameter is the *degeneracy parameter* δ [265] which represents the average number of photons in the same state of polarization in the volume of coherence, i.e. it represents the average number of photons in the same state of polarization which traverse the area of coherence per coherence time. If E_ν is the average number of photons emitted per unit area of the source, per unit frequency interval, per unit solid angle around the direction normal to the source per unit time, then

$$\delta = \tfrac{1}{2} E_\nu S \,\Delta\nu\, \Delta\Omega\, \Delta t\,. \tag{2.61}$$

The factor 1/2 on the right-hand side arises from the fact that the light is assumed to be generated by a chaotic source and so it is unpolarized, i.e. both orthogonal polarizations are present with the same weight. From (2.45) and (2.58) we obtain

$$\delta \approx \frac{c^2}{2\bar{\nu}^2} E_\nu\,, \tag{2.62}$$

which is independent of the geometry. For black-body radiation emerging from an equilibrium enclosure

$$E_\nu = \frac{2\nu^2}{c^2}\left(\exp\frac{h\nu}{KT} - 1\right)^{-1}, \tag{2.63}$$

where K is the Boltzmann constant, T is the absolute temperature of the radiation, h is Planck's constant. Equation (2.62) gives

$$\delta \approx \left(\exp\frac{h\nu}{KT} - 1\right)^{-1}, \tag{2.64}$$

which is the expression first derived by Einstein [139] in connection with his study of radiation in a cavity in thermal equilibrium with the walls of the cavity. This quantity describes the average number of photons in a cell of phase space and in quantum statistics is called the degeneracy parameter of the radiation.

Of course, the term degeneracy parameter can also be defined for non-thermal light. It should be mentioned that for non-thermal (laser) light $\delta \gg 1$ while for thermal light $\delta \ll 1$ [265, 160].

The concept of the volume of coherence may be shown to correspond to a quantum-mechanically defined cell of phase space, i.e. to $\Delta q_x\, \Delta q_y\, \Delta q_z = h^3(\Delta p_x\, \Delta p_y\, \Delta p_z)^{-1} \equiv \Delta V$, where q and p are canonical coordinates and momenta respectively [285].

2.4 Quasi-monochromatic approximation

In cases of practical interest so-called *quasi-monochromatic light* is frequently used. If the frequency bandwidth of such light is $\Delta\nu$ and the mean frequency is $\bar{\nu}$, then for quasi-monochromatic light the condition

$$\Delta\nu \ll \bar{\nu} \tag{2.65}$$

holds. Writing the correlation function in the form

$$\Gamma^{(n,n)}(\boldsymbol{x}_1, \ldots, \boldsymbol{x}_{2n}; \boldsymbol{\tau}) = \\ = \left|\Gamma^{(n,n)}(\boldsymbol{x}_1, \ldots, \boldsymbol{x}_{2n}; \boldsymbol{\tau})\right| \exp\left[i\alpha(\boldsymbol{x}_1, \ldots, \boldsymbol{x}_{2n}; \boldsymbol{\tau}) + i2\pi\bar{\nu}\sum_{j=2}^{2n} \varepsilon_j\tau_j\right], \tag{2.66}$$

where $\alpha = \arg \Gamma^{(n,n)} - 2\pi\bar{\nu}\sum_{j=2}^{2n} \varepsilon_j\tau_j$ and using (2.19) we obtain

$$\Gamma^{(n,n)}(\boldsymbol{x}_1, \ldots, \boldsymbol{x}_{2n}; \boldsymbol{\tau}) = \exp\left(i2\pi\bar{\nu}\sum_{j=2}^{2n} \varepsilon_j\tau_j\right) \int_0^\infty \ldots \int G_H^{(n,n)}(\boldsymbol{x}_1, \ldots, \boldsymbol{x}_{2n}; \boldsymbol{\nu}) \,. \\ . \exp\left[i2\pi\sum_{j=2}^{2n} \varepsilon_j(\nu_j - \bar{\nu})\,\tau_j\right] \prod_{j=2}^{2n} d\nu_j = \exp\left(i2\pi\bar{\nu}\sum_{j=2}^{2n} \varepsilon_j\tau_j\right) . \\ . \int_{-\bar{\nu}}^\infty \ldots \int G^{(n,n)}\left(\boldsymbol{x}_1, \ldots, \boldsymbol{x}_{2n}; \bar{\nu} - \sum_{j=2}^{2n} \varepsilon_j\nu_j, \bar{\nu} + \nu_2, \ldots, -\bar{\nu} - \nu_{2n}\right) . \\ . \exp\left(i2\pi\sum_{j=2}^{2n} \varepsilon_j\nu_j\tau_j\right) \prod_{j=2}^{2n} d\nu_j \,, \tag{2.67}$$

where we have performed the substitutions $\nu_j - \bar{\nu} \to \nu_j$ $\left(-\sum_{j=2}^{2n} \varepsilon_j\nu_j \to \bar{\nu} - \sum_{j=2}^{2n} \varepsilon_j\nu_j\right)$. As we consider quasi-monochromatic light the function G in (2.67) will be effectively non-zero only in a region $(\bar{\nu} - \Delta\nu/2, \bar{\nu} + \Delta\nu/2) \otimes \ldots \otimes (\bar{\nu} - \Delta\nu/2, \bar{\nu} + \Delta\nu/2)$. Thus since $\Delta\nu/\bar{\nu} \ll 1$, the function $\left|\Gamma^{(n,n)}\right| . \exp i\alpha$ contains only low-frequency components and hence will vary slowly with τ_j compared with variations arising from the periodic term $\exp\left(i2\pi\bar{\nu}\sum_{j=2}^{2n} \varepsilon_j\tau_j\right)$. Considering $\boldsymbol{\tau}$ so that

$$|\tau_j| \ll \frac{1}{\Delta\nu}, \quad j = 2, 3, \ldots, 2n \tag{2.68}$$

we see from (2.67) that $\left|\Gamma^{(n,n)}\right| \exp i\alpha$ is independent of $\boldsymbol{\tau}$ and so

$$\Gamma^{(n,n)}(\boldsymbol{x}_1, \ldots, \boldsymbol{x}_{2n}; \boldsymbol{\tau}) = \exp\left(i2\pi\bar{\nu}\sum_{j=2}^{2n} \varepsilon_j\tau_j\right) \Gamma^{(n,n)}(\boldsymbol{x}_1, \ldots, \boldsymbol{x}_{2n}; \boldsymbol{0}) \,. \tag{2.69}$$

Considering two vectors τ_1 and τ_2 for which

$$\left|\tau_{1j} - \tau_{2j}\right| \ll \frac{1}{\Delta\nu}, \quad j = 2, 3, \ldots, 2n \tag{2.70}$$

we obtain

$$\Gamma^{(n,n)}(\boldsymbol{x}_1, \ldots, \boldsymbol{x}_{2n}; \boldsymbol{\tau}_1) = \exp\left[i2\pi\bar{\nu} \sum_{j=2}^{2n} \varepsilon_j(\tau_{1j} - \tau_{2j})\right] \Gamma^{(n,n)}(\boldsymbol{x}_1, \ldots, \boldsymbol{x}_{2n}; \boldsymbol{\tau}_2). \tag{2.71}$$

Similar results apply to the quantum correlation function $\Gamma_{\mathcal{N}}^{(m,n)}$ as well as to the quantum correlation functions involving other operator orderings (Chapter 16).

Chapter 3

SECOND-ORDER COHERENCE

In this chapter we study the second-order coherence effects. We demonstrate the second-order coherence by elementary considerations based on Young's interference experiment and we derive the propagation laws of partial coherence. The description given here is classical. However, we shall see (section 12.2) that the quantum description is formally identical to the classical one and the results given here also apply to the quantum correlation function.

3.1 Interference law for two partially coherent beams

In this chapter we assume a stationary and ergodic optical field. We also assume light to be linearly polarized for simplicity so that it may be represented by a scalar function $V(\boldsymbol{x}, t)$. Further we consider quasi-monochromatic beams.

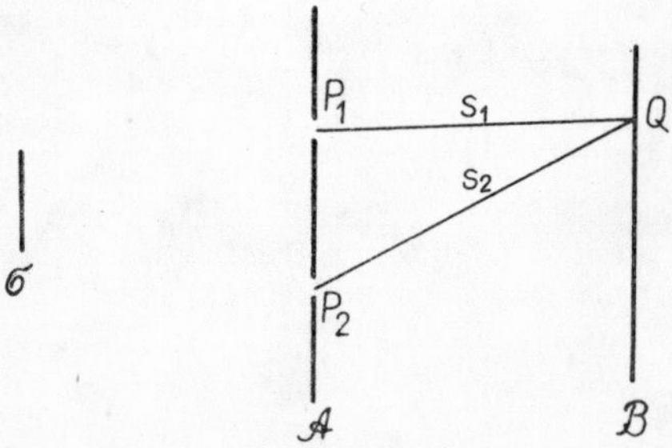

Fig. 3.1. Significance of second-order coherence in the two-beam interference experiment; σ is an extended quasi-monochromatic source, $\mathscr{A}$ is a screen with two pinholes P_1 and P_2 and $\mathscr{B}$ is the screen of observation.

In the Maxwell theory of the electromagnetic field one assumes the electric field $\boldsymbol{E}$ and magnetic field $\boldsymbol{H}$ to be measurable quantities. However, as has already been mentioned, optical vibrations are so rapid that no real detector can follow them and, moreover, the field represents a statistical dynamic system. Therefore it is necessary to introduce an averaging process for physical quantities. Since we have restricted ourselves to the study of stationary and ergodic fields, the time average can be adopted for the definition of the mutual coherence function.

Consider the Young's interference experiment with the arrangement illustrated in Fig. 3.1, where σ represents an extended quasi-monochromatic source of light. Light vibrations from the pinholes $P_1(\boldsymbol{x}_1)$ and $P_2(\boldsymbol{x}_2)$ of the screen $\mathscr{A}$ interfere at the point $Q(\boldsymbol{x})$ on the screen $\mathscr{B}$; the point Q is at a distance s_1 from P_1 and s_2 from P_2 respectively. As we consider linearly polarized light (or we disregard the polarization phenomena in which the vectorial properties of the electromagnetic field manifest themselves) we can describe this phenomenon by the scalar complex function $V(\boldsymbol{x}, t)$ for which at the point $Q(\boldsymbol{x})$

$$V(\boldsymbol{x}, t) = a_1 V(\boldsymbol{x}_1, t - t_1) + a_2 V(\boldsymbol{x}_2, t - t_2), \tag{3.1}$$

where $t_1 = s_1/c$ and $t_2 = s_2/c$, c is velocity of light (for simplicity an air medium is assumed). The imaginary numbers a_1 and a_2 are transmission factors (propagators) between P_1 and Q and P_2 and Q, defined for the mean frequency and they depend on the size of pinholes and on the geometry.†

For the intensity at the point $Q(\boldsymbol{x})$ we obtain

$$\begin{aligned} I(\boldsymbol{x}) &= \langle I(\boldsymbol{x}, t)\rangle = \langle V^*(\boldsymbol{x}, t)\, V(\boldsymbol{x}, t)\rangle = \\ &= |a_1|^2 I(\boldsymbol{x}_1) + |a_2|^2 I(\boldsymbol{x}_2) + 2|a_1 a_2| \operatorname{Re} \Gamma(\boldsymbol{x}_1, \boldsymbol{x}_2; \tau), \end{aligned} \tag{3.2}$$

where $\tau = t_1 - t_2 = (s_1 - s_2)/c$ and

$$\Gamma_{12}(\tau) \equiv \Gamma(\boldsymbol{x}_1, \boldsymbol{x}_2; \tau) = \langle V^*(\boldsymbol{x}_1, t)\, V(\boldsymbol{x}_2, t + \tau)\rangle \tag{3.3}$$

is the mutual coherence function. The intensities at $\boldsymbol{x}_1$ and at $\boldsymbol{x}_2$ are related to the mutual coherence function by

$$I(\boldsymbol{x}_j) = \Gamma(\boldsymbol{x}_j, \boldsymbol{x}_j; 0), \quad j = 1, 2. \tag{3.4}$$

In (3.2) we have also used the stationary condition

$$\langle V^*(\boldsymbol{x}_1, t - t_1)\, V(\boldsymbol{x}_2, t - t_2)\rangle = \langle V^*(\boldsymbol{x}_1, t)\, V(\boldsymbol{x}_2, t + t_1 - t_2)\rangle, \tag{3.5}$$

which expresses the fact that the correlation function depends on the time difference $t_1 - t_2$.

---

† For example their form follows from (3.36) applied to the arrangement: $a = \sqrt{[I(P)]}\, \bar{\Lambda}/2\pi s = i\bar{k} \sqrt{[I(P)]} \cos \vartheta/2\pi s$.

We introduce the intensities

$$I^{(j)}(x) = |a_j|^2 I(x_j), \quad j = 1, 2 \tag{3.6}$$

at Q arising from P_1 and P_2 alone (if only one pinhole were open) and the degree of coherence according to (2.42a) for $n = 1$ (all degrees of coherence $\gamma^{(1,1)}$, ${}^{(G)}\gamma^{(1,1)}$ and ${}^{(S)}\gamma^{(1,1)}$ given by (2.42a – c) are identical in this case)

$$\gamma(x_1, x_2; \tau) = \frac{\Gamma(x_1, x_2; \tau)}{\{\Gamma(x_1, x_1; 0)\, \Gamma(x_2, x_2; 0)\}^{1/2}}, \tag{3.7}$$

for which $|\gamma(x_1, x_2; \tau)| \leqq 1$ as follows from (2.43a) for $n = 1$. We then obtain the final form of the general interference law for stationary quasi-monochromatic optical fields:

$$I(x) = I^{(1)}(x) + I^{(2)}(x) + 2[I^{(1)}(x)\, I^{(2)}(x)]^{1/2} \operatorname{Re} \gamma(x_1, x_2; \tau) = \tag{3.8a}$$

$$= I^{(1)}(x) + I^{(2)}(x) + 2[I^{(1)}(x)\, I^{(2)}(x)]^{1/2} |\gamma(x_1, x_2; \tau)| \,. $$

$$. \cos [\alpha(x_1, x_2; \tau) - 2\pi\bar{\nu}\tau], \tag{3.8b}$$

where (2.66) for $n = 1$ has also been used. Measuring the total intensity $I(x)$ at Q and the intensities $I^{(1)}(x)$ and $I^{(2)}(x)$ at Q arising from single pinholes one can determine $\operatorname{Re} \gamma_{12}(\tau)$ and hence $\operatorname{Re} \Gamma_{12}(\tau)$. Consequently $\gamma_{12}(\tau)$ and $\Gamma_{12}(\tau)$ can be found using the dispersion relations. Hence these are *measurable quantities*. The third term in (3.8) represents an interference term arising if both pinholes are open.

We can define the usual measure of the sharpness of interference fringes, which is called the *visibility*, introduced by Michelson,†

$$\mathscr{V}(x) = \frac{I_{\max} - I_{\min}}{I_{\max} + I_{\min}}, \tag{3.9}$$

where

$$I_{\substack{\max \\ \min}} = I^{(1)}(x) + I^{(2)}(x) \pm 2[I^{(1)}(x)\, I^{(2)}(x)]^{1/2} |\gamma(x_1, x_2; \tau)|, \tag{3.10}$$

which follows from (3.8b) since $\operatorname{Max} \cos x = 1$ $(\alpha - 2\pi\bar{\nu}\tau = 2\pi m,$ $m = 0, \pm 1, \ldots)$ and $\operatorname{Min} \cos x = -1$ $(\alpha - 2\pi\bar{\nu}\tau = (2m + 1)\, \pi)$. Thus for the visibility (3.9) we have

$$\mathscr{V}(x) = \frac{2[I^{(1)}(x)\, I^{(2)}(x)]^{1/2}}{I^{(1)}(x) + I^{(2)}(x)} |\gamma(x_1, x_2; \tau)| =$$

† Another definition of the visibility is adopted by Françon and Slansky [11]: $\mathscr{V}(x) = (I_{\max} - I_{\min})/I_{\max}$.

$$= 2\left\{\left[\frac{I^{(1)}(x)}{I^{(2)}(x)}\right]^{1/2} + \left[\frac{I^{(2)}(x)}{I^{(1)}(x)}\right]^{1/2}\right\}^{-1} |\gamma(x_1, x_2; \tau)| . \quad (3.11)$$

In particular, if the intensities of the two beams are equal, then

$$\mathscr{V}(x) = |\gamma(x_1, x_2; \tau)| , \quad (3.12)$$

i.e. the visibility of fringes is equal to the modulus of the degree of coherence in this case. Hence $|\gamma|$ can be determined from simple measurements.

The phase of γ also has a simple operational significance. From (3.8b) the positions of the maxima of the intensity in the fringe pattern are given by

$$\alpha\left(x_1, x_2; \frac{s_1 - s_2}{c}\right) - \frac{2\pi\bar{\nu}}{c}(s_1 - s_2) =$$

$$= \alpha\left(x_1, x_2; \frac{s_1 - s_2}{c}\right) - \frac{2\pi}{\bar{\lambda}}(s_1 - s_2) =$$

$$= 2m\pi , \quad m = 0, \pm 1, \ldots . \quad (3.13)$$

The positions of the maxima given by (3.13) coincide with those which would be obtained if the two pinholes were illuminated by monochromatic light of wavelength $\bar{\lambda}$ and the phase of the vibrations at P_1 was retarded with respect to that at P_2 by $\alpha(x_1, x_2; (s_1 - s_2)/c)$. Hence $\alpha(x_1, x_2; (s_1 - s_2)/c)$ may be regarded as representing the "effective retardation" of the light at P_1 with respect to the light at P_2. From (3.13) it follows that the phase of γ may in principle be determined from measurements of the positions of the maxima of fringes.

The determination of γ from measurements enables us to calculate the spectrum of radiation if the inverse of the Wiener - Khintchine theorem is used. Unfortunately the positional measurements determining the phase of γ are very difficult to perform. And so we shall use another method of determining the phase of γ based on the analytic properties of γ (Sec. 4.4).

If we restrict ourselves to such time delays τ that (2.68) holds, i. e. $|\tau| \ll (\Delta\nu)^{-1}$ ($\Delta s = |s_1 - s_2| \ll c(\Delta\nu)^{-1}$), it follows from (2.69) for $n = 1$ that $|\Gamma(x_1, x_2; \tau)| = |\Gamma(x_1, x_2; 0)|$, $\arg \Gamma(x_1, x_2; \tau) = \alpha(x_1, x_2; 0) - 2\pi\bar{\nu}\tau$ and so the quantities $|\gamma(x_1, x_2; \tau)|$ and $\alpha(x_1, x_2; \tau)$ appearing in (3.8a) and in the following equations can be replaced by $|\gamma(x_1, x_2; 0)|$ and $\alpha(x_1, x_2; 0)$. The quantity $\Gamma(x_1, x_2; 0) \equiv \Gamma(x_1, x_2)$ will be called the *mutual intensity*.

Returning to (3.8b) we see that no interference fringes are formed if $\gamma = 0$ ($|\gamma| = 0$, $\mathscr{V} = 0$); this is the case when the two beams reaching the point Q from P_1 and P_2 are mutually *incoherent*. If $|\gamma| = 1$ the interference fringes have the maximum possible visibility and the beams are completely *coherent*. The intermediate cases $0 < |\gamma| < 1$ characterize *partial coherence*.

Rewriting the interference law in the form [8]

$$I(x) = |\gamma(x_1, x_2; \tau)| \{I^{(1)}(x) + I^{(2)}(x) + 2[I^{(1)}(x) I^{(2)}(x)]^{1/2} \cdot$$
$$\cdot \cos [\alpha(x_1, x_2; \tau) - 2\pi\bar{\nu}\tau]\} + [1 - |\gamma(x_1, x_2; \tau)|] \{I^{(1)}(x) + I^{(2)}(x)\}, \quad (3.14)$$

we have the interesting result that light which reaches Q from both pinholes P_1 and P_2 may be regarded as a mixture of coherent and incoherent light. The first term in (3.14) may be regarded as arising from the coherent superposition of two beams of intensities $|\gamma(x_1, x_2; \tau)| I^{(1)}(x)$ and $|\gamma(x_1, x_2; \tau)| I^{(2)}(x)$ and of relative phase difference $\alpha(x_1, x_2; \tau) - 2\pi\bar{\nu}\tau$ while the second term may be regarded as arising from the incoherent superposition of two beams of intensities $(1 - |\gamma(x_1, x_2; \tau)|) I^{(1)}(x)$ and $(1 - |\gamma(x_1, x_2; \tau)|) I^{(2)}(x)$.

3.2 Propagation laws of partial coherence

We assume that the mutual coherence function $\Gamma_{12}(\tau) \equiv \Gamma(x_1, x_2; \tau)$ is given over a certain area $\mathscr{A}$ of the space. We have to determine this

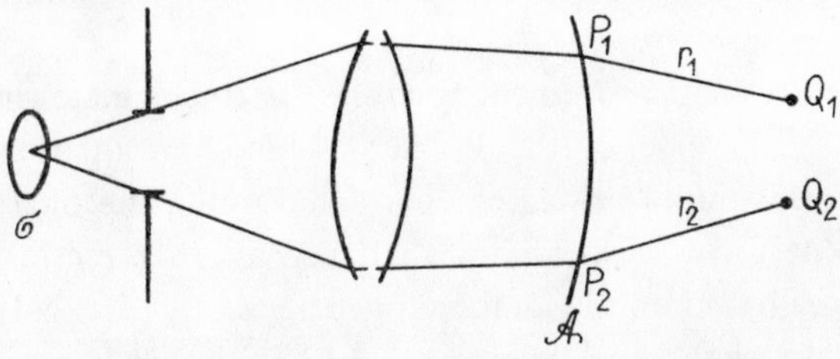

Fig. 3.2. Propagation laws of partial coherence.

function at pairs of points Q_1, Q_2 which do not lie on $\mathscr{A}$ (Fig. 3.2). We denote $r_1 \equiv \overline{P_1Q_1}$ and $r_2 \equiv \overline{P_2Q_2}$. To solve this problem we use the Green's function technique for solving the wave equations (2.36) (with $m = n = 1$) for the mutual coherence function [334, 3]. From (2.36) we have the wave equations for $\Gamma_{12}(\tau)$

$$\nabla_j^2 \, \Gamma_{12}(\tau) = \frac{1}{c^2} \frac{\partial^2 \Gamma_{12}(\tau)}{\partial \tau^2}, \quad j = 1, 2, \quad (3.15a)$$

which can be combined into one equation

$$\nabla_1^2 \nabla_2^2 \Gamma_{12}(\tau) = \frac{1}{c^4} \frac{\partial^4 \Gamma_{12}(\tau)}{\partial \tau^4} . \tag{3.15b}$$

The corresponding equations in the spectral region are

$$\nabla_j^2 G_{12}(\nu) + k^2 G_{12}(\nu) = 0 , \quad j = 1, 2 \quad (\text{for all } \nu) , \tag{3.16}$$

where $k = 2\pi\nu/c$. Let the function $G_{12}(\nu)$ be given on the area $\mathscr{A}$ and let $\mathscr{G}_1$ and $\mathscr{G}_2$ be the Green functions of the equations (3.16), i.e.

$$[\nabla_j^2 + k^2] \mathscr{G}_j(Q_j, P_j; \nu) = -\delta(Q_j - P_j) , \quad j = 1, 2 , \tag{3.17}$$

(δ is the Dirac function) with the boundary conditions

$$\mathscr{G}_j(Q_j, P_j'; \nu)\big|_{P_j' \equiv P_j} = 0 , \quad P_j \in \mathscr{A} , \quad j = 1, 2 . \tag{3.18}$$

Using repeatedly the Green formula [3] we arrive at the propagation law for the mutual spectral density

$$G(Q_1, Q_2; \nu) = \iint_{\mathscr{A}} \frac{\partial \mathscr{G}_1(Q_1, P_1; \nu)}{\partial n_1} \frac{\partial \mathscr{G}_2(Q_2, P_2; \nu)}{\partial n_2} G(P_1, P_2; \nu) \, dP_1 \, dP_2 , \tag{3.19}$$

where $\partial/\partial n_j$ denotes the derivative with respect to the normal to the surface $\mathscr{A}$ at the point P_j ($j = 1, 2$). Since $G^*(Q_1, Q_2; \nu) = G(Q_2, Q_1; \nu)$, (2.32b), it follows that

$$\mathscr{G}_1 = \mathscr{G}_2^* . \tag{3.20}$$

Defining $K(Q, P; \nu) \equiv \partial \mathscr{G}(Q, P; \nu)/\partial n$ in (3.19) we obtain

$$G(Q_1, Q_2; \nu) = \iint_{\mathscr{A}} K^*(Q_1, P_1; \nu) K(Q_2, P_2; \nu) G(P_1, P_2; \nu) \, dP_1 \, dP_2 , \tag{3.21}$$

where $K(Q, P; \nu)$ represents the *diffraction function* (it is the complex amplitude at Q caused by a unit point source of frequency ν situated at P). For example, in applying (3.21) to all the surfaces of an optical system such a function $K(Q, P; \nu)$ may again be introduced as a consequence of the linearity of (3.21) and this function will serve as the kernel of the same integral relation between an object and its image. This result will be used in Chapter 9 in an analysis of optical imaging and this analysis will serve as an example of the application of the propagation laws of partial coherence.

From (3.21) one can determine the *power spectrum* $G(Q, \nu)$ by putting $Q_1 \equiv Q_2 \equiv Q$. This quantity is not determined by the same quantity over the surface $\mathscr{A}$ but by the mutual spectral density $G(P_1, P_2; \nu)$.

The mutual coherence function is obtained by a Fourier transform (2.24)

$$\Gamma(Q_1, Q_2; \tau) = \int_0^\infty F(Q_1, Q_2; \nu) \exp(-i2\pi\nu\tau)\, d\nu\,, \tag{3.22}$$

where

$$F(Q_1, Q_2; \nu) = \iint_{\mathscr{A}} K^*(Q_1, P_1; \nu)\, K(Q_2, P_2; \nu)\, G(P_1, P_2; \nu)\, dP_1\, dP_2\,.$$

If $K(Q, P; \nu)$ depends only slightly on the frequency, this function may be replaced by its value $K(Q, P; \bar{\nu})$ at some mean frequency $\bar{\nu}$ and we obtain the following propagation law for the mutual coherence function

$$\Gamma(Q_1, Q_2; \tau) = \iint_{\mathscr{A}} K^*(Q_1, P_1; \bar{\nu})\, K(Q_2, P_2; \bar{\nu})\, \Gamma(P_1, P_2; \tau)\, dP_1\, dP_2\,. \tag{3.23a}$$

From this we obtain for the mutual intensity, by putting $\tau = 0$,

$$\Gamma(Q_1, Q_2) = \iint_{\mathscr{A}} K^*(Q_1, P_1; \bar{\nu})\, K(Q_2, P_2; \bar{\nu})\, \Gamma(P_1, P_2)\, dP_1\, dP_2\,. \tag{3.23b}$$

The intensity $I(Q)$ is given as $I(Q) \equiv \Gamma(Q, Q; 0)$ and the degree of coherence is determined by (3.7) where $\Gamma(Q_j, Q_j; 0) \equiv I(Q_j)$.

We now consider propagation from a finite plane surface – a problem having very important applications [334,3]. The geometry is shown in Fig. 3.3.

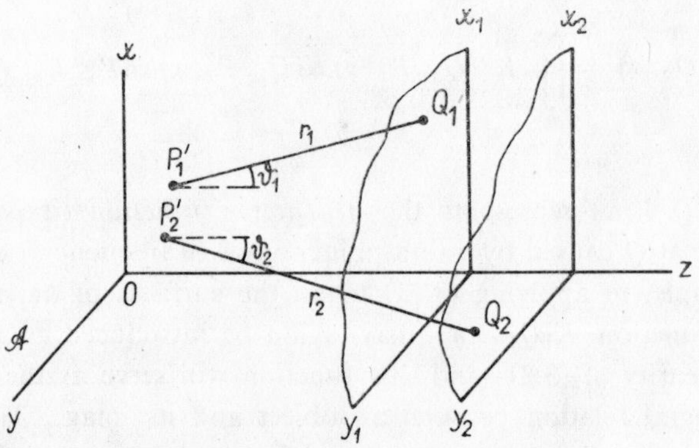

Fig. 3.3. Geometry for describing propagation of partial coherence from a plane surface.

The Green functions of this problem fulfilling the corresponding Sommerfeld radiation condition at infinity (asymptotic behaviour of the form $\exp(ikr)/r$) and having the property (3.18) are

$$\mathcal{G}_j = \frac{\exp(ikr_j)}{4\pi r_j} - \frac{\exp(ikr''_j)}{4\pi r''_j}, \quad j = 1, 2. \tag{3.24}$$

Here $r_j \equiv \overline{P_j Q_j}$ and r''_j is the distance between P_j on $\mathcal{A}$ and the mirror point of Q_j, i.e.

$$r_j = [(x_j - x'_j)^2 + (y_j - y'_j)^2 + (z_j - z'_j)^2]^{1/2},$$
$$r''_j = [(x_j - x'_j)^2 + (y_j - y'_j)^2 + (z_j + z'_j)^2]^{1/2}. \tag{3.25}$$

Hence the normal derivatives give us

$$\frac{\partial \mathcal{G}_j}{\partial n_j} = \left(ik - \frac{1}{r_j}\right)\frac{\exp(ikr_j)}{4\pi r_j}\frac{\partial r_j}{\partial n_j} - \left(ik - \frac{1}{r''_j}\right)\frac{\exp(ikr''_j)}{4\pi r''_j}\frac{\partial r''_j}{\partial n_j}. \tag{3.26}$$

However the following equalities hold,

$$\frac{\partial r_j}{\partial n_j} = -\frac{\partial r''_j}{\partial n_j} = -\left.\frac{\partial r_j}{\partial z'_j}\right|_{\mathcal{A}} = \left.\frac{\partial r''_j}{\partial z'_j}\right|_{\mathcal{A}} = \left.\frac{z_j}{r_j}\right|_{\mathcal{A}} = \cos \vartheta_j,$$
$$r_j|_{\mathcal{A}} = r''_j|_{\mathcal{A}} \tag{3.27}$$

and so from (3.26)

$$\frac{\partial \mathcal{G}_j}{\partial n_j} = 2\left(ik - \frac{1}{r_j}\right)\frac{\exp(ikr_j)}{4\pi r_j}\cos \vartheta_j, \tag{3.28}$$

giving, using (3.19),

$$G(Q_1, Q_2; \nu) = \frac{1}{4\pi^2}\iint_{\mathcal{A}} (1 + ikr_1)(1 - ikr_2) \,.$$
$$. \frac{\exp[-ik(r_1 - r_2)]}{r_1^2 r_2^2} \cos \vartheta_1 \cos \vartheta_2 \, G(P_1, P_2; \nu) \, dP_1 \, dP_2 \,. \tag{3.29}$$

The mutual coherence function is obtained (using the Fourier transform) as

$$\Gamma(Q_1, Q_2; \tau) =$$
$$= \frac{1}{4\pi^2}\int_0^\infty \iint_{\mathcal{A}} H(r_1, r_2; \nu, \tau) \cos \vartheta_1 \cos \vartheta_2 \, G(P_1, P_2; \nu) \, dP_1 \, dP_2 \, d\nu, \tag{3.30}$$

where

$$H(r_1, r_2; \nu, \tau) =$$

$$= \exp\left[-i2\pi\nu\left(\tau + \frac{r_1 - r_2}{c}\right)\right]\left\{\frac{1}{r_1^2 r_2^2} + \frac{ik(r_1 - r_2)}{r_1^2 r_2^2} + \frac{k^2}{r_1 r_2}\right\}.$$

This can be written in a concise form

$$\Gamma(Q_1, Q_2; \tau) = \frac{1}{4\pi^2}\iint_{\mathscr{A}} \frac{\cos\vartheta_1 \cos\vartheta_2}{r_1^2 r_2^2}\,\hat{\mathscr{D}}\Gamma\left(P_1, P_2; \tau + \frac{r_1 - r_2}{c}\right) dP_1\, dP_2\,, \tag{3.31}$$

where the operator $\hat{\mathscr{D}}$ is given by

$$\hat{\mathscr{D}} \equiv 1 - \frac{r_1 - r_2}{c}\frac{\partial}{\partial\tau} - \frac{r_1 r_2}{c^2}\left(\frac{\partial}{\partial\tau}\right)^2. \tag{3.32}$$

Equation (3.31) represents the general propagation law for the mutual coherence of a field produced by a plane polychromatic partially coherent source (primary or secondary).

In practice it is usual that

$$k = \frac{2\pi}{\lambda} \gg \frac{1}{r} \tag{3.33}$$

and then we have from (3.31)

$$\Gamma(Q_1, Q_2; \tau) = \frac{1}{4\pi^2}\iint_{\mathscr{A}} \frac{\Gamma\left(P_1, P_2; \tau + \dfrac{r_1 - r_2}{c}\right)}{r_1 r_2}\,\bar{\Lambda}_1^* \bar{\Lambda}_2\, dP_1\, dP_2\,, \tag{3.34}$$

where the Λ_j are inclination factors defined by $\Lambda_j = ik\cos\vartheta_j$, describing the change of direction of secondary emitted radiation on $\mathscr{A}$ and $\bar{\Lambda}_j$ is a mean value of Λ_j over the frequency region.

Of course, the same propagation law can easily be obtained by using the Huygens-Fresnel principle for the spectral component $\tilde{V}(\nu)$,

$$\tilde{V}(Q_j, \nu) = \int_{\mathscr{A}} \tilde{V}(P_j, \nu)\,\frac{\exp ikr_j}{2\pi r_j}\,\Lambda_j\, dP_j\,, \tag{3.35}$$

if it is substituted into (2.24).

The intensity at Q can be calculated from (3.34) by putting $Q_1 \equiv Q_2 \equiv Q$ and $\tau = 0$

$$I(Q) = \iint_{\mathscr{A}} \frac{[I(P_1)\,I(P_2)]^{1/2}}{4\pi^2 r_1 r_2}\,\gamma\left(P_1, P_2; \frac{r_1 - r_2}{c}\right) \bar{\Lambda}_1^* \bar{\Lambda}_2\, dP_1\, dP_2\,, \tag{3.36}$$

where the degree of coherence γ has been introduced from (3.7).

One can see that only measurable quantities appear in the propagation laws, rather than the non-measurable vectors of Maxwell theory, and this is a characteristic feature of this formulation.

Identifying the area $\mathscr{A}$ with the unilluminated side of the screen $\mathscr{A}$ in Fig. 3.1 we obtain from (3.36) the interference law (3.8) again.

Identifying the area $\mathscr{A}$ with the plane of an extended primary plane source σ, the elementary radiators of which are mutually incoherent, then only pairs of points $P_1 \equiv P_2 \equiv P$ will contribute to Γ in (3.34) and we may write

$$G(P_1, P_2; \nu) = G(P_1, \nu) \qquad \delta(P_2 - P_1)\,, \tag{3.37a}$$

$$\Gamma(P_1, P_2; \tau) = \Gamma(P_1, P_1; \tau)\,\delta(P_2 - P_1)\,, \tag{3.37b}$$

where $G(P, \nu) \equiv G(P, P; \nu)$ is the power spectrum. Thus we have from (3.34)

$$\Gamma(Q_1, Q_2; \tau) =$$

$$= \frac{1}{4\pi^2} \int_0^\infty \exp(-i2\pi\nu\tau)\, d\nu \int_\sigma G(P, \nu) \frac{\exp[-ik(r_1 - r_2)]}{r_1 r_2} \bar{\Lambda}_1^* \bar{\Lambda}_2\, dP\,. \tag{3.38}$$

The degree of coherence $\gamma(Q_1, Q_2; \tau)$ is determined by (3.7), where

$$\Gamma(Q_j, Q_j; 0) \equiv I(Q_j) = \frac{1}{4\pi^2} \int_0^\infty d\nu \int_\sigma \frac{G(P, \nu)}{r_j^2} |\bar{\Lambda}_j|^2\, dP =$$

$$= \frac{1}{4\pi^2} \int_\sigma \frac{I(P)}{r_j^2} |\bar{\Lambda}_j|^2\, dP\,. \tag{3.39}$$

Denoting

$$\sqrt{[G(P, \nu)]}\, \frac{\exp ikr_j}{2\pi r_j}\, \bar{\Lambda}_j = U(Q_j, P; \nu)\,, \tag{3.40}$$

we may rewrite (3.38) in the form

$$\Gamma(Q_1, Q_2; \tau) = [I(Q_1)\, I(Q_2)]^{1/2}\, \gamma(Q_1, Q_2; \tau) =$$

$$= \int_0^\infty \exp(-i2\pi\nu\tau)\, d\nu \int_\sigma U^*(Q_1, P; \nu)\, U(Q_2, P; \nu)\, dP\,, \tag{3.41}$$

where

$$I(Q_j) = \int_0^\infty d\nu \int_\sigma |U(Q_j, P; \nu)|^2 \, dP \,, \quad j = 1, 2 \,. \tag{3.42}$$

This formula expresses the mutual coherence function and the degree of coherence in terms of the light distribution arising from an associated fictional source since, according to (3.40), the function $U(Q, P; \nu)$ may be regarded as the complex amplitude at Q due to a monochromatic point source of frequency ν and amplitude $\sqrt{[G(P, \nu)]}$ situated at P.

3.3 Van Cittert-Zernike theorem

In practice quasi-monochromatic light is often used. If moreover the assumption

$$\left| \tau + \frac{r_1 - r_2}{c} \right| \ll \frac{1}{\Delta\nu} \tag{3.43}$$

holds, then (3.38) gives

$$\Gamma(Q_1, Q_2; \tau) = [I(Q_1) \, I(Q_2)]^{1/2} \, \gamma(Q_1, Q_2; \tau) =$$
$$= \left(\frac{\bar{k}}{2\pi}\right)^2 \exp(-i2\pi\bar{\nu}\tau) \int_\sigma I(P) \frac{\exp[-i\bar{k}(r_1 - r_2)]}{r_1 r_2} \, dP \,, \tag{3.44}$$

where only small angles ϑ_j are considered so that $\cos \vartheta_j \approx 1$. The degree of coherence then has the form

$$\gamma(Q_1, Q_2; \tau) \equiv \gamma_{12}(\tau) = \gamma_{12}(0) \exp(-i2\pi\bar{\nu}\tau) \,, \tag{3.45}$$

with the degree of spatial coherence $\gamma_{12}(0)$ given by

$$\gamma_{12}(0) \equiv \gamma_{12} = \frac{1}{[I(Q_1) \, I(Q_2)]^{1/2}} \int_\sigma I(P) \frac{\exp[-i\bar{k}(r_1 - r_2)]}{r_1 r_2} \, dP \,, \tag{3.46}$$

where

$$I(Q_j) = \int_\sigma \frac{I(P)}{r_j^2} \, dP \,, \quad j = 1, 2 \,. \tag{3.47}$$

This is the Van Cittert-Zernike theorem which enables us to calculate (under the above assumptions) the degree of coherence of an extended quasi-monochromatic incoherent source in terms of the intensity distribution over the source.

It is interesting to note that the incoherent source gives rise in general to a partially coherent field and it means that multiple propagation and diffraction can improve the degree of coherence of the light. This may be important in connection with laser light in a cavity as has been pointed out by Wolf [480, 481].

If (3.45) holds, the visibility of fringes is determined by the geometric properties of the source and it is sufficient to take into account only spatial coherence (described by $\gamma_{12}(0)$) which is independent of the spectral properties of the radiation. However, from (3.38), it is obvious that spatial coherence described by the degree of coherence $\gamma_{12}(0)$ ($\tau = 0$) and temporal coherence described by the degree of coherence $\gamma_{11}(\tau)$ ($P_1 \equiv P_2$) are generally dependent on both the geometrical properties of the source and the spectral properties of the radiation. In general, it is not possible to express the degree of coherence $\gamma_{12}(\tau)$ as a product of the degree of spatial coherence $\gamma_{12}(0)$ and the degree of temporal coherence $\gamma_{11}(\tau)$ due to the fact that spatial and temporal coherence are mutually connected. This follows from the fact that $\Gamma_{12}(\tau)$ obeys the wave equations (3.15a) *in vacuo* (section 2.2). In some cases of practical interest however such a decomposition is possible under certain restrictions. For example, assuming that the power spectrum $G(P, \nu)$ is independent of P and that

$$\frac{|r_1 - r_2|}{c} \ll \frac{1}{\Delta\nu} \tag{3.48}$$

we obtain from (3.38)

$$\begin{aligned}\Gamma(Q_1, Q_2; \tau) &= \frac{1}{4\pi^2}\,\Gamma(\tau)\int_\sigma \frac{\exp\left[-i\bar{k}(r_1 - r_2)\right]}{r_1 r_2}\,\bar{\Lambda}_1^*\bar{\Lambda}_2\,dP = \\ &= \Gamma(\tau)\,\Gamma(Q_1, Q_2; 0)\,,\end{aligned} \tag{3.49}$$

where $\Gamma(\tau)$ is the Fourier transform of $G(\nu)$ and thus

$$\gamma(Q_1, Q_2; \tau) = \gamma(\tau)\,\gamma(Q_1, Q_2; 0)\,. \tag{3.50}$$

Here $\gamma(\tau) = \Gamma(\tau)/\Gamma(0)$ and $\gamma(Q_1, Q_2; 0) = \Gamma(Q_1, Q_2; 0)/[\Gamma(Q_1, Q_1; 0)\,\Gamma(Q_2, Q_2; 0)]^{1/2}$. In this case temporal coherence is connected with the spectral properties of the radiation and spatial coherence is connected with the geometric properties of the source. This result is a special case of more general considerations about the cross-spectral purity of light [266] (section 4.3).

The propagation laws of the mutual coherence function (mutual intensity) for quasi-monochromatic light may be obtained from (3.34) (under the assumption (3.43) and for small angles ϑ_j) in the form

$$\Gamma(Q_1, Q_2; \tau) = \left(\frac{\bar{k}}{2\pi}\right)^2 \exp(-i2\pi\bar{\nu}\tau) \iint_\sigma \frac{\exp[-i\bar{k}(r_1 - r_2)]}{r_1 r_2} \cdot$$

$$\cdot \Gamma(P_1, P_2; 0)\, dP_1\, dP_2 \,. \tag{3.51}$$

Putting $\tau = 0$ it becomes

$$\Gamma(Q_1, Q_2) = [I(Q_1)\, I(Q_2)]^{1/2}\, \gamma(Q_1, Q_2) =$$

$$= \left(\frac{\bar{k}}{2\pi}\right)^2 \iint_\sigma \frac{\exp[-i\bar{k}(r_1 - r_2)]}{r_1 r_2}\, \Gamma(P_1, P_2)\, dP_1\, dP_2 \,, \tag{3.52}$$

which is a special case of (3.23b) when $K(Q, P; \bar{\nu}) \to (\bar{k}/2\pi) \exp(i\bar{k}r)/r$.

Using the analogous substitution to (3.40) in (3.44) we obtain the Hopkins formula

$$\Gamma(Q_1, Q_2) = [I(Q_1)\, I(Q_2)]^{1/2}\, \gamma(Q_1, Q_2) = \int_\sigma U^*(Q_1, P)\, U(Q_2, P)\, dP \,, \tag{3.53}$$

which is particularly useful in instrument optics. This formula gives the degree of spatial coherence without explicit use of an averaging process.

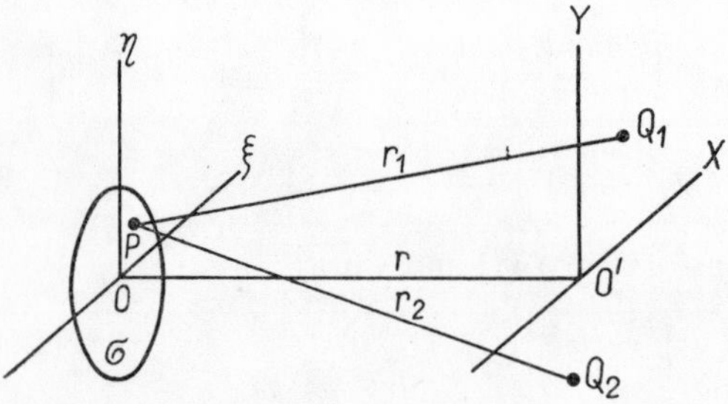

Fig. 3.4. Van Cittert - Zernike theorem.

Restricting ourselves to small linear dimensions of the source and considering only small angular distances of the points Q_1 and Q_2 (Fig. 3.4) we have

$$r_j = [(X_j - \xi)^2 + (Y_j - \eta)^2 + r^2]^{1/2} \approx r + \frac{(X_j - \xi)^2 + (Y_j - \eta)^2}{2r} \tag{3.54}$$

assuming that the linear dimensions of the source are small compared with r and that only small angular distances between OO' and PQ_j are considered; hence

$$r_1 - r_2 = \frac{(X_1^2 + Y_1^2) - (X_2^2 + Y_2^2)}{2r} - \frac{(X_1 - X_2)\,\xi + (Y_1 - Y_2)\,\eta}{r}. \tag{3.55}$$

Denoting

$$\psi = -\frac{\bar{k}[(X_1^2 + Y_1^2) - (X_2^2 + Y_2^2)]}{2r}, \quad p = \frac{X_1 - X_2}{r}, \quad q = \frac{Y_1 - Y_2}{r} \tag{3.56}$$

we obtain from (3.46)

$$\gamma(Q_1, Q_2) = \exp(i\psi)\,\frac{\int_\sigma I(\xi, \eta) \exp[i\bar{k}(p\xi + q\eta)]\, d\xi\, d\eta}{\int_\sigma I(\xi, \eta)\, d\xi\, d\eta}, \tag{3.57}$$

where we have replaced $r_1 r_2$ by r^2. In this case the degree of spatial coherence equals, apart from a phase factor, the *normalized Fourier transform* of the intensity function of the source.

For a central, quasi-monochromatic, uniform, spatially incoherent, circular source of a radius ϱ we obtain from (3.57)

$$\gamma(Q_1, Q_2) = \frac{2\,J_1(v)}{v} \exp i\psi, \tag{3.58}$$

where $v = \bar{k}\varrho[p^2 + q^2]^{1/2}$ and J_1 is the Bessel function. For a rectangular source with sides $2a$ and $2b$ we obtain

$$\gamma(Q_1, Q_2) = \frac{\sin \bar{k}pa}{\bar{k}pa}\,\frac{\sin \bar{k}qb}{\bar{k}qb} \exp i\psi. \tag{3.59}$$

The typical behaviour of the modulus of the degree of spatial coherence (i.e. of the visibility) is shown in Fig. 3.5 where the result obtained in (3.58) has been used.

Equation (3.58) may be used to determine the angular radius of a star by means of the Michelson stellar interferometer the scheme of which is shown in Fig. 3.6. If we change the separation d of the mirrors M_1 and M_2, the visibility of fringes at Q and its neighbourhood will vary according to Fig. 3.5. The first vanishing of the visibility occurs for $v = \bar{k}\varrho d_0/r = 3.83$, i.e. $d_0 = 0{\cdot}61\bar{\lambda}/\alpha$ (for this case $\psi \approx 0$), $\alpha = \varrho/r$ is the angular radius of the star, and is given by $\alpha = 0{\cdot}61\bar{\lambda}/d_0$.

If, further, the effect of temporal coherence is to be considered and if the degree of coherence reduces to the product of the degrees of temporal and

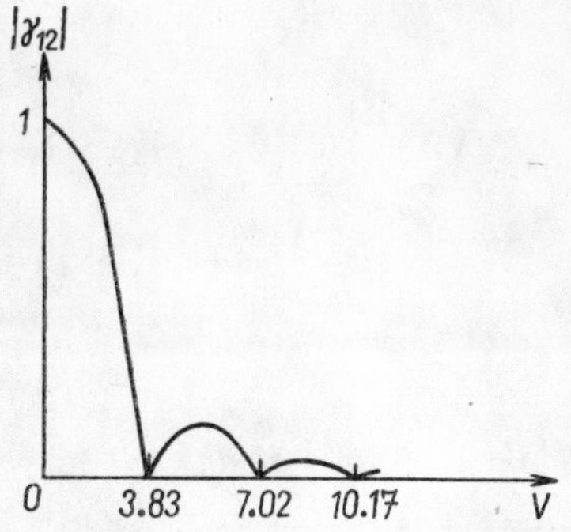

Fig. 3.5. The behaviour of $|\gamma_{12}|$ as a function of $v = (\bar{k}\varrho/r)\,[(X_1 - X_2)^2 + (Y_1 - Y_2)^2]^{1/2}$ for light from a quasi-monochromatic uniform incoherent circular source.

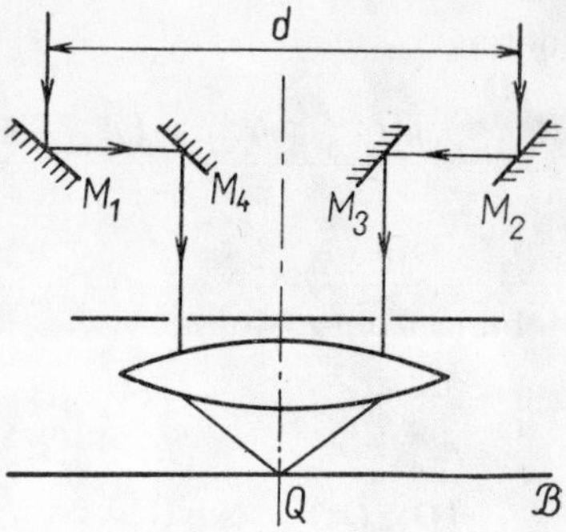

Fig. 3.6. Scheme of the Michelson stellar interferometer; $M_1 - M_4$ are mirrors, the interference is observed on the screen $\mathcal{B}$.

spatial coherence, the resultant visibility of fringes for the Young arrangement is given by the product of the visibilities caused by spatial and temporal coherence $(|\gamma_{12}(\tau)| = |\gamma_{11}(\tau)|\,|\gamma_{12}(0)|)$.

Questions of measurement of the second-order degree of coherence have been discussed in [151].

Chapter 4

SOME GENERAL PROBLEMS OF THE SECOND-ORDER THEORY OF COHERENCE

In this chapter we deal with some interesting general problems of the second-order theory of coherence such as the form of the mutual coherence function for incoherent radiation [71, 3], properties of the coherent radiation as a limiting case of partially coherent radiation [334, 335, 282, 3, 428, 311, 54, 356], the concept of cross-spectral purity of light [266, 267] and also the phase problem of interference spectroscopy. A complete solution of the last problem would make it possible to determine the spectrum of radiation from measurements of the visibility of interference fringes [478, 406, 375, 297, 300, 129, 126, 328].

4.1 Mutual coherence function for incoherent radiation

We return to the propagation law (3.19) and consider the incoherent radiation in the whole space including the surface $\mathscr{A}$, i.e.

$$G(P_1, P_2; \nu) = \begin{cases} G(P, P; \nu) \equiv G(P, \nu), & P_1 \equiv P_2 \equiv P, \\ 0 & , \ P_1 \not\equiv P_2 . \end{cases} \tag{4.1}$$

The integral in (3.19) may be interpreted as an integral in a four-dimensional space while the function under the integral sign is non-vanishing and finite on a two-dimensional hypersurface. Therefore the integral is zero in all cases including $P_1 \equiv P_2$. This is in contradiction to (4.1) and consequently such an incoherent surface does not radiate. As the propagation law (3.19) is a direct consequence of the Helmholtz equations (3.16) we conclude that an incoherent field cannot exist in a free space regardless of the spectral properties of the radiation. However, in agreement with Blanc-Lapierre and Dumontet [83] we may define an incoherent source by (3.37a, b), i.e.

$$G_{12}(\nu) = G_{11}(\nu)\, \delta(P_1 - P_2) \tag{4.2}$$

for all pairs of points P_1, P_2 of the source. Thus (3.19) gives a non-vanishing result

$$G(Q, \nu) = \int_{\mathscr{A}} \left| \frac{\partial \mathscr{G}(Q, P; \nu)}{\partial n} \right|^2 G(P, \nu)\, dP\,, \tag{4.3}$$

where $G(Q, \nu) \equiv G(Q, Q; \nu)$ but the mutual spectral density is non-physical for $P_1 \equiv P_2$ in this case. This difficulty was removed in [71, 3] where it has been shown that the δ-function in (4.2) may be replaced by $2J_1(k|P_1 - P_2|)/k|P_1 - P_2|$, where $k = 2\pi/\lambda$, J_1 is the Bessel function and $|P_1 - P_2| = [(x_1 - x_2)^2 + (y_1 - y_2)^2]^{1/2}$ $(P_j \equiv P_j(x_j, y_j), j = 1, 2)$. Of course such a function expressing the "smooth" δ-function is not unique. For some further details we refer the reader to the papers by Beran and Parrent [71, 3].

4.2 Properties of coherent radiation

A fully coherent field (in second-order) may be defined by

$$|\gamma(x_1, x_2)| \equiv 1\,. \tag{4.4a}$$

If it is stationary, we have

$$|\gamma(\boldsymbol{x}_1, \boldsymbol{x}_2; \tau)| \equiv 1\,. \tag{4.4b}$$

Equations (4.4a, b) must hold for all $\boldsymbol{x}_1, \boldsymbol{x}_2, t_1, t_2$ (τ). This implies

$$|\Gamma(x_1, x_2)|^2 = \Gamma(x_1, x_1)\, \Gamma(x_2, x_2) \tag{4.5a}$$

and

$$|\Gamma(\boldsymbol{x}_1, \boldsymbol{x}_2; \tau)|^2 = \Gamma(\boldsymbol{x}_1, \boldsymbol{x}_1; 0)\, \Gamma(\boldsymbol{x}_2, \boldsymbol{x}_2; 0)\,. \tag{4.5b}$$

Therefore the correlation function $\Gamma(x_1, x_2)$ has the form

$$\Gamma(x_1, x_2) = A(x_1)\, A(x_2) \exp i\varphi(x_1, x_2)\,, \tag{4.6}$$

where $A(x_j)$ $(= [\Gamma(x_j, x_j)]^{1/2})$ and $\varphi(x_1, x_2)$ $(= -\varphi(x_2, x_1)$ since $\Gamma(x_1, x_2) = \Gamma^*(x_2, x_1))$ are real-valued functions. Defining a function $F = \sum_j c_j V(x_j)$ we obtain

$$0 \leqq \langle |F|^2 \rangle = \sum_{j,k} c_j^* c_k \langle V^*(x_j)\, V(x_k) \rangle = \sum_{j,k} c_j^* c_k\, \Gamma(x_j, x_k)\,, \tag{4.7a}$$

i.e.

$$\sum_{j,k} d_j^* d_k\, \gamma(x_j, x_k) \geqq 0\,, \tag{4.7b}$$

which is the *non-negative definiteness condition* for the degree of coherence with arbitrary complex constants d_k $(= c_k [\Gamma(x_k, x_k)]^{1/2})$. This condition gives us successively for $n = 1, 2, 3$

$$\gamma(x, x) = 1 \geqq 0 , \tag{4.8a}$$

$$\begin{vmatrix} 1 & \gamma(x_1, x_2) \\ \gamma(x_2, x_1) & 1 \end{vmatrix} = 1 - |\gamma(x_1, x_2)|^2 \geqq 0 , \tag{4.8b}$$

$$\begin{vmatrix} 1 & \gamma(x_1, x_2) & \gamma(x_1, x_3) \\ \gamma(x_2, x_1) & 1 & \gamma(x_2, x_3) \\ \gamma(x_3, x_1) & \gamma(x_3, x_2) & 1 \end{vmatrix} =$$

$$= (1 - |\gamma(x_1, x_2)|^2)(1 - |\gamma(x_3, x_1)|^2) -$$

$$- |\gamma(x_3, x_2) - \gamma(x_1, x_2)\gamma(x_3, x_1)|^2 \geqq 0 . \tag{4.8c}$$

The first condition is obvious, the second one gives the normalization condition for $\gamma(x_1, x_2)$. From the third condition it follows that

$$|\gamma(x_3, x_2) - \gamma(x_1, x_2)\gamma(x_3, x_1)|^2$$

$$\leqq (1 - |\gamma(x_1, x_2)|^2)(1 - |\gamma(x_3, x_1)|^2) . \tag{4.9}$$

Since (4.4a) holds we have

$$\gamma(x_3, x_2) = \gamma(x_1, x_2)\gamma(x_3, x_1) , \tag{4.10a}$$

i.e.

$$\varphi(x_3, x_2) = \varphi(x_1, x_2) + \varphi(x_3, x_1) = \varphi(x_1, x_2) - \varphi(x_1, x_3) \tag{4.10b}$$

and so putting $x_1 = 0$ we have $\varphi(x_3, x_2) - \varphi(x_3, 0) = \varphi(0, x_2)$ which is independent of x_3. Thus the functional equation (4.10b) defines, up to an additive constant, the function $\alpha(x) \equiv \varphi(0, x)$ and so

$$\varphi(x_1, x_2) = \alpha(x_2) - \alpha(x_1) . \tag{4.11}$$

Substituting (4.11) into (4.6) the mutual coherence function $\Gamma(x_1, x_2)$ takes the factorized form

$$\Gamma(x_1, x_2) = V^*(x_1) V(x_2) , \tag{4.12}$$

where $V(x) = A(x) \exp i\alpha(x)$. Hence, the conditions of full second-order coherence (4.4a) [(4.5a)] and (4.12) are equivalent because (4.12) also implies (4.4a) [(4.5a)]. Some consequences of this factorization property of the second-order correlation function for the higher-order correlation functions will be dealt with in Chapter 12.

Let us note here that a field coherent in all orders may be defined by means of analogous factorization properties imposed upon the correlation functions $\Gamma^{(m,n)}$. The field fulfilling such conditions is obviously a deterministic field without any fluctuations. These questions will be discussed in section 12.2.

A detailed analysis of the factorization property of the mutual coherence function for coherent stationary fields can be found in [311]. Equation (4.12) gives us in this case

$$\Gamma(\boldsymbol{x}_1, \boldsymbol{x}_2; \tau) = U^*(\boldsymbol{x}_1)\, U(\boldsymbol{x}_2) \exp\left(-i2\pi\nu_0\tau\right), \tag{4.13}$$

where $\tau = t_2 - t_1$, ν_0 is a positive constant and $U(\boldsymbol{x}_j)$ is a function depending on $\boldsymbol{x}_j$ only. The factorization property (4.13) of the mutual coherence function was derived using other methods in [334, 335, 3, 282] for quasi-monochromatic light.

One can see that the condition (4.4b) of the second-order fully coherent field leads to the result that the coherent field in this sense must be monochromatic with frequency ν_0. On the other hand, substituting the monochromatic field

$$V(\boldsymbol{x}_j, t) = U(\boldsymbol{x}_j) \exp\left(-i2\pi\nu_0 t\right), \quad j = 1, 2 \tag{4.14}$$

into the definition of $\Gamma(\boldsymbol{x}_1, \boldsymbol{x}_2; \tau)$ we arrive at (4.13) again. Hence, the stationary field is completely coherent (in second order) in the sense of (4.4) if and only if it is monochromatic. This result follows from restricting ourselves to the class of stationary fields; indeed in this case $\Gamma(t_1, t_2) = \Gamma(t_1 - t_2) = V^*(t_1)\, V(t_2)$, and this is a functional equation which may be fulfilled only by an exponential function, i.e. $V(t) \sim \exp\left(-i2\pi\nu_0 t\right)$. However, considering a class of general non-stationary fields the factorization condition (4.12) and analogous conditions for $\Gamma^{(n,n)}$ (and therefore complete coherence) do not restrict the spectral properties of the field V. This may be of arbitrary spectral composition but the statistical properties of the field are rather restricted.

The condition (4.4) is too strong to be fulfilled by a physical field. Therefore Mandel and Wolf [282] considered a weaker, but more realistic, definition of the second-order fully coherent field:

$$\operatorname*{Max}_{\tau} \left|\gamma(\boldsymbol{x}_1, \boldsymbol{x}_2; \tau)\right| \equiv 1 \tag{4.15}$$

for all $\boldsymbol{x}_1, \boldsymbol{x}_2$. They have shown that the mutual coherence function is then factorized in the form

$$\Gamma(\boldsymbol{x}_1, \boldsymbol{x}_2; \tau) = U^*(\boldsymbol{x}_1)\, U(\boldsymbol{x}_2)\, \gamma(\tau), \tag{4.16}$$

where $\gamma(\tau)$ is a unimodular function. Thus a field coherent in this sense need not be monochromatic and so the class of coherent fields in the sense

(4.16) is broader than the class of monochromatic fields. Equation (4.16) reduces to (4.13) if

$$|\tau| \ll \frac{1}{\Delta\nu}. \tag{4.17}$$

It is possible to introduce other generalizations of the definition of the coherent field [351, 352], with respect to which the coherent field need not be monochromatic and also the monochromatic field need not be even coherent. Such definitions are based on the average using functional integrals in a functional space. The fact that coherence is a manifestation of the statistical properties of the field and that it does not restrict generally the spectrum will be more evident in the quantum description, where the formalism of the so-called coherent states will be used.

It should be mentioned here that (4.13) shows that for the stationary coherent field the degree of coherence $\gamma(\boldsymbol{x}_1, \boldsymbol{x}_2; \tau)$ has no zeros in the complex τ-plane. This is an interesting result in connection with the phase problem of the coherence theory (Sec. 4.4); finding conditions under which the knowledge of the modulus of $\gamma(\boldsymbol{x}_1, \boldsymbol{x}_2; \tau)$ – the visibility – for all τ allows a unique determination of its phase and thus the complete function γ from which the spectrum g (its Fourier transform) can be found. It was shown in [136, 459] that the most general unimodular function, analytic in the lower half of the complex τ-plane, has the form

$$\exp\left[i\varphi_{12}(\tau)\right] = \exp\left[i(\beta_{12} - 2\pi\nu_0\tau)\right] \prod_{j=1}^{\infty} \frac{\tau_j^*}{\tau_j} \frac{\tau_j - \tau}{\tau_j^* - \tau}, \tag{4.18}$$

where β_{12} and $\nu_0 > 0$ are real constants and the τ_j are complex numbers ($\operatorname{Im} \tau_j < 0$) determining the positions of the zeros of the function $\exp[i\varphi_{12}(\tau)]$ in the lower half of the complex τ-plane. For stationary fields (4.6) reduces to

$$\Gamma(\boldsymbol{x}_1, \boldsymbol{x}_2; \tau) = A(\boldsymbol{x}_1)\, A(\boldsymbol{x}_2) \exp\left[i\varphi(\boldsymbol{x}_1, \boldsymbol{x}_2; \tau)\right]. \tag{4.19}$$

By using (4.13) we obtain $\varphi(\boldsymbol{x}_1, \boldsymbol{x}_2; \tau) = \beta(\boldsymbol{x}_1, \boldsymbol{x}_2) - 2\pi\nu_0\tau = \alpha(\boldsymbol{x}_2) - \alpha(\boldsymbol{x}_1) - 2\pi\nu_0\tau$, where $\alpha(\boldsymbol{x})$ is a function of $\boldsymbol{x}$ only and $U(\boldsymbol{x}) = A(\boldsymbol{x}) \,.\, \exp i\alpha(\boldsymbol{x})$. The so-called *Blaschke factors* $(\tau - \tau_j)/(\tau - \tau_j^*)$ contained in (4.18) in which zeros are included are not present in (4.13) and consequently $\gamma(\boldsymbol{x}_1, \boldsymbol{x}_2; \tau)$ does not contain any zeros. An elimination of zeros was performed in [335, 3] by using the real part of the cross-symmetry condition $\gamma(\tau) = \gamma^*(-\tau)$ in an incorrect way [311, 54, 356], since this condition leads to a symmetric distribution of zeros with respect to the imaginary axis and not to elimination of them (Sec. 4.4). A possible elimination of zeros follows using the fact that $\Gamma(\boldsymbol{x}_1, \boldsymbol{x}_2; \tau)$ must satisfy the wave equations (3.15a) [356]. The function (4.19), with (4.18) substituted, is of the factorized form $\Gamma(\boldsymbol{x}_1, \boldsymbol{x}_2; \tau) = U^*(\boldsymbol{x}_1)\, U(\boldsymbol{x}_2)\, f(\tau)$, with $f(\tau) = \exp(-i2\pi\nu_0\tau) \prod_j \tau_j^*(\tau_j - \tau)/\tau_j(\tau_j^* - \tau)$ and U independent of τ. It follows that $\Gamma(\boldsymbol{x}_1, \boldsymbol{x}_2; \tau)$ cannot be a solution of the wave equations unless the zeros are absent and $f(\tau)$ equals $\exp(-i2\pi\nu_0\tau)$.

Finally let us note that the property of full coherence stays unchanged during the propagation, i.e. if a field is coherent on a surface $\mathscr{A}$, then it is coherent in the whole space. This follows by substituting

$$G(\boldsymbol{x}_1, \boldsymbol{x}_2; \nu) = U^*(\boldsymbol{x}_1)\, U(\boldsymbol{x}_2)\, \delta(\nu - \nu_0) \tag{4.20}$$

(which is a Fourier transform of (4.13)) into (3.21). We obtain

$$G(Q_1, Q_2; \nu) =$$

$$= \iint_{\mathscr{A}} K^*(Q_1, P_1; \nu)\, K(Q_2, P_2; \nu)\, U^*(P_1)\, U(P_2)\, \delta(\nu - \nu_0)\, dP_1\, dP_2 \tag{4.21}$$

and consequently

$$\Gamma(Q_1, Q_2; \tau) = U^*(Q_1)\, U(Q_2) \exp(-i2\pi\nu_0\tau), \tag{4.22}$$

where

$$U(Q) = \int_{\mathscr{A}} K(Q, P; \nu_0)\, U(P)\, dP, \tag{4.23}$$

which is again of the form (4.13). By using (4.16) and (3.34) one can also prove [282], under certain conditions, that the function $\Gamma(Q_1, Q_2; \tau)$ is of the form (4.16). Therefore the relation (4.23), expressing a more general form of the Huygens - Fresnel principle, is valid for a broader class of fields than the class of monochromatic fields, i.e. it is valid for the class of coherent fields (in the sense of (4.15)). All quantities appearing in (4.23) are measurable.

4.3 Concept of cross-spectral purity of light

As was already mentioned, spatial and temporal coherence are connected in the degree of coherence $\gamma_{12}(\tau)$ and they cannot be separated in general. Nevertheless, it may be shown [266] that in cases of practical interest $\gamma_{12}(\tau)$ may be expressed, at least to a good approximation, as the product of the degrees of spatial and temporal coherence. This property is closely related to the spectral properties of light. The term, cross-spectral purity of light, introduced by means of this factorization property leads to an important classification of optical fields.

We restrict ourselves to linearly polarized light and consider two points $P_1(\boldsymbol{x}_1)$ and $P_2(\boldsymbol{x}_2)$ at which the complex field amplitudes are $V(\boldsymbol{x}_1, t)$ and $V(\boldsymbol{x}_2, t)$ respectively. Let the normalized mutual spectral density of $\gamma(\boldsymbol{x}_1, \boldsymbol{x}_2; \tau)$ be $g(\boldsymbol{x}_1, \boldsymbol{x}_2; \nu) \equiv g_{12}(\nu)$ and let $g_{11}(\nu) \equiv g_{22}(\nu)$. Further let the mean frequency be $\bar{\nu}$ and the width of the spectrum be $\Delta\nu$. Suppose that the light from P_1 and P_2 is superimposed at a point $P_3(\boldsymbol{x}_3)$ with the corresponding normalized spectral density $g_{33}(\nu)$. In general $g_{33}(\nu)$ will not be simply related to $g_{11}(\nu)$ and $g_{22}(\nu)$, even if $g_{11}(\nu)$ and $g_{22}(\nu)$ are identical. If $g_{11}(\nu) \equiv g_{22}(\nu)$ and a region of neighbouring points P_3 exists such that $g_{33}(\nu) \equiv g_{11}(\nu)$, then $V_1(t)$ and $V_2(t)$ are said to be *cross-spectrally pure.*

Performing the superposition of $V_1(t)$ and $V_2(t)$ at P_3 with path differences $c\eta_1$ and $c\eta_2$ and the corresponding propagators a_1 and a_2 between P_1, P_3 and P_2, P_3 respectively we arrive at

$$V(\boldsymbol{x}_3, t) = a_1\, V(\boldsymbol{x}_1, t - \eta_1) + a_2\, V(\boldsymbol{x}_2, t - \eta_2) \tag{4.24}$$

and we obtain for the *autocorrelation function* $\Gamma_{33}(\tau)$

$$\begin{aligned}\Gamma_{33}(\tau) &= |a_1|^2\, \Gamma_{11}(\tau) + |a_2|^2\, \Gamma_{22}(\tau) + \\ &+ a_1^* a_2\, \Gamma_{12}(\tau + \eta_1 - \eta_2) + a_1 a_2^*\, \Gamma_{21}(\tau + \eta_2 - \eta_1)\,.\end{aligned} \tag{4.25}$$

The normalization gives us

$$\gamma_{33}(\tau) = \frac{\gamma_{11}(\tau) + \tfrac{1}{2}K_{12}\, \gamma_{12}(\tau + \eta_1 - \eta_2) + \tfrac{1}{2}K_{12}^*\, \gamma_{21}(\tau + \eta_2 - \eta_1)}{1 + \operatorname{Re}\{K_{12}\, \gamma_{12}(\eta_1 - \eta_2)\}}\,, \tag{4.26}$$

since $g_{11}(\nu) \equiv g_{22}(\nu)$ and so $\gamma_{11}(\tau) \equiv \gamma_{22}(\tau)$, where

$$K_{12} = \frac{2\sqrt{(I_1 I_2)}\, a_1^* a_2}{|a_1|^2 I_1 + |a_2|^2 I_2}\,; \tag{4.27}$$

here $I_j \equiv \Gamma_{jj}(0)$ and obviously $|K_{12}| \leq 1$. Performing the Fourier transformation we obtain

$$g_{33}(\nu) = \frac{g_{11}(\nu) + \operatorname{Re}\{K_{12}\, g_{12}(\nu) \exp\left[-i2\pi\nu(\eta_1 - \eta_2)\right]\}}{1 + \operatorname{Re}\{K_{12}\, \gamma_{12}(\eta_1 - \eta_2)\}} \tag{4.28}$$

and so

$$g_{33}(\nu) - g_{11}(\nu) =$$
$$= \frac{\mathrm{Re}\,\{K_{12}[g_{12}(\nu)\exp[-i2\pi\nu(\eta_1-\eta_2)] - g_{11}(\nu)\,\gamma_{12}(\eta_1-\eta_2)]\}}{1+\mathrm{Re}\,\{K_{12}\,\gamma_{12}(\eta_1-\eta_2)\}}. \quad (4.29)$$

We can now try to find conditions under which the light will be cross-spectrally pure, i.e. $g_{11}(\nu) \equiv g_{22}(\nu) \equiv g_{33}(\nu)$. For this the right-hand side of (4.29) must vanish but this cannot generally be so (except for incoherent beams for which $\gamma_{12}(\tau) \equiv g_{12}(\nu) \equiv 0$). This is readily seen for sufficiently large $|\eta_1 - \eta_2|$ when the function $\gamma_{12}(\eta_1 - \eta_2)$ in (4.29) vanishes and the spectrum is modulated by a cosine function through the first term on the right-hand side. However, if the path difference $|\eta_1 - \eta_2|$ is sufficiently small in the sense that $\eta_1 - \eta_2 = \eta_0 + \Delta\eta$, where $|\Delta\eta| \ll 1/\Delta\nu$, we obtain

$$\gamma_{12}(\eta_1 - \eta_2) = \gamma_{12}(\eta_0 + \Delta\eta) = \gamma_{12}(\eta_0)\exp(-i2\pi\bar{\nu}\,\Delta\eta)\,. \quad (4.30)$$

Substituting this into (4.29) we obtain

$$g_{12}(\nu)\exp(-i2\pi\nu\eta_0) = g_{11}(\nu)\,\gamma_{12}(\eta_0) \quad (4.31)$$

if $g_{33}(\nu) \equiv g_{11}(\nu)$ for all K_{12} and $|\Delta\eta| \ll 1/\Delta\nu$. From this we have, making a Fourier transform, the required factorization property

$$\gamma_{12}(\tau + \eta_0) = \gamma_{11}(\tau)\,\gamma_{12}(\eta_0) \quad (4.32a)$$

or

$$\gamma_{12}(\tau) = \gamma_{11}(\tau - \eta_0)\,\gamma_{12}(\eta_0)\,. \quad (4.32b)$$

As $|\gamma_{11}(\tau)| \leqq |\gamma_{11}(0)| = 1$, it follows that $|\gamma_{12}(\tau + \eta_0)| \leqq |\gamma_{12}(\eta_0)|$, i.e. $c\eta_0$ is the path difference which must be introduced between the light beams from P_1 and P_2 to reach the maximum of $|\gamma_{12}|$. In practice η_0 equals $\eta_1 - \eta_2 = (\overline{P_3P_1} - \overline{P_3P_2})/c$. The first factor on the right of (4.32) represents temporal coherence, the second one represents spatial coherence. Summarizing we have shown (under the above assumptions) that the degree of coherence between points P_1 and P_2 in an optical stationary field is reducible if and only if the corresponding light beams are cross-spectrally pure.

It may be shown [266] with the aid of the propagation law for the mutual coherence function that the reduction formula (4.32), i.e. the cross-spectral purity of light, is conserved during propagation provided that the path differences involved are sufficiently small and that the normalized mutual spectral density is constant on the surface from which it propagates.

From (4.2) it can be seen that a spatially incoherent source, for which

$$\gamma(\boldsymbol{x}_1, \boldsymbol{x}_2; \tau) = \gamma(\boldsymbol{x}_1, \boldsymbol{x}_1; \tau)\, \delta(\boldsymbol{x}_1 - \boldsymbol{x}_2) \tag{4.33}$$

holds, produces a field which is always cross-spectrally pure provided the normalized mutual spectral density is the same for all points of the source (under the restrictions on path differences involved).

The behaviour of cross-spectrally pure beams has been further investigated in [267].

4.4 Determination of the spectrum from measurements of the visibility of interference fringes – phase problem of coherence theory

We now consider the problem in interference spectroscopy of determining the spectrum from measurements of the visibility of interference fringes. This was first considered by Michelson, who showed that one may obtain information about the energy spectrum of the light from the measured visibility $\mathscr{V} = |\gamma(\boldsymbol{x}, \boldsymbol{x}; \tau)| \equiv |\gamma(\tau)|$ as a function of τ.

This can be seen very simply for quasi-monochromatic light obtained from a spectral line of mean frequency $\bar{\nu}$, bandwidth $\Delta\nu$ ($\ll \bar{\nu}$) and symmetric profile, that is the normalized spectral density g satisfies $g(\bar{\nu} + \nu) = g(\bar{\nu} - \nu)$. As (2.49) holds,

$$\gamma(\tau) = \int_0^\infty g(\nu) \exp(-i2\pi\nu\tau)\, d\nu\,, \tag{4.34}$$

the spectrum $g(\nu)$ may be determined by an inverse Fourier transform:

$$\begin{aligned} g(\nu) &= \int_{-\infty}^{+\infty} \gamma(\tau) \exp i2\pi\nu\tau\, d\tau\,, \quad \nu \geqq 0\,, \\ &= 0 \qquad\qquad\qquad\qquad\quad\;\, , \quad \nu < 0\,. \end{aligned} \tag{4.35}$$

In order to determine $g(\nu)$ we must know not only the modulus of $\gamma(\tau)$ but also the phase $\varphi(\tau)$ of $\gamma(\tau)$. However, in the case of a symmetric spectral profile we obtain from (4.34) by substitution $\nu \to \bar{\nu} + \nu$

$$\gamma(\tau) = \exp(-i2\pi\bar{\nu}\tau) \int_{-\Delta\nu/2}^{\Delta\nu/2} g(\bar{\nu} + \nu) \exp(-i2\pi\nu\tau)\, d\nu\,. \tag{4.36}$$

But this integral is real since $g(\bar{\nu} + \nu)$ is symmetric with respect to $\bar{\nu}$ ($g(\bar{\nu} + \nu) = g(\bar{\nu} - \nu)$). Taking the absolute value of (4.36) we see that the integral equals $|\gamma(\tau)|$ and we finally have

$$\gamma(\tau) = \exp(-i2\pi\bar{\nu}\tau)\,|\gamma(\tau)|\,, \tag{4.37}$$

the phase $\varphi(\tau)$ of $\gamma(\tau)$ equalling $(-2\pi\bar{\nu}\tau)$ in this case. The spectrum is determined by (4.35), i.e.

$$g(\nu) = \int_{-\infty}^{+\infty} \mathscr{V}(\tau) \exp[i2\pi\tau(\nu - \bar{\nu})]\, d\tau = $$
$$= 2\int_{0}^{\infty} \mathscr{V}(\tau) \cos[2\pi\tau(\nu - \bar{\nu})]\, d\tau\,, \tag{4.38}$$

since $\mathscr{V}(\tau) \equiv |\gamma(\tau)|$ is an even function of τ.

If the spectrum is non-symmetric the phase of $\gamma(\tau)$ does not equal $(-2\pi\bar{\nu}\tau)$ but we have mentioned in section 3.1 that this phase can in principle be determined from shifts of interference fringes in the Young experiment. Unfortunately, such measurements are difficult to perform. However it has been pointed out by Wolf [478] that information about the phase of γ may be deduced from the analytic properties of $\gamma(\tau)$ if the positions of the zeros of γ in the region of analyticity are known or if they are absent.

As the function $\gamma(\tau)$ given by (4.34) is analytic and regular in the lower half, $\Pi^{(-)}$, of the complex τ-plane, the function

$$\log\gamma(\tau) = \log|\gamma(\tau)| + i\,\varphi(\tau) \tag{4.39}$$

will also be analytic in $\Pi^{(-)}$ but it will have logarithmic branch points at zeros of $\gamma(\tau)$. First we assume that there are no zeros of $\gamma(\tau)$ in $\Pi^{(-)}$ and, applying the dispersion relation of the form (2.26), we can obtain a relation between the modulus $|\gamma(\tau)|$ and the phase $\varphi(\tau)$ [478, 285, 406]. In deriving the dispersion relations for the function $\gamma(\tau)$ we require [6] that $|\gamma(\tau)| \to 0$ as $|\tau| \to \infty$ at least as $|\tau|^{-1}$ and that $|\gamma(\tau)|$ is square integrable (for this case $\int_{-\infty}^{+\infty} |\gamma(\tau)|^2\, d\tau = \int_0^\infty g^2(\nu)\, d\nu < \infty$). One can then write for an analytic and regular function $\gamma(\tau)$ the Cauchy integral with the contour of integration composed of the real axis and the semi-circle of infinite radius lying in $\Pi^{(-)}$. The integral vanishes over the semi-circle as a consequence of the above assumptions so that

$$\gamma(\tau) = -\frac{1}{2\pi i}\int_{-\infty}^{+\infty} \frac{\gamma(\tau')}{\tau' - \tau}\, d\tau'\,, \quad \operatorname{Im}\tau < 0\,. \tag{4.40}$$

Putting $\operatorname{Im}\tau \to 0$ and using the identity

$$\lim_{\substack{\varepsilon\to 0\\ \varepsilon>0}} \frac{1}{\tau \mp i\varepsilon} = P\frac{1}{\tau} \pm \pi i\,\delta(\tau)\,, \tag{4.41}$$

(where P denotes the principal value in the Cauchy sense) and separating the real and the imaginary parts we arrive at dispersion relations of the type (2.26):

$$\operatorname{Im} \gamma(\tau) = \frac{1}{\pi} P \int_{-\infty}^{+\infty} \frac{\operatorname{Re} \gamma(\tau')}{\tau' - \tau} \, d\tau' , \tag{4.42a}$$

$$\operatorname{Re} \gamma(\tau) = -\frac{1}{\pi} P \int_{-\infty}^{+\infty} \frac{\operatorname{Im} \gamma(\tau')}{\tau' - \tau} \, d\tau' . \tag{4.42b}$$

The situation is slightly different if the function $\log \gamma(\tau)$ is considered. Since $|\gamma(\tau)| \approx |\tau|^{-1}$ for $|\tau| \to \infty$, $\log |\gamma(\tau)| \approx \log |\tau|$ and the assumption necessary for the vanishing of the Cauchy integral over the semi-circle is not fulfilled. In this case one must consider the function $\log \gamma(\tau)/\tau$ for which $|\log \gamma(\tau)/\tau| = |\gamma'(\tau)/\gamma(\tau)| = |\tau|^{-1}$ for $|\tau| \to \infty$. Here we adopt another method of solving the phase problem based on the solution of a singular integral equation of the Cauchy type using the method of Muskhelishvili [23]; this has been useful in many branches of physics. We wish also to point out the elegance of this method. We assume the validity of the Lipschitz condition for all the functions used ($|\gamma(\tau_1) - \gamma(\tau_2)| \leqq \alpha |\tau_1 - \tau_2|^{\beta}$, where α and β are constants). The modulus and the phase of $\gamma(\tau)$ may be expressed as

$$|\gamma(\tau)| = [(\operatorname{Re} \gamma(\tau))^2 + (\operatorname{Im} \gamma(\tau))^2]^{1/2} , \tag{4.43a}$$

$$\tan \varphi(\tau) = \frac{\operatorname{Im} \gamma(\tau)}{\operatorname{Re} \gamma(\tau)} . \tag{4.43b}$$

First we assume that $\tan \varphi(\tau) \neq 0$ and $\operatorname{Re} \gamma(\tau) \neq 0$ for all real τ. This is equivalent to the assumption that $\gamma(\tau)$ has no zeros in $\Pi^{(-)}$ since then $\operatorname{Im} \gamma(\tau) \neq 0$ for all τ and the function $\gamma(\tau)$ has no zeros for real τ. Applying a theorem contained in [18] (§ 125) the function $\gamma(\tau)$ indeed has no zeros in $\Pi^{(-)}$. Denoting $\operatorname{Im} \gamma(\tau) = f(\tau)$ and substituting (4.43b) into (4.42b) we obtain a singular integral equation of the type

$$\frac{f(\tau)}{\tan \varphi(\tau)} = -\frac{1}{\pi} P \int_{-\infty}^{+\infty} \frac{f(\tau')}{\tau' - \tau} \, d\tau' . \tag{4.44}$$

To solve this equation we use the above mentioned Muskhelishvili method.

We first derive the formulae of Sochocky and Plemelj. Let $\psi(\tau)$ be defined as the Cauchy integral

$$\psi(\tau) = \frac{1}{2\pi i} \int_{-\infty}^{+\infty} \frac{f(\tau')}{\tau' - \tau} \, d\tau' , \quad \operatorname{Im} \tau \neq 0 . \tag{4.45}$$

Using (4.41) we arrive at the formulae of Sochocky and Plemelj:

$$\psi^+(\tau) - \psi^-(\tau) = f(\tau), \tag{4.46a}$$

$$\psi^+(\tau) + \psi^-(\tau) = \frac{1}{\pi i} P \int_{-\infty}^{+\infty} \frac{f(\tau')}{\tau' - \tau} d\tau', \tag{4.46b}$$

where $\psi^+(\tau)$ and $\psi^-(\tau)$ are boundary values of $\psi(\tau)$ on the real τ-axis from the upper and lower half planes respectively.

Denoting $A(\tau) = i/\tan \varphi(\tau)$ we obtain from (4.44)

$$A(\tau) f(\tau) - \frac{1}{\pi i} P \int_{-\infty}^{+\infty} \frac{f(\tau')}{\tau' - \tau} d\tau' = 0, \tag{4.47}$$

and using (4.46a, b) we have

$$A(\tau)\{\psi^+(\tau) - \psi^-(\tau)\} - \{\psi^+(\tau) + \psi^-(\tau)\} = 0. \tag{4.48}$$

Since $\gamma(\tau)$ is assumed to have no zeros in $\Pi^{(-)}$ (i.e. $\log \gamma(\tau)/2\pi i|_C = \varphi(\tau)/2\pi = 0$ for any closed contour C in $\Pi^{(-)}$) the relation

$$[\log \psi(\tau)]^+ - [\log \psi(\tau)]^- = \log \frac{A(\tau) + 1}{A(\tau) - 1} = -2i\, \varphi(\tau) \tag{4.49}$$

following from (4.48) determines the unique function $\log \psi(\tau)$ for which, according to (4.46a) and (4.45),

$$\log \psi(\tau) = \frac{1}{2\pi i} \int_{-\infty}^{+\infty} \frac{-i2\, \varphi(\tau')}{\tau' - \tau} d\tau' = -\frac{1}{\pi} \int_{-\infty}^{+\infty} \frac{\varphi(\tau')}{\tau' - \tau} d\tau'. \tag{4.50}$$

That is

$$\psi(\tau) = \exp\left[-\frac{1}{\pi} \int_{-\infty}^{+\infty} \frac{\varphi(\tau')}{\tau' - \tau} d\tau'\right]. \tag{4.51}$$

Using (4.46a) again we have

$$f(\tau) = \operatorname{Im} \gamma(\tau) = \psi^+(\tau) - \psi^-(\tau) =$$

$$= 2i \sin \varphi(\tau) \exp\left[-\frac{1}{\pi} P \int_{-\infty}^{+\infty} \frac{\varphi(\tau')}{\tau'' - \tau} d\tau'\right]. \tag{4.52}$$

Making use of the fact that the solution of a homogeneous equation is determined up to a multiplicative constant and omitting the constant $2i$ in (4.52) ($f(\tau) \equiv 1$ for $\varphi = \pi/2$) we obtain from (4.43b)

$$\operatorname{Re} \gamma(\tau) = \cos \varphi(\tau) \exp\left[-\frac{1}{\pi} P \int_{-\infty}^{+\infty} \frac{\varphi(\tau')}{\tau' - \tau} d\tau'\right] \tag{4.53}$$

and from (4.43a)

$$|\gamma(\tau)| = \exp\left[-\frac{1}{\pi} P \int_{-\infty}^{+\infty} \frac{\varphi(\tau')}{\tau' - \tau} d\tau'\right]. \tag{4.54}$$

From this relation the modulus of $\gamma(\tau)$ may be calculated if the phase $\varphi(\tau)$ is known. However, by using the Sochocky - Plemelj formulae one can prove that if two functions G and H are related by the Cauchy integral

$$G(\tau) = \frac{1}{\pi i} P \int_{-\infty}^{+\infty} \frac{H(\tau')}{\tau' - \tau} d\tau', \tag{4.55a}$$

then

$$H(\tau) = \frac{1}{\pi i} P \int_{-\infty}^{+\infty} \frac{G(\tau')}{\tau' - \tau} d\tau'. \tag{4.55b}$$

Hence, from (4.54),

$$-i \log |\gamma(\tau)| = -\frac{1}{\pi i} P \int_{-\infty}^{+\infty} \frac{\varphi(\tau')}{\tau' - \tau} d\tau', \tag{4.56}$$

and finally we obtain

$$\varphi(\tau) = \frac{1}{\pi} P \int_{-\infty}^{+\infty} \frac{\log |\gamma(\tau')|}{\tau' - \tau} d\tau', \tag{4.57}$$

which is the required relation relating the phase $\varphi(\tau)$ to $|\gamma(\tau)|$.

We can now consider the general case when $\gamma(\tau)$ has zeros in $\Pi^{(-)}$ at points τ_j $(j = 1, 2, \ldots)$. Making use of the most general form of an analytic unimodular function (4.18) we can write

$$\gamma(\tau) = \gamma_0(\tau) \exp(-i2\pi\nu_0\tau) \prod_j \frac{\tau - \tau_j}{\tau - \tau_j^*}, \tag{4.58}$$

where $\nu_0 > 0$ and $\operatorname{Im} \tau_j < 0$. The zeros of $\gamma(\tau)$ are described by the Blaschke factors $(\tau - \tau_j)/(\tau - \tau_j^*)$ and $|\gamma(\tau)| \equiv |\gamma_0(\tau)|$ on the real τ-axis; the function $\gamma_0(\tau)$ does not contain any zeros in $\Pi^{(-)}$. Therefore the complete phase $\varphi(\tau)$ of $\gamma(\tau)$ is obtained from (4.58) [443]

$$\varphi(\tau) = \frac{1}{\pi} P \int_{-\infty}^{+\infty} \frac{\log |\gamma(\tau')|}{\tau' - \tau} d\tau' + \sum_j \arg \frac{\tau - \tau_j}{\tau - \tau_j^*} - 2\pi\nu_0\tau. \tag{4.59}$$

The physical significance of the last term is clear – it represents a term causing the shift of the spectrum to ν_0 and so it can be omitted in the following. Further it can be shown that the second term in (4.59) – expressing the so-called *Blaschke phase* – is non-negative, so that the first term given by means of the dispersion relations may be called the *minimal phase.*

Thus, for the unique determination of the phase $\varphi(\tau)$ of $\gamma(\tau)$ and hence for the unique determination of the spectrum from the visibility one must know the positions of the zeros of $\gamma(\tau)$ in $\Pi^{(-)}$; in general the knowledge of only the visibility $\mathscr{V}(\tau) \equiv |\gamma(\tau)|$ is not sufficient to determine $g(\nu)$ uniquely. By using the residuum theorem we can calculate $g(\nu)$ from (4.58) with the explicit contribution of the Blaschke factors to the spectrum $g(\nu)$ [375]:

$$g(\nu) = g_0(\nu) =$$

$$2\,\pi i \sum_j (\tau_j^* - \tau_j) \exp{(i2\pi\tau_j^*\nu)} \prod_{k\neq j} \frac{\tau_j^* - \tau_k}{\tau_j^* - \tau_k^*} \int_0^\nu g_0(\mu) \exp{(-i2\pi\tau_j^*\mu)}\, d\mu\,, \tag{4.60}$$

where the minimum spectrum $g_0(\nu)$ is the Fourier transform of $\gamma_0(\tau)$.

Unfortunately at present very little is known about the physical significance of the zeros and their location, but some restrictions follow from the cross-symmetry condition $\gamma(\tau) = \gamma^*(-\tau)$ expressing the reality of the spectrum $g(\nu)$, $g^*(\nu) = g(\nu)$, and from the non-negativeness of $g(\nu)$. It can easily be shown [406, 375] that the cross-symmetry condition for $\gamma(\tau)$ which has the form (4.58) leads to the symmetrical distribution of zeros with respect to the imaginary τ-axis, i.e. with every factor $(\tau - \tau_j)/(\tau - \tau_j^*)$ the factor $(\tau + \tau_k^*)/(\tau + \tau_k)$ occurs (for every j a k exists such that $\tau_j = -\tau_k^*$). Then (4.59) may be rewritten in the form

$$\varphi(\tau) = \frac{2\tau}{\pi} P \int_0^\infty \frac{\log|\gamma(\tau')|}{\tau'^2 - \tau^2}\, d\tau' + \sum_j \left[\arg\frac{\tau - \tau_j}{\tau - \tau_j^*} + \arg\frac{\tau + \tau_j^*}{\tau + \tau_j}\right], \tag{4.61}$$

since $|\gamma(\tau)|$ is an even function ($\gamma^*(\tau) = \gamma(-\tau)$ implies that $|\gamma(-\tau)| = |\gamma(+\tau)|$ and $\varphi(-\tau) = -\varphi(+\tau)$). The condition $g(\nu) \geqq 0$ leads to the elimination of the zeros from the imaginary τ-axis, since if $\tau = ia$ ($a < 0$ and real) then $\gamma(ia) = \int_0^\infty g(\nu) \exp 2\pi\nu a\, d\nu > 0$, i.e. $\gamma(ia) \neq 0$.

Some physical conditions may be found under which $\gamma(\tau)$ cannot have zeros in $\Pi^{(-)}$ at all. Consequently $\varphi(\tau)$ and $g(\nu)$ may be uniquely determined from the dispersion relations. It was shown in [375] that

$$\frac{1}{\pi} \int_0^\infty \frac{\log|\gamma(\tau)|}{\tau^{2n}}\, d\tau = (-1)^n \frac{\mu_{2n-1}}{2(2n-1)!} - \frac{1}{2n-1} \sum_j \frac{\operatorname{Im} \tau_j^{2n-1}}{|\tau_j|^{2(2n-1)}}\,, \tag{4.62a}$$

$$\sum_j \frac{\operatorname{Re} \tau_j^{2n-1}}{|\tau_j|^{2(2n-1)}} = 0\,, \tag{4.62b}$$

where $n \geqq 1$ and the conditions $|\log \gamma(\tau)| \leqq A|\tau|^{l_1}$, $l_1 < 2n - 1$ for $|\tau| \to \infty$ and $|\log \gamma(\tau)| \leqq B|\tau|^{l_2}$, $l_2 > 2n - 1$ for $|\tau| \to 0$ hold (A and B are real positive constants). As the zeros are distributed symmetrically with respect to the imaginary axis the condition (4.62b) is an identity. The condition (4.62a) relates the behaviour of $|\gamma(\tau)|$ in the neighbourhood of zero and infinity to the moments μ_{2n-1} of the spectrum of the function $\log \gamma(\tau)$ and the distribution of zeros in $\Pi^{(-)}$. For $n = 1$ we obtain [233]

$$\frac{1}{\pi} \int_0^\infty \frac{\log |\gamma(\tau)|}{\tau^2} d\tau = -\frac{\omega}{2} - \sum_j \frac{\operatorname{Im} \tau_j}{|\tau_j|^2}, \tag{4.63}$$

where $\omega \equiv \mu_1$ is the moment of the spectrum of $\gamma(\tau)$ (i.e. of $g(\nu)$). As $0 \leqq |\gamma(\tau)| \leqq 1$, $\gamma(0) = 1$ and $\operatorname{Im} \tau_j < 0$, the minimum value of ω is

$$\omega_{\min} = -\frac{2}{\pi} \int_0^\infty \frac{\log |\gamma(\tau)|}{\tau^2} d\tau > 0 \tag{4.64}$$

and

$$\omega = -\frac{2}{\pi} \int_0^\infty \frac{\log |\gamma(\tau)|}{\tau^2} d\tau - 2 \sum_j \frac{\operatorname{Im} \tau_j}{|\tau_j|^2} \geqq \omega_{\min}. \tag{4.65}$$

Hence, the reconstruction of the phase under the additional assumption of the minimum of the moment of the spectrum is unique. Another kind of elimination of zeros for a fully coherent stationary field has been discussed in section 4.2. As was shown in [226] $\gamma(\tau)$ has no zeros in $\Pi^{(-)}$ for black body radiation so that its phase as well as the spectrum may be reconstructed from measurements of the visibility, $\mathscr{V}(\tau)$ $(\equiv |\gamma(\tau)|)$, of the interference fringes in spite of the fact that the spectrum of this radiation is not symmetric. Some interesting theorems for quasi-monochromatic spectra, particularly of Gaussian and Lorentzian forms, concerning the position of zeros were derived in [328]. In particular it was shown that the contribution of zeros to the phase $\varphi(\tau)$ is significant so that the minimal phase cannot be considered as a good approximation to the real phase.

Recently some alternative experimental approaches to the recovery of the phase $\varphi(\tau)$ from $|\gamma(\tau)|$ have been suggested. Gamo [163] (see also [269]) proposed that the phase can be measured by means of triple intensity correlation measurements. Another method was proposed by Mehta [297] on the basis of the Cauchy - Riemann conditions

$$\frac{\partial |\gamma(\tau_r, \tau_i)|}{\partial \tau_r} = -|\gamma(\tau_r, \tau_i)| \frac{\partial \varphi(\tau_r, \tau_i)}{\partial \tau_i}, \tag{4.66a}$$

$$\frac{\partial |\gamma(\tau_r, \tau_i)|}{\partial \tau_i} = |\gamma(\tau_r, \tau_i)| \frac{\partial \varphi(\tau_r, \tau_i)}{\partial \tau_r}, \tag{4.66b}$$

which must hold in $\Pi^{(-)}$; here $\tau = \tau_r + i\tau_i$. Integrating (4.66b) we may determine the phase φ. To determine $\partial\varphi(\tau_r, \tau_i)/\partial\tau_r|_{\tau_i=0}$ we must also know $|\gamma(\tau)|$ in $\Pi^{(-)}$ for $\tau_i < 0$, but the physical significance of $\gamma(\tau_r, \tau_i)$ is clear as can be seen from (4.34):

$$\gamma(\tau_r, \tau_i) = \int_0^\infty g(\nu) \exp 2\pi\nu\tau_i \exp(-i2\pi\nu\tau_r)\, d\nu\,, \tag{4.67}$$

i.e. $\gamma(\tau_r, \tau_i)$ for $\tau_i < 0$ may be obtained from the spectrum $g(\nu)$ by using the exponential filter $\exp[-2\pi\nu(-\tau_i)]$. Yet another method proposed by Mehta [300] is based on using a reference beam with the known degree of coherence $\gamma_r(\tau)$ and spectrum $g_r(\nu)$. If we superimpose the light from the reference source on light with an unknown spectrum $g(\nu)$, we obtain light with the spectrum

$$\bar{g}(\nu) = g(\nu) + g_r(\nu)\,, \tag{4.68}$$

since the sources are statistically independent. Thus we have

$$\bar{\gamma}(\tau) = \gamma(\tau) + \gamma_r(\tau) \tag{4.69}$$

for the corresponding degree of coherence. From two separate experiments with light of the unknown spectrum and of the superimposed spectrum we obtain $|\gamma(\tau)|$ and $|\bar{\gamma}(\tau)|$ respectively. Taking the squared modulus of (4.69) we arrive at

$$|\bar{\gamma}(\tau)|^2 = |\gamma(\tau)|^2 + |\gamma_r(\tau)|^2 + 2|\gamma(\tau)|\,|\gamma_r(\tau)| \cos[\varphi(\tau) - \varphi_r(\tau)]\,. \tag{4.70}$$

Here all the quantities are known except $\varphi(\tau)$ which can be determined.

The problem of finding zeros of $\gamma(\tau)$ was transferred by Roman and Marathay [406] to a non-linear eigenvalue problem and by Dialetis and Wolf [129] and Dialetis [126] to a certain inhomogeneous eigenvalue problem of the Sturm - Liouville type. It was shown that this eigenvalue problem is equivalent to a certain stability problem in mechanics.

Hence in the optical region, where the function Re $\gamma(\tau)$ is a rapidly oscillating quantity and so cannot be measured directly, there are some unsolved problems concerning the determination of the spectrum $g(\nu)$ from the visibility $|\gamma(\tau)|$. Such measurements of Re $\gamma(\tau)$ are possible in the far infra-red region [434, 209]. As Im $\gamma(\tau)$ is determined uniquely from Re $\gamma(\tau)$ by means of the dispersion relations, the spectrum $g(\nu)$ can in principle be determined uniquely in this case.

Finally let us note that the phase problem arises in other branches of physics too; we can mention the quantum theory of decay [233], scattering theory [183] and the diffraction theory of image formation [330, 460, 347], etc.

Chapter 5

MATRIX DESCRIPTION OF PARTIAL COHERENCE

5.1 Sampling theorem

The theory of coherence we have developed previously may be regarded as a functional theory. Now we follow another approach to the formulation of the coherence properties of the electromagnetic field using the "quantization" of field functions (e.g. of object and image functions) [159] based on the so-called *sampling theorem.*

If a square integrable function $f(x)$ has a spectrum non-vanishing only in the finite interval $(-W, +W)$, then this function is fully determined by its values on a denumerably infinite set of points and it is not difficult to prove that

$$f(x) = \sum_{n=-\infty}^{+\infty} f\left(\frac{n}{2W}\right) u_n(2\pi Wx) , \tag{5.1}$$

where

$$u_n(2\pi Wx) = \frac{\sin \pi(2Wx - n)}{\pi(2Wx - n)} . \tag{5.2}$$

Such a function may be written as

$$f(x) = \int_{-W}^{+W} g(v) \exp(-i2\pi vx) \, dv , \tag{5.3}$$

where $g(v)$ is the spectrum of $f(x)$. Further

$$g(v) = \sigma(v, W) \sum_{n=-\infty}^{+\infty} a_n \exp\left(i\pi n \frac{v}{W}\right), \tag{5.4}$$

where

$$a_n = \frac{1}{2W} \int_{-W}^{+W} g(v) \exp\left(-i\pi n \frac{v}{W}\right) dv \tag{5.5}$$

and

$$\sigma(\nu, W) = \frac{\operatorname{sgn}(W - \nu) + \operatorname{sgn}(W + \nu)}{2}; \tag{5.6}$$

here sgn is the sign function (sgn $x = 1$ for $x > 0$ and sgn $x = -1$ for $x < 0$). Comparing (5.3) and (5.5) we see that

$$a_n = \frac{1}{2W} f\left(\frac{n}{2W}\right) \tag{5.7}$$

and substituting (5.7) into (5.4) and performing the Fourier transformation we finally obtain

$$f(x) = \sum_{n=-\infty}^{+\infty} f\left(\frac{n}{2W}\right) \frac{1}{2W} \int_{-W}^{+W} \exp\left(i\pi n \frac{\nu}{W} - i2\pi\nu x\right) d\nu =$$

$$= \sum_{n=-\infty}^{+\infty} f\left(\frac{n}{2W}\right) \frac{\sin(\pi n - 2\pi x W)}{\pi n - 2\pi x W}, \tag{5.8}$$

which is just (5.1). The information theory worked out in papers by Shannon [28], Gabor [159, 160], Linfoot [255] and Toraldo di Francia [444] (see also [16]) among others is based upon this theorem. A characteristic feature of the theorem is the "quantization" of the function $f(x)$ and we now use it to obtain a *matrix formulation* of partial coherence, the relation of which to the usual functional theory is analogous to the relation between the wave (Schrödinger) and the matrix (Heisenberg - Dirac) formulations of quantum mechanics. Of course, these two formulations are equivalent as a consequence of the isomorphism of the spaces L_2 (the space of functions which are square integrable in the sense of Lebesgue) and l_2 (the space of generalized Fourier coefficients of functions from L_2).

We restrict ourselves to quasi-monochromatic light since this will be sufficient to demonstrate the matrix method. For a more complete treatment of the matrix formulation we refer the reader to the paper by Gamo [165] and also by Gabor [159].

First let us summarize some properties of the functions u_n given by (5.2). Obviously

$$u_n(m\pi) = \delta_{mn} = \begin{cases} 1, & m = n, \\ 0, & m \neq n. \end{cases} \tag{5.9}$$

Further we obtain

$$\int_{-\infty}^{+\infty} u_n(2\pi W x)\, u_m(2\pi W x)\, dx = \frac{1}{2W} \delta_{mn}, \tag{5.10}$$

which is the orthogonality condition for the complete set of functions u_n.

Inverting (5.1) using this condition we obtain

$$f\left(\frac{n}{2W}\right) = 2W \int_{-\infty}^{+\infty} f(x)\, u_n(2\pi Wx)\, dx \tag{5.11}$$

and further we readily see that

$$\int_{-\infty}^{+\infty} |f(x)|^2\, dx = \frac{1}{2W} \sum_{n=-\infty}^{+\infty} \left| f\left(\frac{n}{2W}\right)\right|^2 . \tag{5.12}$$

From (5.1) we obtain, putting $f(x) \equiv 1$ $(z = 2\pi Wx)$,

$$\sum_{n=-\infty}^{+\infty} u_n(z) = \sum_{n=-\infty}^{+\infty} \frac{\sin(z - \pi n)}{z - \pi n} = 1 \tag{5.13}$$

and putting $f(x) = \sin(z - z_2)/(z - z_2)$

$$\sum_{n=-\infty}^{+\infty} \frac{\sin(z_1 - \pi n)}{z_1 - \pi n} \frac{\sin(z_2 - \pi n)}{z_2 - \pi n} = \frac{\sin(z_1 - z_2)}{z_1 - z_2} . \tag{5.14}$$

If $z_1 = z_2 = z$ then

$$\sum_{n=-\infty}^{+\infty} u_n^2(z) = 1 . \tag{5.15}$$

5.2 Interference law in matrix form

As we have restricted ourselves to quasi-monochromatic light we may consider the mutual intensity $\Gamma(\boldsymbol{x}_1, \boldsymbol{x}_2) \equiv J_{12}$ in describing the coherence properties of light. Denoting $\Gamma_{11}(0) \equiv J_{11}$ and $\Gamma_{22}(0) \equiv J_{22}$ we can write the interference law in the form

$$I = |a_1|^2 J_{11} + |a_2|^2 J_{22} + a_1^* a_2 J_{12} + a_1 a_2^* J_{21} , \tag{5.16}$$

where the a_j are propagators introduced in section 3.1. This is a positive semidefinite quadratic form, since $I \geqq 0$. The elements of the matrix $\hat{J} \equiv (J_{ij})$ can be expressed in terms of the eigenvalues λ_1 and λ_2 of the matrix $\hat{J}$, defined by

$$\begin{vmatrix} J_{11} - \lambda & J_{12} \\ J_{21} & J_{22} - \lambda \end{vmatrix} = \lambda^2 - (J_{11} + J_{22})\,\lambda + J_{11}J_{22} - |J_{12}|^2 = 0 , \tag{5.17}$$

and eigenstates

$$\varphi^{(1)} = \begin{pmatrix} U_{11} \\ U_{21} \end{pmatrix}, \quad \varphi^{(2)} = \begin{pmatrix} U_{12} \\ U_{22} \end{pmatrix} \tag{5.18}$$

in the form

$$J_{ij} = \lambda_1 U_{i1} U_{j1}^* + \lambda_2 U_{i2} U_{j2}^*, \quad i, j = 1, 2. \tag{5.19}$$

The elements of the matrix $\hat{U} \equiv (\varphi^{(1)}, \varphi^{(2)})$ satisfy the homogeneous equations

$$\begin{pmatrix} J_{11} - \lambda_j & J_{12} \\ J_{21} & J_{22} - \lambda_j \end{pmatrix} \begin{pmatrix} U_{1j} \\ U_{2j} \end{pmatrix} = 0. \tag{5.20}$$

As $\hat{U}$ is a unitary matrix ($\hat{U}^+ = \hat{U}^{-1}$, $\hat{U}^+$ is the Hermitian conjugate to $\hat{U}$) we have

$$U_{1i}^* U_{1j} + U_{2i}^* U_{2j} = U_{i1} U_{j1}^* + U_{i2} U_{j2}^* = \delta_{ij}, \quad i, j = 1, 2. \tag{5.21}$$

Since $|\gamma_{12}| \leqq 1$, $|J_{12}|^2 \leqq J_{11} J_{22}$ and the equality occurs for coherent light; in this case $\text{Det}\,\hat{J} = J_{11} J_{22} - |J_{12}|^2 = 0$ (Det denotes the determinant) and so, from (5.17), $\lambda_2 = 0$. Then (5.19) gives $J_{ij} = \lambda_1 U_{i1} U_{j1}^*$, i.e. the elements of the matrix $\hat{J}$ are factorized in agreement with the result of section 4.2.

5.3 Intensity matrix and its properties

We now start a more general treatment of the matrix formulation of partial coherence. First we introduce the *intensity matrix* $\hat{A}$ corresponding to the mutual intensity. Consider an object plane which is imaged by an optical system with the numerical apertures $\alpha = n_1 \sin \vartheta_1$ and $\beta = n_2 \sin \vartheta_2$

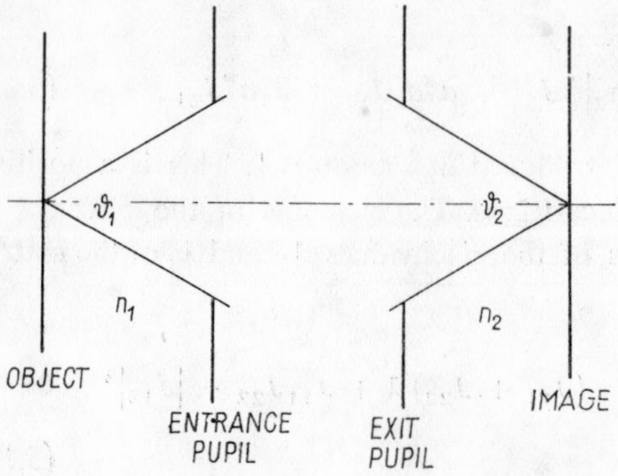

Fig. 5.1. Numerical apertures for entrance and exit pupils.

(n_1, ϑ_1 and n_2, ϑ_2 are the refractive indices and aperture angles in the object and image spaces respectively – Fig. 5.1); we can show by elementary diffraction considerations that the diffraction function $K(x, \xi)$ (we consider a one-dimensional case for simplicity) is a spatial frequency limited function within the domains $(-W, +W) \equiv (-k\alpha/2\pi, +k\alpha/2\pi)$ and $(-k\beta/2\pi, +k\beta/2\pi)$ ([8], p. 480), where k is the mean wave number.

The mutual intensity $J(x_1, x_2)$ may be decomposed into the complete set of functions u_n in the form

$$J(x_1, x_2) = \sum_{n=-\infty}^{+\infty} \sum_{m=-\infty}^{+\infty} A_{nm}\, u_n(k\alpha x_1)\, u_m(k\alpha x_2)\,, \tag{5.22}$$

where, using the orthogonality condition (5.10),

$$A_{nm} = \left(\frac{k\alpha}{\pi}\right)^2 \iint_{-\infty}^{+\infty} J(x_1, x_2)\, u_n(k\alpha x_1)\, u_m(k\alpha x_2)\, dx_1\, dx_2\,. \tag{5.23}$$

As $J^*(x_1, x_2) = J(x_2, x_1)$ the matrix $\hat{A} \equiv (A_{nm})$ is Hermitian, i.e. $A^*_{nm} = A^+_{mn} = A_{mn}$.

The condition of positive semi-definiteness for $J(x_1, x_2)$,

$$\iint_{-\infty}^{+\infty} J(x_1, x_2)\, f^*(x_1)\, f(x_2)\, dx_1\, dx_2 \geqq 0\,, \tag{5.24a}$$

can be transformed to the matrix form

$$\sum_{n=-\infty}^{+\infty} \sum_{m=-\infty}^{+\infty} A_{nm} z^*_n z_m \geqq 0\,, \tag{5.24b}$$

where $z_n = \int_{-\infty}^{+\infty} f(x)\, u_n(k\alpha x)\, dx$ and $f(x)$ is an arbitrary function. Writing the mutual intensity with the help of the Hopkins formula (3.53),†

$$J(x_1, x_2) = \int_\sigma U^*(x_1, \xi)\, U(x_2, \xi)\, d\xi\,, \tag{5.25}$$

where the integration is taken over the surface of the source of light, and using the sampling theorem (5.1) we obtain another form of A_{nm}

$$A_{nm} = \int_\sigma U^*\left(\frac{n\pi}{k\alpha}, \xi\right) U\left(\frac{m\pi}{k\alpha}, \xi\right) d\xi\,. \tag{5.26}$$

† Note that this form of the mutual intensity closely corresponds to the form of the density matrix $\varrho(x_1, x_2)$ in quantum mechanics defined by means of the wave function $\psi(x, q)$ as $\varrho(x_1, x_2) = \int \psi^*(x_1, q)\, \psi(x_2, q)\, dq$.

From this we see that $A_{nm} = J(n\pi/k\alpha,\ m\pi/k\alpha)$ is the mutual intensity between the wave amplitudes at the nth and mth sampling points; $A_{nn} = J(n\pi/k\alpha,\ n\pi/k\alpha)$ is the intensity at nth sampling point. Note that both these quantities are measurable. From $|J_{12}|^2 \leqq J_{11}J_{22}$ it follows that

$$|A_{nm}|^2 \leqq A_{nn}A_{mm}\,. \tag{5.27}$$

From (5.27), summing over n and m,

$$\sum_{n=-\infty}^{+\infty}\sum_{m=-\infty}^{+\infty} A_{nm}A_{mn} \leqq \Big(\sum_{n=-\infty}^{+\infty} A_{nn}\Big)^2\,, \tag{5.28a}$$

where we have used the Hermiticity of $\hat{A}$, $A_{nm}^* = A_{mn}$, and so

$$\mathrm{Tr}\,\hat{A}^2 \leqq (\mathrm{Tr}\,\hat{A})^2\,, \tag{5.28b}$$

where Tr denotes the trace of the matrix. The equality occurs for the coherent field.

The *total intensity* may be obtained from (5.22) by putting $x_1 \equiv x_2 \equiv x$, which gives the intensity $J(x, x)$ as a Hermitian real and non-negative form†, and integrating over x; we obtain

$$I_{\mathrm{tot}} = \int_{-\infty}^{+\infty} J(x, x)\,dx = \frac{\pi}{k\alpha}\,\mathrm{Tr}\,\hat{A}\,, \tag{5.29}$$

where the orthogonality condition (5.10) has also been used. We can also show that

$$\iint_{-\infty}^{+\infty} |J(x_1, x_2)|^2\,dx_1\,dx_2 = \left(\frac{\pi}{k\alpha}\right)^2 \mathrm{Tr}\,\hat{A}^2\,. \tag{5.30}$$

By putting $f(x) \equiv 1$ in (5.24a), with (5.22) substituted, we have

$$\iint_{-\infty}^{+\infty} J(x_1, x_2)\,dx_1\,dx_2 = \left(\frac{\pi}{k\alpha}\right)^2 \sum_{n=-\infty}^{+\infty}\sum_{m=-\infty}^{+\infty} A_{nm} \geqq 0\,, \tag{5.31}$$

where the well-known integral $\int_{-\infty}^{+\infty} \sin(ax)/x\,dx = \pi\,\mathrm{sgn}\,a$ has been used.

Equation (5.22) may be written in the matrix form as a scalar product

$$J(x_1, x_2) = (\phi(x_1), \hat{A}\,\phi(x_2))\,, \tag{5.32}$$

where ϕ is a column vector in the Hilbert space created from u_n. Using the

† Note that the intensity matrix $\hat{A}$ contains the phase information while the intensity $J(x, x)$ does not. From this point of view the intensity matrix $\hat{A} \equiv (A_{nm})$ is a more physical quantity than the intensity $J(x, x)$.

Schwarz inequality we obtain

$$|(\phi(x_1), \hat{A}\,\phi(x_2))|^2 \leqq \|\phi(x_1)\|^2\,\|\hat{A}\,\phi(x_2)\|^2 =$$

$$= \|\hat{A}\,\phi(x_2)\|^2 \leqq \sum_{n=-\infty}^{+\infty}\sum_{m=-\infty}^{+\infty}|A_{nm}|^2 = \mathrm{Tr}\,\hat{A}^2 =$$

$$= \left(\frac{k\alpha}{\pi}\right)^2 \iint_{-\infty}^{+\infty} |J(x_1, x_2)|^2\,dx_1\,dx_2\,, \tag{5.33}$$

where the norms $\|\phi\|$ and $\|\hat{A}\phi\|$ of ϕ and $\hat{A}\phi$ fulfil the following equations:

$$\|\phi\|^2 = (\phi, \phi) = \sum_{n=-\infty}^{+\infty} u_n^2 = 1 \tag{5.34}$$

according to (5.15) and

$$\|\hat{A}\phi\|^2 \leqq \|A\|^2\,\|\phi\|^2 = \sum_{n=-\infty}^{+\infty}\sum_{m=-\infty}^{+\infty}|A_{nm}|^2 = \mathrm{Tr}\,\hat{A}^2\,. \tag{5.35}$$

In (5.33) equation (5.30) has also been used. Equation (5.33) shows that the form (5.22) is bounded and uniformly convergent if

$$\iint_{-\infty}^{+\infty} |J(x_1, x_2)|^2\,dx_1\,dx_2 < \infty\,. \tag{5.36}$$

The quadratic form (5.22) may be transformed to the normal (diagonal) form using standard methods. Let λ_n be the eigenvalues of the matrix $\hat{A}$ determined from the condition

$$\mathrm{Det}\,(\hat{A} - \lambda\hat{1}) = 0\,, \tag{5.37}$$

where $\hat{1}$ is the unit matrix (the λ_n are real and non-negative since $\hat{A}$ is Hermitian and non-negative), and let $\varphi^{(n)}$ be the eigenvectors of the matrix $\hat{A}$ obeying the homogeneous equations

$$\hat{A}\varphi^{(n)} = \lambda_n\varphi^{(n)}\,. \tag{5.38}$$

Define a matrix $\hat{U}$ such that U_{mn} is the mth component of the vector $\varphi^{(n)}$, i.e.

$$\hat{U} \equiv (\varphi^{(1)}, \varphi^{(2)}, \ldots)\,.$$

The matrix $\hat{U}$ is unitary

$$\hat{U}^+\hat{U} = \hat{U}\hat{U}^+ = \hat{1}\,, \tag{5.39}$$

that is

$$\sum_{j=-\infty}^{+\infty} U_{jm}^* U_{jn} = \sum_{j=-\infty}^{+\infty} U_{mj}U_{nj}^* = \delta_{mn}\,, \tag{5.40}$$

which is a more general form of (5.21). Introducing a new vector

$$\psi = \hat{U}^+\phi\,, \quad \phi = \hat{U}\psi \tag{5.41}$$

in (5.32) we arrive at the diagonal form of $J(x_1, x_2)$ in terms of ψ

$$J(x_1, x_2) = (\hat{U}\,\psi(x_1), \hat{A}\hat{U}\,\psi(x_2)) = (\psi(x_1), \hat{U}^+\hat{A}\hat{U}\,\psi(x_2))\,. \tag{5.42}$$

As $\hat{U}^+\hat{A}\hat{U} = \hat{\Lambda}$, where $\hat{\Lambda}$ is the diagonal matrix with elements λ_n, we obtain from (5.42)

$$J(x_1, x_2) = \sum_{n=-\infty}^{+\infty} \lambda_n\,\psi_n^*(x_1)\,\psi_n(x_2)\,, \tag{5.43}$$

which is the required diagonal form of the mutual intensity $J(x_1, x_2)$.

Further we obtain

$$\hat{A} = \hat{U}\hat{\Lambda}\hat{U}^+\,, \tag{5.44a}$$

that is

$$A_{jl} = \sum_{n=-\infty}^{+\infty} \lambda_n U_{jn} U_{ln}^*\,, \tag{5.44b}$$

and from (5.41)

$$\psi_n(x) = \sum_{j=-\infty}^{+\infty} U_{jn}^*\,u_j(k\alpha x)\,,$$

$$u_n(k\alpha x) = \sum_{j=-\infty}^{+\infty} U_{nj}\,\psi_j(x)\,. \tag{5.45}$$

It is obvious that

$$\|\psi\|^2 = (\psi, \psi) = \sum_{n=-\infty}^{+\infty}\sum_{j=-\infty}^{+\infty}\sum_{l=-\infty}^{+\infty} U_{jn}U_{ln}^*\,u_j(k\alpha x)\,u_l(k\alpha x)$$

$$= \sum_{j=-\infty}^{+\infty} u_j^2(k\alpha x) = \|\phi\|^2 = 1\,, \tag{5.46}$$

where we have used (5.40) and (5.34).

Equation (5.44b) may be written in the form

$$\hat{A} = \sum_{n=-\infty}^{+\infty} \lambda_n\,\hat{P}^{(n)}\,, \tag{5.47}$$

where

$$(\hat{P}^{(n)})_{jl} = U_{jn}U_{ln}^* \tag{5.48}$$

is the matrix element of the *projection operator* $\hat{P}^{(n)}$ for which

$$(\hat{P}^{(n)}\hat{P}^{(m)})_{jl} = \sum_r (\hat{P}^{(n)})_{jr}\,(\hat{P}^{(m)})_{rl} = \sum_r U_{jn}U_{rn}^*U_{rm}U_{lm}^* =$$

$$= U_{jn}U_{lm}^*\delta_{nm} = (\hat{P}^{(n)})_{jl}\,\delta_{nm}\,, \tag{5.49a}$$

that is

$$\hat{P}^{(n)}\hat{P}^{(m)} = \hat{P}^{(n)}\delta_{nm}. \tag{5.49b}$$

Further

$$\mathrm{Tr}\,\hat{P}^{(n)} = 1 \tag{5.50}$$

using (5.40).

Some further properties of the mutual intensity may be obtained if it is considered as a kernel of a homogeneous integral equation instead of considering it as a quadratic form in the Hilbert space. Writing $\phi_n(x) = u_n(k\alpha x)$ we obtain, if (5.10), (5.41) [(5.45)] and (5.39) [(5.40)] are used,

$$(\phi_n(x), \phi_m(x)) = \int_{-\infty}^{+\infty} \phi_n(x)\, \phi_m(x)\, dx =$$

$$= (\psi_n(x), \psi_m(x)) = \int_{-\infty}^{+\infty} \psi_n^*(x)\, \psi_m(x)\, dx = \frac{\pi}{k\alpha}\, \delta_{nm}. \tag{5.51}$$

Multiplying (5.43) by $\psi_n(x_1)$ and integrating over x_1 yields

$$\int_{-\infty}^{+\infty} J(x_1, x_2)\, \psi_n(x_1)\, dx_1 = \frac{\pi}{k\alpha}\, \lambda_n\, \psi_n(x_2). \tag{5.52}$$

Therefore $\psi_n(x)$ are eigenfunctions of the kernel $J(x_1, x_2)$ corresponding to the eigenvalues $\pi\lambda_n/k\alpha$. Equation (5.43) represents the decomposition of the mutual intensity – the kernel of the integral equation (5.52) – in terms of the eigenfunctions $\psi_n(x)$ (the Mercer theorem).

If $\lambda_j = 0$ for all $j \neq n$ and $\lambda_n > 0$ (all λ_j must be real, since $J^*(x_1, x_2) = J(x_2, x_1)$, and non-negative, since (5.24a) holds), then we obtain from (5.43)

$$J(x_1, x_2) = \lambda_n\, \psi_n^*(x_1)\, \psi_n(x_2). \tag{5.53}$$

The total intensity (5.29) equals $\pi\lambda_n/k\alpha$ in this case. Thus the mutual intensity is factorized and the field is coherent. A partially coherent field may be regarded, according to (5.43), as the superposition of elementary coherent fields emitted by discrete statistically independent elementary sources. Equation (5.43) can be written in terms of the projection operator

$$P^{(n)}(x_1, x_2) = \psi_n^*(x_1)\, \psi_n(x_2) \tag{5.54}$$

as

$$J(x_1, x_2) = \sum_{n=-\infty}^{+\infty} \lambda_n\, P^{(n)}(x_1, x_2) \tag{5.55}$$

for which, in analogy to (5.49b) and (5.50),

$$\frac{k\alpha}{\pi} \int_{-\infty}^{+\infty} P^{(n)}(x_1, x_2)\, P^{(m)}(x_2, x_3)\, dx_2 = P^{(n)}(x_1, x_3)\, \delta_{nm} \tag{5.56a}$$

and

$$\frac{k\alpha}{\pi}\int_{-\infty}^{+\infty} P^{(n)}(x, x)\, dx = 1\,. \tag{5.56b}$$

5.4 Propagation law of partial coherence in matrix form

It is interesting to formulate the propagation law (3.23b) in the matrix form. We can write

$$J'(x_1, x_2) = \iint_{\mathscr{A}_1} K^*(x_1, \xi_1)\, K(x_2, \xi_2)\, J(\xi_1, \xi_2)\, d\xi_1\, d\xi_2\,, \tag{5.57}$$

where $\Gamma(Q_1, Q_2)$ and $\Gamma(P_1, P_2)$ are denoted as $J'(x_1, x_2)$ and $J(\xi_1, \xi_2)$ respectively and $K(x, \xi) \equiv K(Q, P; \bar{\nu})$; $\mathscr{A}_1$ is an object region in which $J(\xi_1, \xi_2)$ is defined while the image region in which the mutual intensity is $J'(x_1, x_2)$ will be denoted as $\mathscr{A}_2$. The numerical aperture in the image space is β so that the interval of spatial frequencies present in the image space will be $(-k\beta/2\pi, +k\beta/2\pi)$ as we have mentioned. The *transmission matrix* $(K_{nm}) \equiv \hat{K}$ may be defined in an identical way to the intensity matrix. We can write

$$K(x, \xi) = \sum_{n=-\infty}^{+\infty} \sum_{m=-\infty}^{+\infty} K_{nm}\, u_m(k\beta x)\, u_n(k\alpha\xi)\,, \tag{5.58}$$

where

$$K_{nm} = \frac{k^2\alpha\beta}{\pi^2} \int_{\mathscr{A}_2}\int_{\mathscr{A}_1} K(x, \xi)\, u_m(k\beta x)\, u_n(k\alpha\xi)\, dx\, d\xi\,. \tag{5.59}$$

The physical significance of the elements K_{nm} is obtained, similarly to the significance of A_{nm}, by applying (5.11): $K_{nm} = K(m\pi/k\beta, n\pi/k\alpha)$ represents the wave amplitude at mth sampling point in the image plane caused by the unit amplitude at nth sampling point of the object plane.

Substituting (5.58) into (5.57), multiplying (5.57) by $u_m(k\beta x_1)\, u_j(k\beta x_2)$, integrating over x_1 and x_2 under the assumptions that $J \equiv 0$ outside $\mathscr{A}_1$ and $J' \equiv 0$ outside $\mathscr{A}_2$ and introducing the elements of the intensity matrix $\hat{A}$ from (5.23), we arrive at

$$A'_{mj} = \left(\frac{\pi}{k\alpha}\right)^2 \sum_{n=-\infty}^{+\infty} \sum_{l=-\infty}^{+\infty} K^*_{nm} A_{nl} K_{lj}\,, \tag{5.60a}$$

that is

$$\hat{A}' = \left(\frac{\pi}{k\alpha}\right)^2 \hat{K}^+ \hat{A} \hat{K}\,. \tag{5.60b}$$

The solution of the problem of the reconstruction of an object from its image by solving the Fredholm integral equation (5.57) corresponds to the solution of the matrix equation (5.60).

We refer the reader for some further details about the matrix formulation to the review paper [165] by Gamo. We note that a simplified matrix formulation of the coherence theory in the isoplanatic region, in which $K(x, \xi) = K(x - \xi)$, was developed by O'Neill and Asakura [329] (see also [24], Chapter 8).

5.5 The entropy of light beams

Interpreting (5.43) as the incoherent superposition of elementary beams we may introduce the probability p_n that a photon belongs to the nth beam

$$p_n = \frac{\lambda_n}{\sum_n \lambda_n}, \qquad \sum_n p_n = 1, \qquad \lambda_n \geqq 0. \tag{5.61}$$

Further we may introduce the *entropy* as the mean value of the quantity $\log(1/p_n) = -\log p_n$, (which varies in the interval $(0, \infty)$ while p_n varies in the interval $(0, 1)$) using the relation [18]†

$$H = -\sum_n p_n \log p_n . \tag{5.62}$$

This quantity is a measure of the disorder in the field and it also reflects the correlation properties of the field. One can easily show that the minimum of the entropy occurs if all $\lambda_j = 0$ except say $\lambda_n\ (>0)$; then $p_j = 0$, $j \neq n$ and $p_n = 1$ so that $H = H_{\min} = 0$. This is the case of the coherent field. It can also be shown by the method of Lagrange multipliers that the maximum of the entropy occurs if $\lambda_j = 1/N$ $(j = 1, 2, \ldots, N)$ when $H_{\max} = \log N$; this corresponds to the incoherent field. For a partially coherent field the entropy lies between these two values.

Some applications of thermodynamics and the concept of the entropy to problems of interference were considered by Laue [245] and Gamo [165].

† This expression corresponds to the expression for the entropy defined by means of the density matrix $\hat{\varrho}$

$$H = -\operatorname{Tr} \hat{\varrho} \log \hat{\varrho}.$$

5.6 Generally covariant formulation of partial coherence

Another kind of matrix formulation which may be called a generally covariant formulation of partial coherence, developed in [346], makes use of the "quantization" of an object and its image [159] to obtain an interpretation of the degree of coherence as the metric tensor in a non-Euclidian space, which is called the *optical space*. This fact allows a study of partial coherence as a property of the optical space. We consider here real functions and quasi-monochromatic light with time delays $|\tau| \ll 1/\Delta\nu$ only.

Introducing the degree of coherence $\gamma(\xi_1, \xi_2) = J(\xi_1, \xi_2)/[I(\xi_1)\, I(\xi_2)]^{1/2}$ into (5.57), denoting the amplitude $\sqrt{[I(\xi)]}$ as $u(\xi)$, considering $K \equiv 1$ for simplicity and performing the "quantization" of the object we obtain the intensity in the quadratic form

$$I = \sum_{i,j}^{n} \gamma_{ij} u^i u^j \,, \tag{5.63}$$

where $u^i \equiv u(\xi_i)$, ξ_i are points of "quantization" and $\gamma_{ij} \equiv \gamma(u^i, u^j)$. We may define a non-Euclidian space with the metric tensor $\gamma_{ij} = \gamma_{ji}$ in the usual way [26] with the use of the quadratic form for the interval $(ds)^2$:

$$(ds)^2 = \sum_{i,j}^{n} \gamma_{ij}\, du^i\, du^j \,. \tag{5.64}$$

This space is called the optical space. As $I \geqq 0$ this quadratic form is positive semi-definite. Thus $(ds)^2$ may be regarded as the interval in a Riemann space of n dimensions with the metric tensor γ_{ij}.

One can see that for incoherent light, since $\gamma_{ij} = \delta_{ij}$, the optical space is Euclidian and

$$I = \sum_{i=1}^{n} (u^i)^2 \,. \tag{5.65}$$

For coherent light $\gamma_{ij} \equiv 1$ for all i, j and we obtain

$$I = \Big(\sum_{i=1}^{n} u^i\Big)^2 \,. \tag{5.66}$$

Therefore for coherent and partially coherent light the optical space is non-Euclidian.

This interpretation of the degree of coherence as the metric tensor in the non-Euclidian (Riemann) space makes it possible to apply the formalism of Riemann spaces and to study the coherence properties of optical fields as properties of these spaces.

We conclude this chapter by summarizing some basic results.

The necessary and sufficient condition for γ_{ij} to describe incoherent light in a finite region is

$$R^i_{jlm} = 0\,, \tag{5.67}$$

where R^i_{jlm} is the Riemann tensor. The equivalence principle of the general theory of relativity leads to the conclusion that one can always change the amplitudes of the light in such a way that the light will be incoherent (locally).

If we introduce the light tensor T_{ij} characterizing sources and their positions we obtain for γ_{ij} the n-dimensional Einstein equation

$$R_{ij} - \tfrac{1}{2}\gamma_{ij}R = \varkappa T_{ij}\,, \tag{5.68}$$

where R_{ij} is the Ricci tensor and $R = \sum_{i,j}\gamma^{ij}R_{ij}$; γ^{ij} is the contravariant metric tensor and $\varkappa$ is a proportionality constant between geometrical and non-geometrical quantities. Equation (5.68) leads to the wave equation in a linear approximation. If $T_{ij} = 0$ it follows that

$$R_{ij} = 0\,, \tag{5.69}$$

which are equations for $n(n+1)/2$ components of γ_{ij} corresponding to the curvature of the optical space created by the pure light (in absence of sources). With respect to the non-linearity of equation (5.68) partial coherence creates partial coherence.

The tensor T_{ij} is not generally determined completely by sources and their position; it is also dependent on γ_{ij}. Therefore, in solving (5.68), we must use a consistent method. The use of the optical space enables T_{ij} to be constructed from the topological properties of the optical space if the sources are interpreted topologically.

The sources of light create the metric of the space according to (5.68). As a consequence of the non-linearity of this equation the metric causes a "movement" of the sources (variations of the amplitude of the light) according to the equation of motion

$$\frac{d^2u^l}{ds^2} + \sum_{i,j}\Gamma^l_{ij}\frac{du^i}{ds}\frac{du^j}{ds} = 0\,, \tag{5.70}$$

where Γ^l_{ij} are the Christoffel symbols.

One can study the question of variations of the amplitudes of the light which do not change the degree of coherence or on the other hand, one can find the degree of coherence which is unchanged by given variations of the amplitudes. These questions may be solved by using Lie groups. The following theorem holds:

A necessary and sufficient condition for the amplitude transformation

$$u'^i = u^i + \xi^i(u)\,\delta t \tag{5.71}$$

(t is a parameter and ξ^i is a direction vector) not to change the metric of the optical space (i.e. the degree of coherence) is the validity of the condition

$$\underset{L}{\delta}\gamma_{ij} = 0 \tag{5.72}$$

in the direction ξ^i, where $\underset{L}{\delta}$ is the so-called Lie differential.

In this case one says that (5.71) is the group of movements. Equation (5.72) is called the Killing equation and it enables both the problems mentioned above to be solved.

The space of maximal homogeneity is invariant for an $n(n+1)/2$ parametric group of movements (n translations and $\binom{n}{2}$ rotations; this is a group of maximal movability). The optical space of maximal homogeneity is the space of ideally incoherent light. The optical space of partially coherent as well as coherent light is invariant for groups with a smaller number of parameters than $n(n+1)/2$.

Some invariants may be constructed in the optical space from the degree of coherence and its derivatives with the help of the Noether theorem.

For further details concerning the present mathematical formalism we refer the reader to the book of Petrov [26].

Chapter 6

POLARIZATION PROPERTIES OF LIGHT

So far we have considered light beams as scalar quantities making either a scalar approximation to the vector electromagnetic field or taking one component of the vector field. In the following we shall take into account the vector properties of the field. In this chapter we consider the correlation of light beams at one space point with particular attention to the correlation between polarization components of the light beams. A more complete theory including correlations between beams at different space-time points and between different polarization components of the field will be dealt with in the next chapter.

Wiener [465–467] was the first to introduce the *coherence matrix* describing the polarization properties of the electromagnetic field. He showed its connection with the density matrix of quantum mechanics and he also showed that the coherence matrix can be decomposed into matrices (which are now called the Pauli matrices) with the so-called Stokes parameters as the expansion coefficients. Further he proposed an operational procedure for the experimental determination of the Stokes parameters. These questions were later studied by Fano [140], Wolf [471, 476] (see also [8]), Parrent and Roman [337] (see also [3]), Barakat [53] and Marathay [289] (see also [24]). While the papers by Wolf, Parrent and Roman and Marathay treat quasi-monochromatic fields only, results obtained in the papers by Wiener and Barakat are correct for an arbitrary spectral bandwidth of the fields. An analysis, by Pancharatnam [331, 332], has also been made for polychromatic light.

6.1 Definitions of coherence matrix and Stokes parameters

First let us note that the completely monochromatic wave is also *completely polarized*, i.e. the end point of the electric or the magnetic vector moves periodically with time, in general, over an ellipse. If the end point moves completely chaotically, one speaks of *unpolarized light*. However, light is generally in intermediate states of polarization and one speaks of *partially polarized light*.

In all considerations here we use the electric vector $\boldsymbol{E}$ only, but there is no great difficulty in taking into account the magnetic vector $\boldsymbol{H}$, too. We assume stationary plane waves propagating in the direction of the positive z-axis. The complex analytic signals of a typical wave will be $E_x(\boldsymbol{x}, t)$ and $E_y(\boldsymbol{x}, t)$. These are the components of $\boldsymbol{E}$ in two mutually orthogonal directions perpendicular to the z-direction, with (x, y, z) forming a right-handed triad; $E_z = 0$ because of the transversality.

We introduce the *coherence matrix* [471, 476]†

$$\mathscr{J}(\tau) = \langle \mathscr{E}^+(t) \otimes \mathscr{E}(t+\tau) \rangle = \left\langle \begin{pmatrix} E_x^*(t) \\ E_y^*(t) \end{pmatrix} \otimes (E_x(t+\tau), E_y(t+\tau)) \right\rangle =$$

$$= \begin{pmatrix} \langle E_x^*(t)\, E_x(t+\tau)\rangle & \langle E_x^*(t)\, E_y(t+\tau)\rangle \\ \langle E_y^*(t)\, E_x(t+\tau)\rangle & \langle E_y^*(t)\, E_y(t+\tau)\rangle \end{pmatrix} = \begin{pmatrix} \mathscr{J}_{xx}(\tau) & \mathscr{J}_{xy}(\tau) \\ \mathscr{J}_{yx}(\tau) & \mathscr{J}_{yy}(\tau) \end{pmatrix}, \tag{6.1}$$

where $\otimes$ is the direct (Kronecker) product and

$$\mathscr{J}_{ij} = \langle E_i^*(t)\, E_j(t+\tau)\rangle\,, \quad i, j = x, y \tag{6.2}$$

are elements of the coherence matrix and $\mathscr{E} \equiv (E_x, E_y)$ is a row matrix. The explicit dependence on $\boldsymbol{x}$ is suppressed here. Further we introduce the *spectral coherence matrix* [466–468]

$$\mathscr{R}(\nu) = \begin{pmatrix} \mathscr{R}_{xx}(\nu) & \mathscr{R}_{xy}(\nu) \\ \mathscr{R}_{yx}(\nu) & \mathscr{R}_{yy}(\nu) \end{pmatrix} = \lim_{T\to\infty} \frac{1}{2T} \left\langle \begin{pmatrix} \tilde{E}_x^*(\nu) \\ \tilde{E}_y^*(\nu) \end{pmatrix} \otimes (\tilde{E}_x(\nu), \tilde{E}_y(\nu)) \right\rangle_e, \tag{6.3}$$

where

$$\mathscr{R}_{ij}(\nu) = \int_{-\infty}^{+\infty} \mathscr{J}_{ij}(\tau) \exp i2\pi\nu\tau\, d\tau = \lim_{T\to\infty} \frac{1}{2T} \langle \tilde{E}_i^*(\nu)\, \tilde{E}_j(\nu)\rangle_e \tag{6.4}$$

and (2.25) has been used.

From these definitions of $\mathscr{J}(\tau)$ and $\mathscr{R}(\nu)$ it is seen that

$$\mathscr{J}^+(\tau) = \mathscr{J}(-\tau) \tag{6.5a}$$

and

$$\mathscr{R}^+(\nu) = \mathscr{R}(\nu)\,, \tag{6.5b}$$

---

† We note that Wolf's coherence matrix differs slightly from $\mathscr{J}(\tau)$ given by (6.1); his coherence matrix equals $\mathscr{J}^*(0)$.

i.e. the spectral coherence matrix $\mathscr{R}(\nu)$ is Hermitian while the coherence matrix $\mathscr{J}(\tau)$ is Hermitian for $\tau = 0$ only; this may be useful for quasi-monochromatic light if time delays τ involved are restricted by $|\tau| \ll 1/\Delta\nu$.

Writing the matrix $\mathscr{R}(\nu)$ in the form

$$\mathscr{R}(\nu) = \frac{1}{2}\begin{pmatrix} s_0 + s_1 & s_2 + is_3 \\ s_2 - is_3 & s_0 - s_1 \end{pmatrix} \tag{6.6}$$

we see that

$$\begin{aligned} s_0 &= \mathscr{R}_{xx} + \mathscr{R}_{yy}\,, \\ s_1 &= \mathscr{R}_{xx} - \mathscr{R}_{yy}\,, \\ s_2 &= \mathscr{R}_{xy} + \mathscr{R}_{yx}\,, \\ s_3 &= i(\mathscr{R}_{yx} - \mathscr{R}_{xy})\,. \end{aligned} \tag{6.7}$$

These quantities are called the *spectral Stokes parameters* [331].

Further we obtain

$$4 \operatorname{Det} \mathscr{R} = 4\{\mathscr{R}_{xx}\mathscr{R}_{yy} - |\mathscr{R}_{xy}|^2\} = s_0^2 - s_1^2 - s_2^2 - s_3^2 \geqq 0\,, \tag{6.8a}$$

and

$$\operatorname{Tr} \mathscr{R} = \mathscr{R}_{xx} + \mathscr{R}_{yy} = s_0 \geqq 0\,. \tag{6.8b}$$

Therefore the trace of $\mathscr{R}$ equals the spectral intensity $\tilde{I}(\nu)$. The property (6.8a) of the Stokes parameters suggests the use of the Lorentz group of transformations to study the polarization properties of light [53].

Introducing the Pauli matrices

$$\sigma_0 = \begin{pmatrix} 1 & 0 \\ 0 & 1 \end{pmatrix} \quad \sigma_1 = \begin{pmatrix} 1 & 0 \\ 0 & -1 \end{pmatrix}, \quad \sigma_2 = \begin{pmatrix} 0 & 1 \\ 1 & 0 \end{pmatrix}, \quad \sigma_3 = \begin{pmatrix} 0 & i \\ -i & 0 \end{pmatrix} \tag{6.9}$$

for which

$$\sigma_\alpha\sigma_\beta = -\sigma_\beta\sigma_\alpha = -i\sigma_\gamma\,, \quad (\alpha, \beta, \gamma) = (1, 2, 3) \text{ and cycl}\,, \tag{6.10a}$$

$$(\sigma_j)^2 = \sigma_0\,, \qquad j = 0, 1, 2, 3\,, \tag{6.10b}$$

$$\sigma_j\sigma_0 = \sigma_0\sigma_j = \sigma_j\,, \qquad j = 0, 1, 2, 3\,, \tag{6.10c}$$

$$\operatorname{Tr}(\sigma_i\sigma_j) = 2\delta_{ij}\,, \qquad i, j = 0, 1, 2, 3\,, \tag{6.10d}$$

we can write the spectral coherence matrix in the form

$$\mathscr{R} = \tfrac{1}{2}\sum_{i=0}^{3} s_i\sigma_i\,. \tag{6.11}$$

Multiplying this relation by σ_j, taking the trace and using (6.10d) we arrive at

$$s_j = \mathrm{Tr}\,(\sigma_j \mathscr{R})\,, \tag{6.12}$$

which is a compact form of (6.7).

An identical decomposition to (6.11) holds for the coherence matrix $\mathscr{J}(0)$ (assuming quasi-monochromatic light and $|\tau| \ll 1/\Delta\nu$) in terms of the *Stokes parameters* $S_j(0) \equiv S_j$ defined by

$$S_j(\tau) = \int_0^\infty s_j(\nu) \exp\,(-i2\pi\nu\tau)\, d\nu\,. \tag{6.13}$$

6.2 Matrix description of non-image-forming optical elements

Some optical devices not forming an image, such as a compensator or polarizer, may be described by a matrix $\mathscr{L}(\nu)$ depending on the properties of the device only; this matrix relates the emerging spectral field $\mathscr{F}'(\nu)$ to the incident spectral field $\mathscr{F}(\nu)$

$$\mathscr{F}' = \mathscr{F}\mathscr{L}\,. \tag{6.14}$$

The spectral coherence matrix behind the device is given, using (6.3), by

$$\mathscr{R}'(\nu) = \lim_{T\to\infty} \frac{1}{2T} \langle \mathscr{F}'^{+}(\nu) \otimes \mathscr{F}'(\nu)\rangle_e =$$

$$= \lim_{T\to\infty} \frac{1}{2T} \langle \mathscr{L}^{+}\mathscr{F}^{+} \otimes \mathscr{F}\mathscr{L}\rangle_e = \mathscr{L}^{+}(\nu)\,\mathscr{R}(\nu)\,\mathscr{L}(\nu)\,. \tag{6.15}$$

The spectral intensity $\tilde{I}'(\nu)$ behind the device is

$$\tilde{I}'(\nu) = \mathrm{Tr}\,\mathscr{R}'(\nu) = \mathrm{Tr}\,\{\mathscr{L}^{+}(\nu)\,\mathscr{R}(\nu)\,\mathscr{L}(\nu)\} = \mathrm{Tr}\,\{\mathscr{R}\mathscr{L}\mathscr{L}^{+}\} \tag{6.16}$$

since a cyclic permutation does not change the trace.

It is not difficult to specialize these general results to quasi-monochromatic light under the assumption that $|\tau| \ll 1/\Delta\nu$. In this case the coherence matrix $\mathscr{J}(0) \equiv \mathscr{J}$ must be used instead of the spectral coherence matrix $\mathscr{R}(\nu)$ [476, 337, 8, 285]. It is seen from (6.5a) that the coherence matrix $\mathscr{J}$ is Hermitian in this case and consequently equations (6.6), (6.7), (6.8), (6.11), (6.12), (6.14), (6.15) and (6.16) hold if the spectral quantities are replaced by their corresponding Fourier transform quantities; if $\tau = 0$

$$\begin{aligned} &\mathscr{R}(\nu) \to \mathscr{J}(0), \quad (\tilde{I}(\nu) \to I = \mathrm{Tr}\,\mathscr{J}), \\ &s_j(\nu) \to S_j(0), \\ &\mathscr{F}(\nu) \to \mathscr{E}(t). \end{aligned} \tag{6.17}$$

The device matrix $\mathscr{L}(\nu)$ must be replaced by $\mathscr{L}(\bar{\nu})$, i.e. this matrix must be taken at the mean frequency. This case was treated for example in [3].

Let us now consider some special kinds of optical devices.

Compensator

When the radiation passes through the compensator, the phases of the $\tilde{E}_x$- and $\tilde{E}_y$-components of the field are changed to $\varepsilon_x(\nu)$ and $\varepsilon_y(\nu)$ respectively. Introducing the difference $\delta(\nu) = \varepsilon_x(\nu) - \varepsilon_y(\nu)$ we may describe this device by the matrix

$$\mathscr{L}_c = \begin{pmatrix} \exp(i\delta/2) & 0 \\ 0 & \exp(-i\delta/2) \end{pmatrix}. \tag{6.18}$$

This matrix is unitary so that from (6.15)

$$\mathscr{R}' = \mathscr{L}_c^{-1}\mathscr{R}\mathscr{L}_c \tag{6.19a}$$

and

$$\tilde{I}'_c = \tilde{I}_c, \tag{6.19b}$$

since a unitary transform does not change the trace (there is no absorption and no reflection).

Absorber

This device is described by the matrix

$$\mathscr{L}_a = \begin{pmatrix} \exp(-\eta_x/2) & 0 \\ 0 & \exp(-\eta_y/2) \end{pmatrix}, \tag{6.20}$$

where η_x and η_y are real non-negative constants. The spectral coherence matrix $\mathscr{R}'$ is given by (6.15) and the spectral intensity equals

$$\tilde{I}'_a = \mathrm{Tr}\,\{\mathscr{R}\mathscr{L}_a^2\}. \tag{6.21}$$

Rotator

A device which rotates the electric vector through an angle α is described by the real unimodular and unitary matrix

$$\mathscr{L}_r = \begin{pmatrix} \cos\alpha & \sin\alpha \\ -\sin\alpha & \cos\alpha \end{pmatrix} \tag{6.22}$$

and we obtain

$$\mathscr{R}' = \mathscr{L}_r^{-1}\mathscr{R}\mathscr{L}_r = \mathscr{L}_r(-\alpha)\,\mathscr{R}\mathscr{L}_r(\alpha) \tag{6.23a}$$

and

$$\tilde{I}'_r = \tilde{I}_r\,. \tag{6.23b}$$

Polarizer

A polarizer is a device projecting the electric field onto a direction ϑ and it is characterized by a projection operator $\mathscr{L}_p^{(+)}(\vartheta)$ fulfilling the idempotency condition

$$\mathscr{L}_p^{(+)}(\vartheta)\,\mathscr{L}_p^{(+)}(\vartheta) = \mathscr{L}_p^{(+)}(\vartheta)\,. \tag{6.24a}$$

The conjugate operator $\mathscr{L}_p^{(-)}(\vartheta)$ projecting the field onto the direction perpendicular to ϑ satisfies a condition of the same form and both these operators fulfil the following conditions

$$\mathscr{L}_p^{(+)}\mathscr{L}_p^{(-)} = \mathscr{L}_p^{(-)}\mathscr{L}_p^{(+)} = 0\,, \tag{6.24b}$$

$$\mathscr{L}_p^{(+)} + \mathscr{L}_p^{(-)} = \hat{1}\,. \tag{6.24c}$$

The projecting operators can be represented by the Hermitian matrices

$$\mathscr{L}_p^{(+)}(\vartheta) = \begin{pmatrix} \cos^2\vartheta & \cos\vartheta\sin\vartheta \\ \sin\vartheta\cos\vartheta & \sin^2\vartheta \end{pmatrix}, \tag{6.25a}$$

$$\mathscr{L}_p^{(-)}(\vartheta) = \begin{pmatrix} \sin^2\vartheta & -\sin\vartheta\cos\vartheta \\ -\cos\vartheta\sin\vartheta & \cos^2\vartheta \end{pmatrix}. \tag{6.25b}$$

It is obvious that $\mathscr{L}_p^{(-)}(\vartheta) = \mathscr{L}_p^{(+)}(\vartheta + \pi/2)$. The spectral coherence matrix transforms according to

$$\mathscr{R}' = \mathscr{L}_p^{(+)}\mathscr{R}\mathscr{L}_p^{(+)} \tag{6.26a}$$

since $\mathscr{L}_p^{(+)+} = \mathscr{L}_p^{(+)}$ and for the spectral intensity we obtain

$$\tilde{I}'_p = \mathrm{Tr}\,\{\mathscr{R}\mathscr{L}_p^{(+)}\} \tag{6.26b}$$

if (6.24a) is taken into account.

6.3 Interference law

Multiplying such matrices successively we may describe cascade systems. As an important example, enabling us to give an operational procedure for defining the elements of the coherence matrix and the Stokes parameters, we consider a combination of a compensator followed by a polarizer. We obtain from (6.18) and (6.25a)

$$\mathscr{L} = \mathscr{L}_c\mathscr{L}_p^{(+)} = \begin{pmatrix} \exp(i\delta/2) & 0 \\ 0 & \exp(-i\delta/2) \end{pmatrix} \cdot \begin{pmatrix} \cos^2\vartheta & \cos\vartheta\sin\vartheta \\ \sin\vartheta\cos\vartheta & \sin^2\vartheta \end{pmatrix} =$$

$$= \begin{pmatrix} \exp(i\delta/2)\cos^2\vartheta, & \exp(i\delta/2)\cos\vartheta\sin\vartheta \\ \exp(-i\delta/2)\sin\vartheta\cos\vartheta, & \exp(-i\delta/2)\sin^2\vartheta \end{pmatrix} \tag{6.27}$$

and so

$$\tilde{I}(\nu) = \mathrm{Tr}\,\{\mathscr{L}_p^{(+)+}\mathscr{L}_c^+\mathscr{R}\mathscr{L}_c\mathscr{L}_p^{(+)}\} = \mathrm{Tr}\,\{\mathscr{R}\mathscr{L}_c\mathscr{L}_p^{(+)}\mathscr{L}_c^+\} =$$

$$= \mathrm{Tr}\left\{\begin{pmatrix} \mathscr{R}_{xx} & \mathscr{R}_{xy} \\ \mathscr{R}_{yx} & \mathscr{R}_{yy} \end{pmatrix} \begin{pmatrix} \exp(i\delta/2)\cos^2\vartheta & \exp(i\delta/2)\cos\vartheta\sin\vartheta \\ \exp(-i\delta/2)\sin\vartheta\cos\vartheta & \exp(-i\delta/2)\sin^2\vartheta \end{pmatrix}\right. .$$

$$\left. \cdot \begin{pmatrix} \exp(-i\delta/2) & 0 \\ 0 & \exp(i\delta/2) \end{pmatrix}\right\} =$$

$$= \mathscr{R}_{xx}\cos^2\vartheta + \mathscr{R}_{yy}\sin^2\vartheta + \mathscr{R}_{xy}\exp(-i\delta)\sin\vartheta\cos\vartheta + \mathscr{R}_{yx}\exp(i\delta)\,.$$

$$.\sin\vartheta\cos\vartheta =$$

$$= \mathscr{R}_{xx}\cos^2\vartheta + \mathscr{R}_{yy}\sin^2\vartheta + 2\sin\vartheta\cos\vartheta\,\mathrm{Re}\,\{\mathscr{R}_{xy}\exp(-i\delta)\}\,. \tag{6.28}$$

Denoting

$$\tilde{I}^{(1)}(\nu) = \mathscr{R}_{xx}(\nu)\cos^2\vartheta\,, \quad \tilde{I}^{(2)}(\nu) = \mathscr{R}_{yy}(\nu)\sin^2\vartheta \tag{6.29a}$$

and

$$\mu_{xy}(\nu) = |\mu_{xy}(\nu)|\exp i\beta_{xy}(\nu) = \frac{\mathscr{R}_{xy}(\nu)}{[\mathscr{R}_{xx}(\nu)\,\mathscr{R}_{yy}(\nu)]^{1/2}} \tag{6.29b}$$

we can rewrite (6.28) in the form [332]

$$\tilde{I}(\nu) = \tilde{I}^{(1)}(\nu) + \tilde{I}^{(2)}(\nu) + 2[\tilde{I}^{(1)}(\nu)\,\tilde{I}^{(2)}(\nu)]^{1/2}\,|\mu_{xy}(\nu)|\cos[\beta_{xy}(\nu) - \delta], \tag{6.30}$$

which is the spectral interference law; its form is very similar to the form of the interference law (3.8b) for partial coherence.

Considering quasi-monochromatic light and time delays restricted by $|\tau| \ll 1/\Delta\nu$ and also taking all device quantities in (6.30) or (6.28) at the mean frequency we obtain, by means of a Fourier transform,

$$I(\vartheta, \delta) = \mathscr{J}_{xx} \cos^2 \vartheta + \mathscr{J}_{yy} \sin^2 \vartheta + 2 \sin \vartheta \cos \vartheta \operatorname{Re} \{\mathscr{J}_{xy} \exp(-i\delta)\} \tag{6.31a}$$

$$= I^{(1)} + I^{(2)} + 2[I^{(1)}I^{(2)}]^{1/2} |\gamma_{xy}| \cos(\beta_{xy} - \delta), \tag{6.31b}$$

where the correlation between the x and y components of the electric field is expressed by

$$\gamma_{xy} = |\gamma_{xy}| \exp i\beta_{xy} = \frac{\mathscr{J}_{xy}}{[\mathscr{J}_{xx}\mathscr{J}_{yy}]^{1/2}}. \tag{6.32}$$

This is Wolf's interference law for partial polarization [476, 8, 285], which is very similar to the interference law (3.8b) for quasi-monochromatic partially coherent light beams. The correlation coefficient γ_{xy} plays the role of the second-order degree of coherence $\gamma(\boldsymbol{x}_1, \boldsymbol{x}_2; 0)$ of the scalar theory and it expresses the correlation between the components of the electric field at a particular space point. From (6.8a) and the corresponding relation between the elements of $\mathscr{J}$ and S_j we obtain

$$0 \leqq |\mu_{xy}(\nu)| \leqq 1, \tag{6.33a}$$

$$0 \leqq |\gamma_{xy}| \quad \leqq 1. \tag{6.33b}$$

These inequalities correspond to $|\gamma(\boldsymbol{x}_1, \boldsymbol{x}_2; \tau)| \leqq 1$.

It can be shown from (6.31a) that

$$I_{\substack{\max \\ \min}(\vartheta,\delta)} = \frac{1}{2} \operatorname{Tr} \mathscr{J} \left[1 \pm \sqrt{\left(1 - \frac{4 \operatorname{Det} \mathscr{J}}{(\operatorname{Tr} \mathscr{J})^2}\right)}\right] \tag{6.34}$$

and the visibility of fringes equals

$$\frac{I_{\max} - I_{\min}}{I_{\max} + I_{\min}} = \sqrt{\left(1 - \frac{4 \operatorname{Det} \mathscr{J}}{(\operatorname{Tr} \mathscr{J})^2}\right)}. \tag{6.35}$$

This quantity depends on the rotational invariants Tr $\mathscr{J}$ and Det $\mathscr{J}$ only and so is invariant during a rotation of the (x, y)-system of coordinates.

6.4 Unpolarized and polarized light

For unpolarized (natural) light, $I(\vartheta, \delta)$ is a constant and consequently

$$|\gamma_{xy}| = 0\,, \quad \mathscr{J}_{xx} = \mathscr{J}_{yy}\,, \tag{6.36}$$

i.e. E_x and E_y are mutually incoherent. Therefore $\mathscr{J}_{xy} = \mathscr{J}_{yx} = 0$ and the coherence matrix has the form

$$\mathscr{J} = \tfrac{1}{2}I \begin{pmatrix} 1 & 0 \\ 0 & 1 \end{pmatrix}. \tag{6.37}$$

Consider next coherent light; in this case we have

$$\mathscr{J} = \begin{pmatrix} |E_x|^2 & |E_x E_y| \exp i\delta \\ |E_x E_y| \exp(-i\delta) & |E_y|^2 \end{pmatrix}, \tag{6.38}$$

where $\delta = \arg E_y - \arg E_x$. Therefore

$$\operatorname{Det} \mathscr{J} = 0 \tag{6.39}$$

and

$$\gamma_{xy} = \exp i\delta\,, \tag{6.40a}$$

that is

$$|\gamma_{xy}| = 1 \tag{6.40b}$$

and so the light is also completely polarized. The matrix (6.38) can also be specialized for linearly and circularly polarized light [8].

Let us now show that every coherence matrix may be decomposed into a sum of the coherence matrices for completely polarized and unpolarized light:

$$\mathscr{J} = \mathscr{J}_p + \mathscr{J}_u\,, \tag{6.41}$$

where

$$\mathscr{J}_u = \begin{pmatrix} A & 0 \\ 0 & A \end{pmatrix}, \quad \mathscr{J}_p = \begin{pmatrix} B & D \\ D^* & C \end{pmatrix} \tag{6.42}$$

with A, B, C real and non-negative and

$$BC - |D|^2 = 0\,. \tag{6.43}$$

It follows that

$$A + B = \mathscr{J}_{xx}\,, \qquad D = \mathscr{J}_{xy}\,,$$
$$D^* = \mathscr{J}_{yx}\,, \quad A + C = \mathscr{J}_{yy} \tag{6.44}$$

and from (6.43) we obtain

$$(\mathcal{J}_{xx} - A)(\mathcal{J}_{yy} - A) - \mathcal{J}_{xy}\mathcal{J}_{yx} = \begin{vmatrix} \mathcal{J}_{xx} - A & \mathcal{J}_{xy} \\ \mathcal{J}_{yx} & \mathcal{J}_{yy} - A \end{vmatrix} = 0, \quad (6.45)$$

i.e. A is the eigenvalue of the coherence matrix $\mathcal{J}$ and so

$$A_{1,2} = \frac{1}{2} \operatorname{Tr} \mathcal{J} \left[1 \pm \sqrt{\left(1 - \frac{4 \operatorname{Det} \mathcal{J}}{(\operatorname{Tr} \mathcal{J})^2}\right)} \right], \quad (6.46)$$

which equals (6.34). As the following inequalities hold

$$\operatorname{Det} \mathcal{J} \leqq \mathcal{J}_{xx}\mathcal{J}_{yy} \leqq \tfrac{1}{4}(\operatorname{Tr} \mathcal{J})^2, \quad (6.47)$$

both these roots are real and non-negative. The quantities B, C, D are determined from (6.44). Taking the negative sign in (6.46) (the positive sign leads to negative values of B and C) we obtain

$$\begin{aligned} A &= \tfrac{1}{2}(\mathcal{J}_{xx} + \mathcal{J}_{yy}) - \tfrac{1}{2}[(\operatorname{Tr} \mathcal{J})^2 - 4 \operatorname{Det} \mathcal{J}]^{1/2}, \\ B &= \tfrac{1}{2}(\mathcal{J}_{xx} - \mathcal{J}_{yy}) + \tfrac{1}{2}[(\operatorname{Tr} \mathcal{J})^2 - 4 \operatorname{Det} \mathcal{J}]^{1/2}, \\ C &= \tfrac{1}{2}(\mathcal{J}_{yy} - \mathcal{J}_{xx}) + \tfrac{1}{2}[(\operatorname{Tr} \mathcal{J})^2 - 4 \operatorname{Det} \mathcal{J}]^{1/2}, \\ D &= \mathcal{J}_{xy}, \quad D^* = \mathcal{J}_{yx}; \end{aligned} \quad (6.48)$$

$B, C \geqq 0$ since

$$[(\operatorname{Tr} \mathcal{J})^2 - 4 \operatorname{Det} \mathcal{J}]^{1/2} = [(\mathcal{J}_{xx} - \mathcal{J}_{yy})^2 + 4|\mathcal{J}_{xy}|^2]^{1/2} \geqq |\mathcal{J}_{xx} - \mathcal{J}_{yy}|. \quad (6.49)$$

In this sense the decomposition (6.41) is unique.

6.5 Degree of polarization

The *degree of polarization* is defined as

$$P = \frac{\operatorname{Tr} \mathcal{J}_p}{\operatorname{Tr} \mathcal{J}_p + \operatorname{Tr} \mathcal{J}_u} = \frac{B + C}{B + C + 2A} = \sqrt{\left(1 - \frac{4 \operatorname{Det} \mathcal{J}}{(\operatorname{Tr} \mathcal{J})^2}\right)} = $$

$$= \left|\frac{A_1 - A_2}{A_1 + A_2}\right| = \frac{\sqrt{(S_1^2 + S_2^2 + S_3^2)}}{S_0} \quad (6.50)$$

and so it equals the visibility (6.35). Here we have used (6.48), (6.46) and the corresponding relations to (6.8a) and (6.8b) for the Stokes parameters S_j.

We see that the degree of polarization is rotationally invariant and from (6.47)

$$0 \leqq P \leqq 1 . \tag{6.51}$$

For $P = 1$, Det $\mathscr{J} = 0$ and $|\gamma_{xy}| = 1$; the light is *completely polarized.* For $P = 0$, $(\mathscr{J}_{xx} - \mathscr{J}_{yy})^2 + 4|\mathscr{J}_{xy}|^2 = 0$ (i.e. $\mathscr{J}_{xx} = \mathscr{J}_{yy}$ and $\mathscr{J}_{xy} = \mathscr{J}_{yx} = 0$) and the light is *completely unpolarized* ($\gamma_{xy} = 0$). In intermediate cases $0 < P < 1$ one speaks of *partially polarized light.*

It may be shown [8, 337] that

$$P \geqq |\gamma_{xy}| \tag{6.52}$$

and that

$$P = \mathrm{Max}\, |\gamma_{xy}| \,. \tag{6.53}$$

The angle of rotation φ of the system (x, y) to reach this maximum is given by

$$\tan 2\varphi = \frac{\mathscr{J}_{yy} - \mathscr{J}_{xx}}{\mathscr{J}_{xy} + \mathscr{J}_{yx}} . \tag{6.54}$$

In this new system of co-ordinates (x', y'), $\mathscr{J}_{x'x'} = \mathscr{J}_{y'y'} = I/2$ holds.

6.6 Probability interpretation of the eigenvalues of the coherence matrix

In analogy to the case of partial coherence and the probability interpretation of the eigenvalues of the intensity matrix (Sec. 5.5) we can give a similar interpretation to the eigenvalues A_1 and A_2 of the coherence matrix. Writing

$$\frac{1}{A_1 + A_2}\begin{pmatrix} A_1 & 0 \\ 0 & A_2 \end{pmatrix} = \frac{1}{A_1 + A_2}\begin{pmatrix} A_1 & 0 \\ 0 & 0 \end{pmatrix} + \frac{1}{A_1 + A_2}\begin{pmatrix} 0 & 0 \\ 0 & A_2 \end{pmatrix} \tag{6.55}$$

we see that the coherence matrix may be written as the incoherent superposition of the coherent matrices of completely polarized light. We may say that the probability p_1 that a photon belongs to the x-direction is $p_1 = A_1/(A_1 + A_2)$ and the probability that it belongs to the y-direction is $p_2 = A_2/(A_1 + A_2)$. The entropy will have its minimum if only one of the numbers A_1 and A_2 is not zero, which is the case of completely polarized light; it will have its maximum if $A_1 = A_2$, which is the case of completely unpolarized light. The entropy of partially polarized light will take on values between these two cases.

The eigenvalues A_1 and A_2 may be expressed in terms of P in the form

$$A_{1,2} = \tfrac{1}{2}I(1 \pm P). \tag{6.56}$$

6.7 Operational definition of coherence matrix and Stokes parameters

The elements of the coherence matrix may be defined operationally from the interference law (6.31a)

$$\mathscr{I}_{xx} = I(0, 0), \quad \mathscr{I}_{yy} = I(\pi/2, 0),$$

$$\mathscr{I}_{xy} = \tfrac{1}{2}\{I(\pi/4, 0) - I(3\pi/4, 0)\} + \frac{i}{2}\{I(\pi/4, \pi/2) - I(3\pi/4, \pi/2)\},$$

$$\mathscr{I}_{yx} = \mathscr{I}^*_{xy}. \tag{6.57}$$

The Stokes parameters S_j may be defined operationally by means of the elements of the coherence matrix, since both these methods are equivalent for the description of the second-order statistical properties of optical fields. They represent a complete description for Gaussian light, for which knowledge of the second-order moment is sufficient. For non-Gaussian light moments of all orders are necessary for a complete description.

6.8 Analogy of coherence matrix and density matrix of quantum mechanics

The above equations show that in general a measurable quantity M may be represented by the trace relation

$$M = \mathrm{Tr}\,\{\hat{M}\mathscr{I}\}, \tag{6.58}$$

where the coherence matrix $\mathscr{I}$ describes the state of the field and the operator $\hat{M}$ describes the system. This equation is completely analogous to the corresponding quantum-mechanical equations giving the expectation values of observable physical quantities (with the density matrix $\hat{\varrho}$ replacing the coherence matrix $\mathscr{I}$). The transformation equations for $\mathscr{R}$ [(6.15)] and $\mathscr{I}$ are also analogous to the corresponding equation for the density matrix. However, even if there is a very close analogy between $\mathscr{I}$ and $\hat{\varrho}$, this analogy is not complete [337]. The similarity between $\mathscr{I}$ and $\hat{\varrho}$ has been investigated on the basis of the interference law in [348]. The analogy between the coherence matrix and the density matrix was extensively investigated by Fano [141], ter Haar [440], Parrent and Roman [337] and Gamo [165] among others.

A generalization of the Stokes parameters for a quasi-monochromatic but not plane waves was considered by Roman [403]. In this case the coherence matrix possesses (3×3) elements and it may be decomposed in terms of nine linearly independent (3×3)-matrices forming a complete system. The coherence matrix for N beams has been considered by Barakat [53].

Chapter 7

FIELD EQUATIONS AND CONSERVATION LAWS FOR CORRELATION TENSORS

The scalar theory of coherence phenomena developed in Chapters 1–5 is applicable to linearly polarized light and it may serve as an approximate description of the coherence properties of light as we have already pointed out. Chapter 6 was devoted to correlations between the different polarization components of the field; in this chapter we took into account the vectorial properties of the electromagnetic field, but the coincidence of the space points was assumed there. We now introduce a system of correlation tensors which provide a mathematical as well as physical framework for a unified treatment of coherence and polarization phenomena.

The correlation tensors were introduced into the theory of coherence by Wolf [471, 473] and they have been extensively studied by Roman and Wolf [407, 408], Roman [404, 405], Kano [221], Beran and Parrent [70, 3] and Horák [204], using a classical non-relativistic formulation, and by Kujawski [239] and Dialetis [128], using a relativistic formulation. Analogous studies of the quantum correlation tensors were carried out by Mehta and Wolf [308, 309] and by Horák [205, 206]. Some applications of this method were proposed by Germey [168] and Karczewski [227, 228] (to problems of interference and diffraction) and by Bourret [88], Mehta and Wolf [306, 307, 310] and Brevik and Suhonen [498, 499] (to the study of the coherence properties of black-body radiation).

7.1 Definition of correlation tensors and their properties

We introduce four *correlation tensors* $\mathscr{E}$, $\mathscr{H}$, $\mathscr{M}$ and $\mathscr{N}$ ordered in the matrix $\mathscr{K}$ as follows

$$\mathscr{K}_{jk}(x_1, x_2) = \left\langle \begin{pmatrix} E_j^*(x_1) \\ H_j^*(x_1) \end{pmatrix} \otimes (E_k(x_2), H_k(x_2)) \right\rangle =$$

$$= \begin{pmatrix} \mathscr{E}_{jk}(x_1, x_2) & \mathscr{M}_{jk}(x_1, x_2) \\ \mathscr{N}_{jk}(x_1, x_2) & \mathscr{H}_{jk}(x_1, x_2) \end{pmatrix}, \tag{7.1}$$

where E_j and H_j are the complex analytic signals of the components of the electric and magnetic field respectively and

$$\begin{aligned} \mathscr{E}_{jk}(x_1, x_2) &= \langle E_j^*(x_1)\ E_k(x_2)\rangle, \quad j, k = 1, 2, 3, \\ \mathscr{H}_{jk}(x_1, x_2) &= \langle H_j^*(x_1)\ H_k(x_2)\rangle, \\ \mathscr{M}_{jk}(x_1, x_2) &= \langle E_j^*(x_1)\ H_k(x_2)\rangle, \\ \mathscr{N}_{jk}(x_1, x_2) &= \langle H_j^*(x_1)\ E_k(x_2)\rangle. \end{aligned} \tag{7.2}$$

Of course, interpreting the brackets here as the average using the density matrix with E_j and H_j represented by their corresponding operators, we may define the same system of quantum correlation tensors (cf. Sec. 12.1). Relations of the type (2.26) and (2.28)–(2.31) hold for the correlation tensors, and the higher-order correlation tensors, which may be ordered in a matrix $\mathscr{K}^{(m,n)}$, can also be defined and some of their properties can be studied [205]; but since their structure is rather complicated it will suffice to demonstrate the method of correlation tensors using the second-order tensors.

First we see from (7.1) that

$$\mathscr{K}_{jk}^{+}(x_1, x_2) = \mathscr{K}_{kj}(x_2, x_1) \tag{7.3}$$

and for later convenience we define the tensors

$$U_{jk}(x_1, x_2) = \mathscr{E}_{jk}(x_1, x_2) + \mathscr{H}_{jk}(x_1, x_2) = \operatorname{Tr} \mathscr{K}_{jk}(x_1, x_2), \tag{7.4a}$$

$$S_{jk}(x_1, x_2) = \mathscr{M}_{jk}(x_1, x_2) - \mathscr{N}_{jk}(x_1, x_2) = \operatorname{Tr}\{\hat{\sigma}\, \mathscr{K}_{jk}(x_1, x_2)\}, \tag{7.4b}$$

where

$$\hat{\sigma} = \begin{pmatrix} 0 & -1 \\ 1 & 0 \end{pmatrix}. \tag{7.5}$$

With the tensor U_{jk} we can associate the scalar U and the vector $\boldsymbol{U}(U_1, U_2, U_3)$ defined as follows

$$U(x_1, x_2) = \sum_{k=1}^{3} U_{kk}(x_1, x_2) = \langle \boldsymbol{E}^*(x_1) \cdot \boldsymbol{E}(x_2)\rangle + \langle \boldsymbol{H}^*(x_1) \cdot \boldsymbol{H}(x_2)\rangle, \tag{7.6a}$$

$$\begin{aligned} U_i(x_1, x_2) &= \sum_{j,k=1}^{3} \varepsilon_{ijk}\, U_{jk}(x_1, x_2) = \\ &= \{\langle \boldsymbol{E}^*(x_1) \times \boldsymbol{E}(x_2)\rangle + \langle \boldsymbol{H}^*(x_1) \times \boldsymbol{H}(x_2)\rangle\}_i, \end{aligned} \tag{7.6b}$$

where . and × denote the scalar and the vector products respectively and ε_{ijk} is the Levi-Civita unit antisymmetric tensor ($\varepsilon_{ijk} = 1\,(-1)$ if (i, j, k) is an even (odd) permutation of (1, 2, 3) and $\varepsilon_{ijk} = 0$ if at least two indices are the same). We may also associate a scalar and a vector with the tensor S_{jk} in the same way

$$S(x_1, x_2) = \sum_{k=1}^{3} S_{kk}(x_1, x_2) = \langle \boldsymbol{E}^*(x_1) \,.\, \boldsymbol{H}(x_2)\rangle - \langle \boldsymbol{H}^*(x_1) \,.\, \boldsymbol{E}(x_2)\rangle\,, \quad (7.7a)$$

$$S_i(x_1, x_2) = \sum_{j,k=1}^{3} \varepsilon_{ijk}\, S_{jk}(x_1, x_2) = $$
$$= \{\langle \boldsymbol{E}^*(x_1) \times \boldsymbol{H}(x_2)\rangle - \langle \boldsymbol{H}^*(x_1) \times \boldsymbol{E}(x_2)\rangle\}_i\,. \quad (7.7b)$$

One can see that there is a simple physical meaning to the quantities (7.6a) and (7.7b). The first one multiplied by $1/4\pi$ and considered for $x_1 \equiv x_2$ ($\boldsymbol{x}_1 \equiv \boldsymbol{x}_2$, $t_1 = t_2$) represents the expectation value of the electromagnetic energy density; the second one multiplied by $c/4\pi$ or $1/4\pi c$ for $x_1 \equiv x_2$ represents the component of the Poynting vector or the component of the field momentum density *in vacuo*.

Some further tensors may be defined as follows

$$T_{jk}(x_1, x_2) = U_{jk}(x_1, x_2) + U_{kj}(x_1, x_2) - \delta_{jk} \sum_{i=1}^{3} U_{ii}(x_1, x_2)\,, \quad (7.8a)$$

$$Q_{jk}(x_1, x_2) = S_{jk}(x_1, x_2) + S_{kj}(x_1, x_2) - \delta_{jk} \sum_{i=1}^{3} S_{ii}(x_1, x_2)\,. \quad (7.8b)$$

The tensor $(1/4\pi)\; T_{jk}(x_1, x_2)|_{x_1 \equiv x_2}$ represents the Maxwell stress tensor of the electromagnetic field. If the quantum tensors are considered a vacuum contribution occurs as an additive constant to these tensors as a consequence of the commutation rules for the field operators [309]. However the renormalized quantum correlation tensors in which this contribution of the vacuum is omitted are identical with the classical correlation tensors.

From the definition of these correlation tensors and from (7.3) it follows that

$$U_{kj}(x_1, x_2) = U^*_{jk}(x_2, x_1)\,, \quad (7.9a)$$

$$S_{kj}(x_1, x_2) = -S^*_{jk}(x_2, x_1)\,, \quad (7.9b)$$

$$T_{kj}(x_1, x_2) = T_{jk}(x_1, x_2) = T^*_{jk}(x_2, x_1)\,, \quad (7.9c)$$

$$Q_{kj}(x_1, x_2) = Q_{jk}(x_1, x_2) = -Q^*_{jk}(x_2, x_1)\,. \quad (7.9d)$$

Therefore we obtain from (7.6a) and (7.9a)

$$\operatorname{Im} U(x, x) = 0\,, \quad (7.10a)$$

from (7.7b) and (7.9b)

$$\mathrm{Im}\ S_i(x, x) = 0\,, \tag{7.10b}$$

and from (7.8a) and (7.9c)

$$\mathrm{Im}\ T_{jk}(x, x) = 0\,. \tag{7.10c}$$

Similarly from (7.7a) and (7.9b)

$$\mathrm{Re}\ S(x, x) = 0\,, \tag{7.11a}$$

from (7.6b) and (7.9a)

$$\mathrm{Re}\ U_i(x, x) = 0\,, \tag{7.11b}$$

and from (7.8b) and (7.9d)

$$\mathrm{Re}\ Q_{jk}(x, x) = 0\,. \tag{7.11c}$$

We can easily modify all these tensors, vectors and scalars for stationary fields and all the quantities will then depend upon the time difference $\tau = t_2 - t_1$ so that the space-time dependence of the quantities is given by $(x_1, x_2) \equiv (\boldsymbol{x}_1, \boldsymbol{x}_2; \tau)$; if x_1 and x_2 coincide then $(x, x) \equiv (\boldsymbol{x}, \boldsymbol{x}; 0)$. The bracket (x_2, x_1) is equivalent to $(\boldsymbol{x}_2, \boldsymbol{x}_1; -\tau)$.

7.2 Dynamical equations for correlation tensors

We begin with Maxwell's equations for the electromagnetic field *in vacuo*. These equations may be written, using the Levi-Civita tensor ε_{jkl}, in the form

$$\sum_{k,l} \varepsilon_{jkl} \frac{\partial}{\partial x_k} \begin{pmatrix} E_l^*(x) \\ H_l^*(x) \end{pmatrix} - \frac{\hat{\sigma}}{c} \frac{\partial}{\partial t} \begin{pmatrix} E_j^*(x) \\ H_j^*(x) \end{pmatrix} = 0\,, \quad j = 1, 2, 3\,, \tag{7.12}$$

where the matrix $\hat{\sigma}$ is given by (7.5) and E_j and H_j are understood to be the analytic signals. The divergence conditions may be written as

$$\sum_j \frac{\partial}{\partial x_j} \begin{pmatrix} E_j^*(x) \\ H_j^*(x) \end{pmatrix} = 0\,. \tag{7.13}$$

Considering equations (7.12) and (7.13) for $x \equiv x_1$ and multiplying them directly by $(E_m(x_2), H_m(x_2))$ from the right and taking the average we arrive at

$$\sum_{k,l} \varepsilon_{jkl} \frac{\partial}{\partial x_{1k}} \mathscr{K}_{lm}(x_1, x_2) - \frac{\hat{\sigma}}{c} \frac{\partial}{\partial t_1} \mathscr{K}_{jm}(x_1, x_2) = 0\,, \tag{7.14}$$

$$\sum_j \frac{\partial}{\partial x_{1j}} \mathscr{K}_{jm}(x_1, x_2) = 0\,, \tag{7.15}$$

which is the first system of *dynamical equations* for the correlation tensors.

Another set of dynamical equations may be obtained in the following way. We consider the Hermitian conjugate to (7.12) with $x \equiv x_2$, multiply this equation directly from the left by

$$\begin{pmatrix} E_m^*(x_1) \\ H_m^*(x_1) \end{pmatrix}$$

and take the average. Performing the same with equation (7.13) we arrive at the equations

$$\sum_{k,l} \varepsilon_{jkl} \frac{\partial}{\partial x_{2k}} \mathscr{K}_{ml}(x_1, x_2) + \frac{1}{c} \frac{\partial}{\partial t_2} \mathscr{K}_{mj}(x_1, x_2)\, \hat{\sigma} = 0\,, \tag{7.16}$$

$$\sum_j \frac{\partial}{\partial x_{2j}} \mathscr{K}_{mj}(x_1, x_2) = 0\,, \tag{7.17}$$

where $\hat{\sigma}^+ = -\hat{\sigma}$ has been used.

Equations (7.14)–(7.17) can be regarded as the basic differential equations of the second-order correlation theory of the electromagnetic field, although the two sets (7.14)–(7.15) and (7.16)–(7.17) of dynamical equations are equivalent in a certain sense. If we use (7.3) we can obtain (7.16) and (7.17) from (7.14) and (7.15) if the first system is Hermitian conjugated and the following substitutions are used: $x_{1k} \to x_{2k}$, $t_1 \to t_2$, $\partial/\partial x_{1k} \to \partial/\partial x_{2k}$, $\partial/\partial t_1 \to \partial/\partial t_2$ ($\hat{\sigma}^+ = -\hat{\sigma}$). In the converse way the first system can be obtained from the second one.

Note that both systems of equations are invariant with respect to the transformations $\mathscr{K} \to \hat{\sigma}\mathscr{K}\hat{\sigma}^{-1} = \hat{\sigma}^{-1}\mathscr{K}\hat{\sigma} = -\hat{\sigma}\mathscr{K}\hat{\sigma}$ ($\hat{\sigma}^{-1} = -\hat{\sigma}$) and also the system (7.14)–(7.15) is invariant with respect to the transformation

$$\mathscr{K} \to \mathscr{K} \begin{pmatrix} 0 & 1 \\ 1 & 0 \end{pmatrix}$$

while the system (7.16)–(7.17) is invariant with respect to the transformation

$$\mathscr{K} \to \begin{pmatrix} 0 & 1 \\ 1 & 0 \end{pmatrix} \mathscr{K}.$$

Considering the first system, applying the operator $\partial/\partial x_{1j}$ to (7.14) and summing over j we obtain

$$\sum_j \frac{\partial}{\partial x_{1j}} \mathscr{K}_{jm}(x_1, x_2) \equiv \text{const}$$

since the first term is zero as a consequence of the antisymmetry of ε_{jkl}. Hence, if this constant is equal to zero for $t_1 = t_0$ (i.e. the divergence condition (7.15) holds) then this condition holds for all time.

7.3 Second-order equations — wave equations — for the correlation tensors

The derived first-order dynamical equations for the correlation tensors can be used to obtain second-order equations for correlation tensors – the wave equations.

Applying the operator $\partial/\partial(ct_1)$ to (7.14) and using (7.14) again to express the first term in the form

$$\frac{1}{c}\sum_{k,l}\varepsilon_{jkl}\frac{\partial}{\partial x_{1k}}\frac{\partial}{\partial t_1}\mathscr{K}_{lm} = \sum_{k,l,i,n}\hat{\sigma}^{-1}\varepsilon_{jkl}\varepsilon_{lin}\frac{\partial}{\partial x_{1k}}\frac{\partial}{\partial x_{1i}}\mathscr{K}_{nm}, \tag{7.18}$$

we obtain

$$\sum_{k,l,i,n}\varepsilon_{jkl}\varepsilon_{lin}\frac{\partial}{\partial x_{1k}}\frac{\partial}{\partial x_{1i}}\mathscr{K}_{nm} = -\frac{1}{c^2}\frac{\partial^2}{\partial t_1^2}\mathscr{K}_{jm}, \tag{7.19}$$

where $\hat{\sigma}^2 = -\hat{1}$ has been used. Using the identity [15]

$$\sum_l \varepsilon_{jkl}\varepsilon_{lin} = \delta_{ji}\delta_{kn} - \delta_{jn}\delta_{ki}, \tag{7.20}$$

where δ_{jk} is the Kronecker symbol, we finally obtain from (7.19)

$$\nabla_1^2\mathscr{K}_{jm} = \frac{1}{c^2}\frac{\partial^2}{\partial t_1^2}\mathscr{K}_{jm}, \tag{7.21a}$$

which is the required wave equation for the correlation tensors. Here $\nabla_1^2 \equiv \sum_i \partial^2/\partial x_{1i}^2$ is the Laplace operator acting with respect to the coordinates of the point $\boldsymbol{x}_1$. (Equation (7.15) has also been used.) One can derive the same type of wave equation with respect to the second space-time point in an identical way

$$\nabla_2^2\mathscr{K}_{jm} = \frac{1}{c^2}\frac{\partial^2}{\partial t_2^2}\mathscr{K}_{jm}. \tag{7.21b}$$

A further second-order equation can be obtained by applying the operator $\partial/\partial(ct_2)$ to (7.14) and using (7.16)

$$\sum_{k,l,i,n}\varepsilon_{jkl}\varepsilon_{\min}\frac{\partial}{\partial x_{1k}}\frac{\partial}{\partial x_{2i}}\mathscr{K}_{ln} = \frac{1}{c^2}\frac{\partial^2}{\partial t_1\,\partial t_2}\hat{\sigma}\mathscr{K}_{jm}\hat{\sigma}^{-1}. \tag{7.22}$$

7.4 Conservation laws for correlation tensors

Taking the trace of (7.14) and (7.15) and making use of (7.4a) and (7.4b) we obtain the conservation laws

$$\sum_{k,l}\varepsilon_{jkl}\frac{\partial}{\partial x_{1k}}U_{lm}-\frac{1}{c}\frac{\partial}{\partial t_1}S_{jm}=0\,, \tag{7.23a}$$

$$\sum_{j}\frac{\partial}{\partial x_{1j}}U_{jm}=0\,. \tag{7.23b}$$

Multiplying (7.14) and (7.15) by $\hat{\sigma}$ from the left (or right) and taking the trace we obtain another system of conservation laws

$$\sum_{k,l}\varepsilon_{jkl}\frac{\partial}{\partial x_{1k}}S_{lm}+\frac{1}{c}\frac{\partial}{\partial t_1}U_{jm}=0\,, \tag{7.24a}$$

$$\sum_{j}\frac{\partial}{\partial x_{1j}}S_{jm}=0\,. \tag{7.24b}$$

Both these systems of conservation laws may be written in the matrix form again

$$\sum_{k,l}\varepsilon_{jkl}\frac{\partial}{\partial x_{1k}}\begin{pmatrix}U\\S\end{pmatrix}_{lm}+\frac{\hat{\sigma}}{c}\frac{\partial}{\partial t_1}\begin{pmatrix}U\\S\end{pmatrix}_{jm}=0\,, \tag{7.25a}$$

$$\sum_{j}\frac{\partial}{\partial x_{1j}}\begin{pmatrix}U\\S\end{pmatrix}_{jm}=0\,. \tag{7.25b}$$

These equations are invariant with respect to the transformation

$$\begin{pmatrix}U\\S\end{pmatrix}\to\hat{\sigma}\begin{pmatrix}U\\S\end{pmatrix}. \tag{7.26}$$

The system of conservation laws with respect to the space-time point $(\boldsymbol{x}_2, t_2)$ may be obtained from (7.25) by Hermitian conjugation if the substitutions $x_1\to x_2$, $\partial/\partial x_{1k}\to\partial/\partial x_{2k}$, $\partial/\partial t_1\to\partial/\partial t_2$ and the equations (7.9a) and (7.9b) are used (or from (7.23) and (7.24) making use of the substitutions $S_{jm}\to -S_{mj}$, $U_{jm}\to U_{mj}$, $x_1\to x_2$, $\partial/\partial x_{1k}\to\partial/\partial x_{2k}$ and $\partial/\partial t_1\to\partial/\partial t_2$).

From (7.21a) and (7.21b) we obtain the wave equations for U_{jm} and S_{jm} taking the trace of (7.21a, b) and the trace of (7.21a, b) multiplied by $\hat{\sigma}$

$$\nabla_i^2\begin{pmatrix}U\\S\end{pmatrix}_{jm}=\frac{1}{c^2}\frac{\partial^2}{\partial t_i^2}\begin{pmatrix}U\\S\end{pmatrix}_{jm}, \qquad i=1,2\,. \tag{7.27}$$

Equations (7.25a, b) may be called the *tensor conservation laws.* Putting $j = m$ and summing over j in (7.23a) and (7.24a) we obtain

$$\nabla_1 \,.\, \boldsymbol{U} = \frac{1}{c}\frac{\partial}{\partial t_1} S\,, \tag{7.28}$$

$$\nabla_1 \,.\, \boldsymbol{S} = -\frac{1}{c}\frac{\partial}{\partial t_1} U\,, \tag{7.29}$$

where U, $\boldsymbol{U}$ and S, $\boldsymbol{S}$ are defined in (7.6) and (7.7). These two equations represent the *scalar conservation laws.*

Multiplying (7.23a) and (7.24a) by ε_{rjm} and summing over j and m we obtain (with the use of the identity (7.20) and (7.25b))

$$\sum_j \frac{\partial}{\partial x_{1j}} T_{rj} = \frac{1}{c}\frac{\partial}{\partial t_1} S_r\,, \tag{7.30}$$

$$\sum_j \frac{\partial}{\partial x_{1j}} Q_{rj} = -\frac{1}{c}\frac{\partial}{\partial t_1} U_r\,, \tag{7.31}$$

where the quantities T_{rj}, S_r, Q_{rj} and U_r are defined by (7.8a), (7.7b), (7.8b) and (7.6b) respectively. Equations (7.30) and (7.31) represent the *vectorial conservation laws.* The second set of the conservation laws may be obtained from (7.28)–(7.31) by the substitutions $\nabla_1 \to \nabla_2$, $\partial/\partial x_{1j} \to \partial/\partial x_{2j}$ and $\partial/\partial t_1 \to \partial/\partial t_2$.

The conservation laws just derived are linear since they contain the correlation tensors linearly. Some nonlinear conservation laws, which involve the correlation tensors quadratically, have been derived in [405] and [204].

7.5 Stationary fields

We have noted at the end of Sec. 7.1 that the corresponding equations for stationary fields may be obtained by substitutions $(x_1, x_2) \to (\boldsymbol{x}_1, \boldsymbol{x}_2; \tau)$, $\tau = t_2 - t_1$ $((x_2, x_1) \to (\boldsymbol{x}_2, \boldsymbol{x}_1; -\tau)$, $(x, x) \to (\boldsymbol{x}, \boldsymbol{x}; 0)$, $-\partial/\partial t_1 = \partial/\partial t_2 = \partial/\partial\tau)$. For example we obtain from (7.14) and (7.16)

$$\sum_{k,l} \varepsilon_{jkl} \frac{\partial}{\partial x_{1k}} \mathscr{K}_{lm}(\boldsymbol{x}_1, \boldsymbol{x}_2; \tau) + \frac{\hat{\sigma}}{c}\frac{\partial}{\partial\tau} \mathscr{K}_{jm}(\boldsymbol{x}_1, \boldsymbol{x}_2; \tau) = 0\,, \tag{7.32}$$

$$\sum_{k,l} \varepsilon_{jkl} \frac{\partial}{\partial x_{2k}} \mathscr{K}_{ml}(\boldsymbol{x}_1, \boldsymbol{x}_2; \tau) + \frac{1}{c}\frac{\partial}{\partial\tau} \mathscr{K}_{mj}(\boldsymbol{x}_1, \boldsymbol{x}_2; \tau)\,\hat{\sigma} = 0\,. \tag{7.33}$$

The wave equations become in this case

$$\nabla_i^2 \mathscr{K}_{jm} = \frac{1}{c^2} \frac{\partial^2}{\partial \tau^2} \mathscr{K}_{jm}, \quad i = 1, 2 \tag{7.34}$$

and from (7.22) we have

$$\sum_{k,l,i,n} \varepsilon_{jkl} \varepsilon_{min} \frac{\partial}{\partial x_{1k}} \frac{\partial}{\partial x_{2i}} \mathscr{K}_{ln} = -\frac{1}{c^2} \frac{\partial^2}{\partial \tau^2} \hat{\sigma} \mathscr{K}_{jm} \hat{\sigma}^{-1} . \tag{7.35}$$

The conservation laws (7.28)–(7.31) reduce to

$$\nabla_i \,.\, \boldsymbol{U} = \mp \frac{1}{c} \frac{\partial}{\partial \tau} \boldsymbol{S} , \tag{7.36}$$

$$\nabla_i \,.\, \boldsymbol{S} = \pm \frac{1}{c} \frac{\partial}{\partial \tau} U , \tag{7.37}$$

$$\sum_j \frac{\partial}{\partial x_{ij}} T_{mj} = \mp \frac{1}{c} \frac{\partial}{\partial \tau} S_m , \tag{7.38}$$

$$\sum_j \frac{\partial}{\partial x_{ij}} Q_{mj} = \pm \frac{1}{c} \frac{\partial}{\partial \tau} U_m . \tag{7.39}$$

Here the upper and lower signs are taken according as i takes the value 1 or 2 respectively.

It can be shown [408, 309] that the standard conservation laws in the averaged form follow from the present conservation laws as special cases. Thus from (7.37) one can obtain the averaged form of the energy conservation law and from (7.38) the averaged form of the momentum conservation law of the electromagnetic field is obtained if $x_1 \to x_2$, $\tau \to 0$ and the real part of (7.37) and (7.38) is taken. In this sense the conservation laws (7.36) and (7.39) have no classical analogue, since they reduce to the identity $0 = 0$ as a consequence of (7.11a, b, c).

7.6 Cross-spectral tensors

The *cross-spectral tensors* $\tilde{\mathscr{E}}_{jk}(\boldsymbol{x}_1, \boldsymbol{x}_2; \nu)$, $\tilde{\mathscr{H}}_{jk}(\boldsymbol{x}_1, \boldsymbol{x}_2; \nu)$, $\tilde{\mathscr{M}}_{jk}(\boldsymbol{x}_1, \boldsymbol{x}_2; \nu)$, $\tilde{\mathscr{N}}_{jk}(\boldsymbol{x}_1, \boldsymbol{x}_2; \nu)$ as well as $\tilde{U}_{jk}(\boldsymbol{x}_1, \boldsymbol{x}_2; \nu)$, $\tilde{S}_{jk}(\boldsymbol{x}_1, \boldsymbol{x}_2; \nu)$, arranged in the spectral matrix $\tilde{\mathscr{K}}_{jk}(\boldsymbol{x}_1, \boldsymbol{x}_2; \nu)$, may be introduced by the relation

$$\tilde{\mathscr{K}}_{jk}(\boldsymbol{x}_1, \boldsymbol{x}_2; \nu) = \int_{-\infty}^{+\infty} \mathscr{K}_{jk}(\boldsymbol{x}_1, \boldsymbol{x}_2; \tau) \exp(i2\pi\nu\tau)\, d\tau . \tag{7.40}$$

In agreement with the spectral analysis of general correlation functions for stationary fields contained in Sec. 2.2 (equation (2.23)) we can obtain

$$\langle \tilde{E}_j^*(x_1, \nu)\, \tilde{E}_k(x_2, \nu')\rangle = \tilde{\mathscr{E}}_{jk}(x_1, x_2; \nu)\, \delta(\nu - \nu'), \tag{7.41}$$

where $\tilde{E}_j(x, \nu)$ is the spectral component of $E_j(x, t)$. Thus we can conclude that two spectral components of the electric field are correlated if and only if the frequencies coincide. Similar results are also valid for the other correlation tensors.

The inverse form of (7.40),

$$\mathscr{K}_{jk}(x_1, x_2; \tau) = \int_0^\infty \tilde{\mathscr{K}}_{jk}(x_1, x_2; \nu) \exp(-i2\pi\nu\tau)\, d\nu, \tag{7.42}$$

expresses the Wiener-Khintchine theorem (cf. (2.24)). If this theorem is used, one can derive the corresponding dynamical equations and the conservation laws for the cross-spectral tensors.

From (7.32) and (7.33) we obtain

$$\sum_{l,k} \varepsilon_{jkl} \frac{\partial}{\partial x_{1k}} \tilde{\mathscr{K}}_{lm}(x_1, x_2; \nu) - \frac{i2\pi\nu\hat{\sigma}}{c} \tilde{\mathscr{K}}_{jm}(x_1, x_2; \nu) = 0, \tag{7.43}$$

$$\sum_{k,l} \varepsilon_{jkl} \frac{\partial}{\partial x_{2k}} \tilde{\mathscr{K}}_{ml}(x_1, x_2; \nu) - \frac{i2\pi\nu}{c} \tilde{\mathscr{K}}_{mj}(x_1, x_2; \nu)\, \hat{\sigma} = 0 \tag{7.44}$$

and the divergence conditions give

$$\sum_j \frac{\partial}{\partial x_{1j}} \tilde{\mathscr{K}}_{jm}(x_1, x_2; \nu) = 0, \tag{7.45}$$

$$\sum_j \frac{\partial}{\partial x_{2j}} \tilde{\mathscr{K}}_{mj}(x_1, x_2; \nu) = 0; \tag{7.46}$$

the wave equations reduce to the Helmholtz equations

$$\nabla_i^2 \tilde{\mathscr{K}}_{jm}(x_1, x_2; \nu) + \frac{4\pi^2\nu^2}{c^2} \tilde{\mathscr{K}}_{jm}(x_1, x_2; \nu) = 0, \quad i = 1, 2. \tag{7.47}$$

7.7 Non-negative definiteness conditions for correlation tensors and cross-spectral tensors

Just as the non-negative definiteness condition (4.7) was derived for the correlation function $\Gamma(x_1, x_2)$, analogous non-negative definiteness conditions may be derived for the correlation and cross-spectral tensors [308, 309, 206].

Introducing the quantity

$$A = \sum_i \int [f_i(x)\, E_i(x) + g_i(x)\, H_i(x)]\, d^4x \tag{7.48}$$

we obtain

$$0 \leqq \langle |A|^2 \rangle = \sum_{i,j} \iint [f_i^*(x_1)\, \mathscr{E}_{ij}(x_1, x_2)\, f_j(x_2) + g_i^*(x_1)\, \mathscr{H}_{ij}(x_1, x_2)\, g_j(x_2) +$$

$$+ f_i^*(x_1)\, \mathscr{M}_{ij}(x_1, x_2)\, g_j(x_2) + g_i^*(x_1)\, \mathscr{N}_{ij}(x_1, x_2)\, f_j(x_2)]\, d^4x_1\, d^4x_2\,, \tag{7.49}$$

where $f_i(x)$ and $g_i(x)$, $i = 1, 2, 3$ are arbitrary functions of space-time points $x \equiv (\boldsymbol{x}, t)$ for which the integrals exist. A number of special conditions may be derived by choosing some special functions f and g. Thus if $g \equiv 0$ we obtain the corresponding non-negative definiteness condition for the correlation tensor $\mathscr{E}_{ij}$; if $f \equiv 0$ we obtain the condition for $\mathscr{H}_{ij}$ only. Putting

$$f_i(x) = \sum_{m=1}^{M} \sum_{k=1}^{N} \alpha_{mki}\, \delta(\boldsymbol{x}_m - \boldsymbol{x})\, \delta(t_k - t)\,, \tag{7.50}$$

$$g_i(x) = \sum_{m=1}^{M} \sum_{k=1}^{N} \beta_{mki}\, \delta(\boldsymbol{x}_m - \boldsymbol{x})\, \delta(t_k - t)\,, \tag{7.51}$$

$(i = 1, 2, 3)$, we obtain from (7.49)

$$\sum_{\substack{m,k,n,l \\ i,j}} [\alpha_{mki}^*\, \mathscr{E}_{ij}(\boldsymbol{x}_m, \boldsymbol{x}_n; t_k, t_l)\, \alpha_{nlj} + \beta_{mki}^*\, \mathscr{H}_{ij}(\boldsymbol{x}_m, \boldsymbol{x}_n; t_k, t_l)\, \beta_{nlj} +$$

$$+ \alpha_{mki}^*\, \mathscr{M}_{ij}(\boldsymbol{x}_m, \boldsymbol{x}_n; t_k, t_l)\, \beta_{nlj} + \beta_{mki}^*\, \mathscr{N}_{ij}(\boldsymbol{x}_m, \boldsymbol{x}_n; t_k, t_l)\, \alpha_{nlj}] \geqq 0\,. \tag{7.52}$$

Corresponding conditions for stationary fields follow by introducing $\tau = t_2 - t_1$ in (7.49) and (7.52).

Introducing the quantity

$$\tilde{A} = \sum_i \int_{\nu - \Delta\nu/2}^{\nu + \Delta\nu/2} d\nu \int d^3\boldsymbol{x} [f_i(\boldsymbol{x})\, \tilde{E}_i(\boldsymbol{x}, \nu) + g_i(\boldsymbol{x})\, \tilde{H}_i(\boldsymbol{x}, \nu)]\,, \tag{7.53}$$

where $\tilde{H}$ has a similar significance to $\tilde{E}$ and $(\nu - \Delta\nu/2, \nu + \Delta\nu/2)$ is an arbitrarily small frequency interval, we obtain in a similar way the non-negative definiteness condition for the cross-spectral tensors

$$\sum_{i,j} \iint d^3x_1 \, d^3x_2 [f_i^*(x_1) \, \tilde{\mathscr{E}}_{ij}(x_1, x_2; \nu) \, f_j(x_2) + \\ + g_i^*(x_1) \, \tilde{\mathscr{H}}_{ij}(x_1, x_2; \nu) \, g_j(x_2) + \\ + f_i^*(x_1) \, \tilde{\mathscr{M}}_{ij}(x_1, x_2; \nu) \, g_j(x_2) + \\ + g_i^*(x_1) \, \tilde{\mathscr{N}}_{ij}(x_1, x_2; \nu) \, f_j(x_2)] \geqq 0 \,. \tag{7.54}$$

Finally let us note that the results derived for the correlation tensors are independent of the kind of averaging (they characterize the space-time behaviour of the field) and consequently they are also valid for the quantum correlation tensors, if the contribution of the physical vacuum is omitted [308, 309]. The present results can be extended to the case of the existence of random sources in the space [404, 70, 3, 204] and from many further results obtained in this field we mention the following. The general propagation laws, including both the interference law for partial coherence and the interference law for partial polarization, have been derived in [204] on the basis of the differential equations for the correlation tensors. A theory of the correlation properties of optical fields using the correlation tensors of arbitrary order has been developed in [205, 206]. All these formulations of the correlation theory were non-relativistic, which is the typical case in quantum optics applications. A relativistic correlation theory based on the use of the anti-symmetric electromagnetic tensor $F_{\alpha\beta}(x)$ has been developed in [239] and [128].

The technique of correlation tensors may be useful in connection with the study of the properties of turbulent fluids such as plasmas.

Chapter 8

GENERAL CLASSICAL STATISTICAL DESCRIPTION OF THE FIELD

In section 2.2 we introduced both the classical and quantum correlation functions of arbitrary order, which are the fundamental quantities in the present treatment of the coherence of light. We now give a more precise classical definition in terms of classical stochastic processes, while a more precise specification of the averaging brackets from a quantum-mechanical point of view will be discussed later (Chapters 12 and 14) together with the correspondence between these two approaches.

All experimental results reached in optical coherence, interference, diffraction and polarization experiments prior to about 1955 can be described in terms of the second-order coherence theory which we have developed in the preceding chapters. Since that time, a number of unconventional experiments have been carried out or proposed. For their description a systematic broadening of the second-order coherence theory is necessary. From these experiments we mention the following: the experiments of Forrester, Gudmundsen and Johnson [150] on beats between light from independent sources; the experiments of Hanbury Brown and Twiss [92—98] on photon correlations and intensity interferometry; the experiments of Magyar and Mandel [261, 262] on transient interference effects with light from independent sources (lasers); an experiment proposed by Gamo [163] on measuring the three-point (sixth-order) correlation function; experiments measuring photon statistics or alternatively the statistics of emitted photoelectrons by a photodetector (see Chapter 10); an experiment to measure the sixth-order correlation function by Davidson and Mandel [125] (see also [503]) and an experiment on the interference of individual photons by Pfleegor and Mandel [378—380] and Radloff [390]. Such a general formulation of the theory of coherence phenomena based on the classical theory of stochastic functions was first proposed by Wolf [479, 481] and Mandel [272], while a theory based on quantum electrodynamics was introduced by Glauber [173, 174, 175].

8.1 Stochastic description of light

Consider an optical field represented by the vector complex analytic wave amplitude $V(\boldsymbol{x}_i, t_i) \equiv V_i$ at the space-time point $(\boldsymbol{x}_i, t_i)$. The stochastic behaviour of the vectorial field $V(\boldsymbol{x}, t)$ can be described by the sequence of *probability densities*

$$P_1(V_1), P_2(V_1, V_2), \ldots, P_n(V_1, V_2, \ldots, V_n), \ldots \tag{8.1}$$

having the following significance. The joint n-fold probability distribution

$$P_n(V_1, V_2, \ldots, V_n)\, d^2V_1\, d^2V_2 \ldots d^2V_n, \tag{8.2}$$

where $d^2V_j = d(\text{Re } V_j)\, d(\text{Im } V_j)$, $j = 1, 2, \ldots, n$, represents the joint n-fold probability that at the space-time point $(\boldsymbol{x}_1, t_1)$ the quantities Re V and Im V lie in the intervals (Re V_1, Re $V_1 + d(\text{Re } V_1)$) and (Im V_1, Im $V_1 + d(\text{Im } V_1)$) respectively, etc., and at the space-time point $(\boldsymbol{x}_n, t_n)$ they lie at the intervals (Re V_n, Re $V_n + d(\text{Re } V_n)$) and (Im V_n, Im $V_n +$ $+ d(\text{Im } V_n)$) respectively.

If $F(V_1, \ldots, V_n)$ represents a function depending on the values of the field V at the points $(\boldsymbol{x}_1, t_1), \ldots, (\boldsymbol{x}_n, t_n)$ we may define the ensemble average of F by the relation†

$$\langle F(V_1, \ldots, V_n)\rangle = \int_{-\infty}^{+\infty} \ldots \int F(V_1, \ldots, V_n)\, P_n(V_1, \ldots, V_n)\, d^2V_1 \ldots d^2V_n. \tag{8.3}$$

Let us note that P_n is a probability distribution of $6n$ variables and the integral (8.3) extends over $3n$ complex planes. The complex representation of the field is very convenient for a description of measurements of the statistical properties of optical fields by means of photoelectric detectors. It represents a bridge between the classical and quantum descriptions of the coherence properties of light, even if, since the specification of the complex field V is equivalent to the simultaneous specifications of the amplitude and the phase of the wave, it might appear that $P_n(V_1, \ldots, V_n)$ has no analogue in the description of the quantized field.

† For a full description of the statistics of the field we must put $n \to \infty$ and the ensemble average of F will be given by the *functional integral*

$$\langle F\rangle = \int_{-\infty}^{+\infty} \ldots \int F\{V(x)\}\, P\{V(x)\} \prod_x d^2V(x),$$

where $x \equiv (\boldsymbol{x}, t)$. A discussion can be found in [3], section 13.3.

In the theory of coherence, some correlation measurements involve the *higher-order moments* – correlation functions or tensors – the classical form of which is

$$\Gamma^{(m,n)}_{\mu_1\ldots\mu_{m+n}}(x_1, \ldots, x_{m+n}) = \langle \prod_{j=1}^{m} V^*_{\mu_j}(x_j) \prod_{k=m+1}^{m+n} V_{\mu_k}(x_k)\rangle , \qquad (8.4)$$

where $x \equiv (\boldsymbol{x}, t)$, as we have seen in Chapter 2. A special case of stationary and ergodic fields, for which the probability distributions are invariant with respect to translations of the time origin so that the ensemble average is equal to the time average, has also been discussed in Chapter 2.

In the terminology of this chapter some experiments determining the fourth - and sixth-order correlation functions have been proposed and carried out. Some further experiments provide the distribution of photons in the field, and this is related to distributions of the form (8.1); in this way one may also calculate the higher-order correlation functions from experimental data. This method is much easier than a direct measurement of the higher-order correlation functions.

8.2 Functional formulation

We will not go into the details of functional theory here. We only introduce the *characteristic functional* by the relation

$$C\{y(x)\} =$$

$$= \int \exp\left\{\int [y(x)\, V^*(x) - y^*(x)\, V(x)]\, d^4x\right\} P\{V(x)\} \prod_x d^2V(x) , \qquad (8.5)$$

where $P\{V(x)\}$ is the *probability functional* and x is a concise notation for (x, μ) $(\int \ldots d^4x \equiv \sum_\mu \int \ldots d^4x)$. The correlation function (8.4) can be obtained as the functional derivative of the characteristic functional

$$\prod_{j=1}^{m} \frac{\delta}{\delta y(x_j)} \prod_{k=m+1}^{m+n} \frac{\delta}{\delta(-y^*(x_k))} C\{y(x)\}\big|_{y(x)\equiv 0} = \langle \prod_{j=1}^{m} V^*(x_j) \prod_{k=m+1}^{m+n} V(x_k)\rangle , \qquad (8.6)$$

where the functional derivative is defined by

$$\frac{\delta F}{\delta y(x_0)} = \lim_{\varepsilon\to 0} \frac{1}{\varepsilon} [F\{y(x) + \varepsilon\, \delta(x - x_0)\} - F\{y(x)\}] ; \qquad (8.7)$$

F is a continuous functional and $\delta(x)$ is the four-dimensional Dirac function.

A quantum analogue of these equations will be given in Sec. 12.3.

Some mathematical details about the functional theory can be found in the book by Volterra [34] (see also [167] and [13], Chapter III). Some physical applications of the functional theory to the study of the coherence properties of light can be found in [3], [350–352] and [207, 208]. Numerous applications exist in field theory [144, 10, 7, 2, 166].

Chapter 9

SOME APPLICATIONS OF THE THEORY OF COHERENCE TO OPTICAL IMAGING

In this chapter we apply some of the results of earlier chapters to the determination of the relation between object and image for optical systems which image extended polychromatic objects. Although only the second-order coherence effects are usually taken into account in such an analysis of optical imaging processes (if coherence effects are included at all) [336, 3, 201, 202, 8, 20], we can, without particular difficulty, perform the analysis including coherence effects of arbitrary order. Such an analysis may in principle lead to a description of optical imaging with partially coherent non-chaotic (e.g. laser) light [363, 373] since the complete specification of the statistics of such light demands the knowledge of moments of all orders, while for chaotic light the second-order moment is sufficient (Sec. 17.1). Special cases of the general relation between an object and its image having practical significance are (i) the reconstruction of the object when the image is given (the image may not resemble the object as a consequence of the diffraction properties of the optical system and of aberrations in the system) and (ii) the similarity between an object and its image [347, 372, 373], which characterizes the imaging process with partially coherent light by means of a typical non-linearity.

9.1 Spatial Fourier analysis of optical imaging — transfer functions for partially coherent light

An important characteristic of an optical system is the *transfer function*, the Fourier transform of the diffraction function of the system for coherent light. This describes the ability of the system to transfer spatial frequencies — the number of lines per unit length on a test object. The transfer function may be introduced under the assumption of a linear and stationary (in space) system and in this case the optical system may be treated as a *spatial frequency filter*, in terms of the theory of electric circuits. Among interesting applications of spatial Fourier analysis the spatial filtering technique proposed by Maréchal and Croce [290] should be mentioned. This was continued in papers by Stroke and his co-workers [433, 430, 432, 29, 431] and the technique enabled pictures very similar to the original objects to be reconstructed from blurred photographs. This method represents an extension of the Maréchal-Croce spatial filtering principle realized in the form of *holograms* by Gabor [158, 161]. Since the relation between an object and its image has the form of a convolution of the object function with the diffraction function, the frequency spectrum of the image $\tilde{g}(\mu)$ is equal to the product of the transfer function $\tilde{K}(\mu)$ and the spatial frequency function of the object $\tilde{f}(\mu)$ (assuming coherent light). The dimensions of apertures in optical systems are finite so the transfer function is zero outside a finite region. Consequently the system filters out higher spatial frequencies and these cannot be present in the image. Therefore the object cannot be reconstructed uniquely from $\tilde{f}(\mu) = \tilde{g}(\mu)/\tilde{K}(\mu)$ by means of a Fourier transform, since $\tilde{f}(\mu) = 0/0$ in some region and consequently the values of $\tilde{f}(\mu)$ for all μ cannot be deduced. In connection with the "deconvolution" holographic method some possibilities of obtaining "super-resolution" (using analytic continuation to determine the uncertainty 0/0) were discussed in [368, 374, 431].

Calculation of the transfer functions usually involves numerical integration [203, 232, 24] and the analysis is normally limited to strictly coherent or incoherent light. The transfer function for partially coherent objects has been computed in some cases in [427, 319]. In the first stage all analysis was performed for quasi-monochromatic light [201, 202, 8, 20]. This restriction was removed in [336, 3] where an analysis applicable to polychromatic light was developed. However, this analysis does not take into account detection of light in the process of imaging. A general analysis of the optical imaging for polychromatic light and coherence effects of arbitrary order was developed in [363] and applied in [373].

An optical system regarded as a spatial filter may be compared with a temporal filter in electronics. If an electronic system is described by the response function $h(t)$ at time t and if the input and output are described by the functions $f(t)$ and $g(t)$ respectively, then

$$g(t') = \hat{L}f(t) = \int_{-\infty}^{+\infty} h(t' - t) f(t)\, dt\,, \tag{9.1}$$

where the integral operator $\hat{L}$ is obviously linear and the system is stationary since the response function depends on the time difference $(t' - t)$. The physical condition of causality is

$$h(t) = 0\,, \quad t < 0 \tag{9.2}$$

and the stability condition demands that

$$\int_{-\infty}^{+\infty} |h(t)|\, dt < \infty\,, \tag{9.3}$$

since we obtain by using this condition

$$|\tilde{h}(\nu)| = \left| \int_{-\infty}^{+\infty} h(t) \exp i2\pi\nu t\, dt \right| \leqq \int_{-\infty}^{+\infty} |h(t)|\, dt < \infty\,,$$

i.e. the response to a finite input is finite.

All these conditions must be realized in the imaging process except the condition (9.2) which has no spatial analogue since the distribution of light exists on both sides of the axis of an optical system (we are dealing with the space variable x instead of the time variable t in this case). The stationary condition holds for optical systems approximately and regions in which the stationary condition holds are said to be the *isoplanatic regions.* A typical relation between the object and the image is two-dimensional and it may be written as

$$g(x', y') = \iint_{-\infty}^{+\infty} K(x' - x, y' - y) f(x, y)\, dx\, dy\,, \tag{9.4}$$

where (x, y) and (x', y') are coordinates in the object and image planes respectively and K is the diffraction function. The condition (9.2) may sometimes be replaced by

$$f(x, y) = 0\,, \quad x, y \notin \mathscr{A}\,, \tag{9.5}$$

where $\mathscr{A}$ is a region of the object plane. Equation (9.5) expresses the fact that objects are usually finite in extension. The transfer function $\tilde{K}(\mu, \nu)$ may be defined by using a Fourier transform as

$$\tilde{g}(\mu, \nu) = \tilde{K}(\mu, \nu) \tilde{f}(\mu, \nu)\,, \tag{9.6}$$

where μ, ν are spatial frequencies and

$$\tilde{K}(\mu, \nu) = \iint_{-\infty}^{+\infty} K(x, y) \exp i2\pi(\mu x + \nu y)\, dx\, dy = \frac{\tilde{g}(\mu, \nu)}{\tilde{f}(\mu, \nu)}\,, \tag{9.7}$$

under the assumption that $\tilde{f}(\mu, \nu) \not\equiv 0$ for all μ, ν.

Such a formulation is useful for coherent light and, if K is replaced by $|K|^2$ and the functions f, g represent the intensities, for incoherent light.

If the light is partially coherent we must begin with the propagation law (3.21) for the mutual spectral density $G(\boldsymbol{x}_1, \boldsymbol{x}_2; \nu)$. More generally, solving a boundary problem for the wave equations (2.35) (or (2.37) if a stationary field is assumed) we obtain instead of (3.29)

$$G^{(n,n)}(Q_1, \ldots, Q_{2n}; \nu) =$$

$$\frac{1}{(2\pi)^{2n}} \int_{\mathscr{A}} \cdots \int G^{(n,n)}(P_1, \ldots, P_{2n}; \nu)\, R(P_1, \ldots, P_{2n}; Q_1, \ldots, Q_{2n}; \nu) \prod_{j=1}^{n} dP_j\, dP_{n+j}\,, \tag{9.8}$$

where

$$R(P_1, \ldots, P_{2n}; Q_1, \ldots, Q_{2n}; \nu) =$$

$$\prod_{j=1}^{n} (1 + ik_j r_j)(1 - ik_{n+j} r_{n+j}) \frac{\exp\left[-i(k_j r_j - k_{n+j} r_{n+j})\right]}{r_j^2 r_{n+j}^2} \cos \vartheta_j \cos \vartheta_{n+j}\,.$$

Considering this propagation law on the hypersurface (2.21) (i.e. $\nu_1 = -\sum_{j=2}^{2n} \varepsilon_j \nu_j$) with $k \gg 1/r$, we obtain using (2.19)

$$\Gamma^{(n,n)}(Q_1, \ldots, Q_{2n}; \tau)$$

$$= \frac{1}{(2\pi)^{2n}} \int_{\mathscr{A}} \cdots \int \Gamma^{(n,n)}(P_1, \ldots, P_{2n}; \tau_2', \ldots, \tau_{2n}') \prod_{j=1}^{n} \frac{\bar{\Lambda}_j^* \bar{\Lambda}_{n+j}}{r_j r_{n+j}}\, dP_j\, dP_{n+j}$$

$$= \frac{1}{(2\pi)^{2n}} \int_{\mathscr{A}} \cdots \int \gamma^{(n,n)}(P_1, \ldots, P_{2n}; \tau_2', \ldots, \tau_{2n}')\,.$$

$$\cdot \prod_{j=1}^{n} \frac{\left[I^{(n)}(P_j)\, I^{(n)}(P_{n+j})\right]^{1/2n}}{r_j r_{n+j}} \bar{\Lambda}_j^* \bar{\Lambda}_{n+j}\, dP_j\, dP_{n+j}\,, \tag{9.9}$$

where

$$\tau_j' = \tau_j + \frac{r_1 - r_j}{c}\,, \quad j = 2, \ldots, 2n\,.$$

Here (2.42a) has been used and $I^{(n)}(P) \equiv \Gamma^{(n,n)}(P, \ldots, P; 0, \ldots, 0)$. Equation (9.9) is a generalization of (3.34). Under the usual assumptions analogous to (3.43) the corresponding equations can be obtained for quasi-monochromatic light.

Linear process of imaging

A still more general propagation law valid in non-homogeneous media may be written in the form

$$ {}^{i}G^{(m,n)}(\boldsymbol{x}'_1, \ldots, \boldsymbol{x}'_{m+n}; \boldsymbol{v}) = \int_{\mathscr{A}} \ldots \int \prod_{j=1}^{m} K^*(\boldsymbol{x}'_j - \boldsymbol{x}_j, v_j) \,. $$

$$ \cdot \prod_{k=m+1}^{m+n} K(\boldsymbol{x}'_k - \boldsymbol{x}_k, v_k)\, {}^{o}G^{(m,n)}(\boldsymbol{x}_1, \ldots, \boldsymbol{x}_{m+n}; \boldsymbol{v}) \prod_{j=1}^{m+n} d\boldsymbol{x}_j \,, \tag{9.10} $$

where ${}^{o}G$ and ${}^{i}G$ are the object and image spectral correlation functions respectively; the equation is considered in the isoplanatic region so that K depends on the difference $(\boldsymbol{x}' - \boldsymbol{x})$. If the field is assumed to be stationary (9.10) must be considered on the hypersurface $\sum_{j=1}^{m+n} \varepsilon_j v_j = 0$ again. The image correlation function ${}^{i}\Gamma^{(m,n)}$ may be obtained by means of a Fourier transform.

The transfer functions may also be introduced to describe the quality of the transfer properties of the optical system for partially coherent polychromatic light of arbitrary statistical behaviour. As (9.10), relating the object and image characteristics, is linear we obtain using a Fourier transform

$$ {}^{i}\tilde{G}^{(m,n)}(\boldsymbol{\mu}_1, \ldots, \boldsymbol{\mu}_{m+n}; \boldsymbol{v}) = $$

$$ = \prod_{j=1}^{m} \tilde{K}^*(-\boldsymbol{\mu}_j, v_j) \prod_{k=m+1}^{m+n} \tilde{K}(\boldsymbol{\mu}_k, v_k)\, {}^{o}\tilde{G}^{(m,n)}(\boldsymbol{\mu}_1, \ldots, \boldsymbol{\mu}_{m+n}; \boldsymbol{v}) \,, \tag{9.11} $$

where $\boldsymbol{\mu}$ is the spatial frequency vector associated with $\boldsymbol{x}$. The quantities

$$ \mathscr{L}^{(m,n)}(\boldsymbol{\mu}_1, \ldots, \boldsymbol{\mu}_{m+n}; \boldsymbol{v}) = \prod_{j=1}^{m} \tilde{K}^*(-\boldsymbol{\mu}_j, v_j) \prod_{k=m+1}^{m+n} \tilde{K}(\boldsymbol{\mu}_k, v_k) \tag{9.12} $$

are the transfer functions.

Non-linear process of imaging

If the field is detected with the use of, say n, quadratic detectors (placed at points $\boldsymbol{x}'_j$) whose outputs are correlated so that we are able in principle to measure the quantities ${}^{i}\Gamma^{(n,n)}(\boldsymbol{x}'_1, \ldots, \boldsymbol{x}'_n, \boldsymbol{x}'_n, \ldots, \boldsymbol{x}'_1)$† we must set $m = n$

† A detailed analysis is given in section 10.6.

and $x'_j \equiv x'_{n+j}$ in (9.10). Thus we arrive at the non-linear relation

$$ {}^{i}G^{(n)}(x'_1, \dots, x'_n; \nu) = \int_{\mathscr{A}} \dots \int \prod_{j=1}^{n} K^*(x'_j - x_j, \nu_j)\, K(x'_{n+j} - x_{n+j}, \nu_{n+j})\,. $$

$$.\, {}^{o}g^{(n,n)}(x_1, \dots, x_{2n}; \nu) \prod_{j=1}^{n} {}^{o}u^*(x_j)\, {}^{o}u(x_{n+j})\, dx_j dx_{n+j}\,, \qquad (9.13) $$

where ${}^{o}g^{(n,n)}$ is the normalized spectral correlation function (in the sense of (2.42a)) (i.e. it is a Fourier transform of ${}^{o}\gamma^{(n,n)}$), ${}^{o}u(x)$ is the object amplitude and ${}^{i}G^{(n)}(x'_1, \dots, x'_n; \nu) \equiv {}^{i}G^{(n,n)}(x'_1, \dots, x'_n, x'_n, \dots, x'_1; \nu)$.

Restricting ${}^{o}g^{(n,n)}$ by the spatial stationary condition

$$ {}^{o}g^{(n,n)}(x_1, \dots, x_{2n}; \nu) = {}^{o}g^{(n,n)}(x_{n+1} - x_1, \dots, x_{2n} - x_n; \nu) \qquad (9.14) $$

we obtain from (9.13), by means of the spatial Fourier analysis, the following equations describing the transfer of spatial frequencies

$$ {}^{i}G^{(n)}(x'_1, \dots, x'_n; \nu) = \int_{-\infty}^{+\infty} \dots \int \mathscr{L}^{(n,n)}(\eta_1, \dots, \eta_{2n}; \nu)\,. $$

$$.\, \prod_{j=1}^{n} {}^{o}\tilde{u}^*(\eta_j)\, {}^{o}\tilde{u}(\eta_{n+j}) \exp\left[i2\pi x'_j \,.\, (\eta_j - \eta_{n+j})\right] d\eta_j\, d\eta_{n+j} \qquad (9.15a) $$

and

$$ {}^{i}\tilde{G}^{(n)}(\mu_1, \dots, \mu_n; \nu) = \int_{-\infty}^{+\infty} \dots \int \mathscr{L}^{(n,n)}(\eta_1, \dots, \eta_n, \eta_1 + \mu_1, \dots, \eta_n + \mu_n; \nu)\,. $$

$$.\, \prod_{j=1}^{n} {}^{o}\tilde{u}^*(\eta_j)\, {}^{o}\tilde{u}(\eta_j + \mu_j)\, d\eta_j\,. \qquad (9.15b) $$

The *generalized transfer functions* $\mathscr{L}^{(n,n)}$, generalizing the *transmission cross-coefficient* of Hopkins [201, 202], are given by

$$ \mathscr{L}^{(n,n)}(\eta_1, \dots, \eta_{2n}; \nu) = \int_{-\infty}^{+\infty} \dots \int \prod_{j=1}^{n} \tilde{K}^*(\mu'_j + \eta_j; \nu_j)\,. $$

$$.\, \tilde{K}(\mu'_j + \eta_{n+j}; \nu_{n+j})\, {}^{o}\tilde{g}^{(n,n)}(\mu'_1, \dots, \mu'_n; \nu) \prod_{k=1}^{n} d\mu'_k\,. \qquad (9.16) $$

These generalized transfer functions are completely determined by the diffraction properties of the system (the function K) and the coherence properties of light (the function g). The integration in (9.16) is taken over the common parts of regions in which both the functions $\tilde{K}$ and $\tilde{g}$ are non-zero (there are finite apertures in the system).

If light is detected with the use of n equivalently placed detectors, the process of imaging will be described by equation (9.15a) with $x'_1 \equiv x'_2 \equiv$

$\equiv \ldots \equiv \boldsymbol{x}'_n \equiv \boldsymbol{x}'$. Denoting ${}^iG^{(n)}(\boldsymbol{x}', \ldots, \boldsymbol{x}'; \nu)$ as ${}^iG^{(n)}(\boldsymbol{x}', \nu)$ we obtain the following equation in the spatial frequency region

$$ {}^i\tilde{G}^{(n)}(\boldsymbol{\mu}; \nu) = \int_{-\infty}^{+\infty} \ldots \int \mathscr{L}^{(n,n)}(\boldsymbol{\eta}_1, \ldots, \boldsymbol{\eta}_{2n}; \nu) \, . $$

$$. \, \delta\left(\sum_{j=1}^{2n} \varepsilon_j \boldsymbol{\eta}_j + \boldsymbol{\mu}\right) \prod_{j=1}^{n} {}^o\tilde{u}^*(\boldsymbol{\eta}_j) \, {}^o\tilde{u}(\boldsymbol{\eta}_{n+j}) \, d\boldsymbol{\eta}_j \, d\boldsymbol{\eta}_{n+j} \, . \tag{9.17} $$

Equation (9.17) shows that the transfer of spatial frequencies is non-trivial if and only if the spatial frequencies $\boldsymbol{\eta}_1, \ldots, \boldsymbol{\eta}_{2n}$ of $2n$ waves going from the object satisfy the relation

$$ \sum_{j=1}^{2n} \varepsilon_j \boldsymbol{\eta}_j + \boldsymbol{\mu} = 0 \, . \tag{9.18} $$

All these results can be specialized for the case of stationary fields if $\nu_1 = -\sum_{j=2}^{m+n} \varepsilon_j \nu_j$ is substituted.

Quasi-monochromatic analysis

In the quasi-monochromatic approximation described in section 2.4 we obtain the following equation for the transmission of spatial frequencies

$$ {}^i\tilde{\Gamma}^{(n)}(\boldsymbol{\mu}_1, \ldots, \boldsymbol{\mu}_n; \tau) = $$

$$ = \exp\left(i2\pi\bar{\nu} \sum_{j=2}^{2n} \varepsilon_j \tau_j\right) \int_{-\infty}^{+\infty} \ldots \int \mathscr{L}^{(n,n)}(\boldsymbol{\eta}_1, \ldots, \boldsymbol{\eta}_n, \boldsymbol{\eta}_1 + \boldsymbol{\mu}_1, \ldots, \boldsymbol{\eta}_n + \boldsymbol{\mu}_n) \, . $$

$$. \prod_{j=1}^{n} {}^o\tilde{u}^*(\boldsymbol{\eta}_j) \, {}^o\tilde{u}(\boldsymbol{\eta}_j + \boldsymbol{\mu}_j) \, d\boldsymbol{\eta}_j \, , \tag{9.19} $$

where ${}^i\Gamma^{(n)}$ corresponded to ${}^iG^{(n)}$ and

$$ \mathscr{L}^{(n,n)}(\boldsymbol{\eta}_1, \ldots, \boldsymbol{\eta}_{2n}) = $$

$$ = \int_{-\infty}^{+\infty} \ldots \int \prod_{j=1}^{n} \tilde{K}^*(\boldsymbol{\mu}'_j + \boldsymbol{\eta}_j, \bar{\nu}) \, \tilde{K}(\boldsymbol{\mu}'_j + \boldsymbol{\eta}_{n+j}, \bar{\nu}) \, {}^o\tilde{\gamma}^{(n,n)}(\boldsymbol{\mu}'_1, \ldots, \boldsymbol{\mu}'_n) \prod_{j=1}^{n} d\boldsymbol{\mu}'_j \, . \tag{9.20} $$

If the size of apertures in the optical system and the size of the source are finite (i.e. the *pupil function* $\tilde{K}$ and the degree of coherence ${}^o\tilde{\gamma}$ over the source are non-zero only in finite regions), then the generalized transfer functions $\mathscr{L}^{(n,n)}$ (as well as the function ${}^i\tilde{\Gamma}^{(n)}$) are also non-zero only in finite regions. Spatial frequencies of the object structure filtered by the system through $\mathscr{L}^{(n,n)}$ do not contribute to the image function ${}^i\Gamma^{(n)}$. This fact means that in general the object cannot be reconstructed uniquely from its image. However, considering finite-sized objects and using the so-called Paley-Wiener theorem on entire functions, square integrable objects may be reconstructed uniquely (section 9.2).

The case $n = 1$

Putting $n = m = 1$ in (9.12) we arrive at the transfer function introduced by Parrent [336] (if it is specialized to stationary fields). Further from (9.15a) we obtain for stationary fields and $n = 1$

$$ {}^iI^{(1)}(\mathbf{x}') = \iint_{-\infty}^{+\infty} \mathscr{L}^{(1,1)}(\boldsymbol{\eta}_1, \boldsymbol{\eta}_2)\, {}^o\tilde{u}^*(\boldsymbol{\eta}_1)\, {}^o\tilde{u}(\boldsymbol{\eta}_2)\, . $$

$$. \exp\left[i2\pi\, \mathbf{x}' . (\boldsymbol{\eta}_1 - \boldsymbol{\eta}_2)\right] d\boldsymbol{\eta}_1\, d\boldsymbol{\eta}_2\,, \tag{9.21} $$

where

$$ \mathscr{L}^{(1,1)}(\boldsymbol{\eta}_1, \boldsymbol{\eta}_2) = $$

$$ = \int_0^{\infty} \left\{ \iint_{-\infty}^{+\infty} \tilde{K}^*(\boldsymbol{\mu}' + \boldsymbol{\eta}_1, \nu)\, \tilde{K}(\boldsymbol{\mu}' + \boldsymbol{\eta}_2, \nu)\, {}^o\tilde{g}^{(1,1)}(\boldsymbol{\mu}', \nu)\, d\boldsymbol{\mu}' \right\} d\nu \tag{9.22} $$

is the transmission cross-coefficient for imaging with polychromatic light; second-order coherence effects are included. For quasi-monochromatic light we get the transmission cross-coefficient introduced by Hopkins [201, 202]

$$ \mathscr{L}^{(1,1)}(\boldsymbol{\eta}_1, \boldsymbol{\eta}_2) = \int_{-\infty}^{+\infty} \tilde{K}^*(\boldsymbol{\mu}' + \boldsymbol{\eta}_1, \bar{\nu})\, \tilde{K}(\boldsymbol{\mu}' + \boldsymbol{\eta}_2, \bar{\nu})\, {}^o\tilde{\gamma}^{(1,1)}(\boldsymbol{\mu}')\, d\boldsymbol{\mu}'\,. \tag{9.23} $$

If the apertures and the source have the form of circles, then the integration in (9.23) is taken over the common part of the circle of ${}^o\tilde{\gamma}$ with the centre at the origin and the circles of $\tilde{K}^*$ and $\tilde{K}$ centred on $\boldsymbol{\eta}_1$ and $\boldsymbol{\eta}_2$ respectively.

Putting $n = 1$ in (9.19) and using (9.20), we obtain [20]

$$ {}^i\tilde{I}^{(1)}(\boldsymbol{\mu}) = \int_{-\infty}^{+\infty} \left\{ \iint_{-\infty}^{+\infty} \tilde{K}^*(\boldsymbol{\mu}' + \boldsymbol{\eta}, \bar{\nu})\, \tilde{K}(\boldsymbol{\mu}' + \boldsymbol{\eta} + \boldsymbol{\mu}, \bar{\nu})\, {}^o\tilde{\gamma}^{(1,1)}(\boldsymbol{\mu}')\, d\boldsymbol{\mu}' \right\} . $$

$$. {}^o\tilde{u}^*(\boldsymbol{\eta})\, {}^o\tilde{u}(\boldsymbol{\eta} + \boldsymbol{\mu})\, d\boldsymbol{\eta} = $$

$$ = \int_{-\infty}^{+\infty} \left\{ \iint_{-\infty}^{+\infty} \tilde{K}^*\left(\boldsymbol{\mu}' - \frac{\boldsymbol{\mu}}{2}, \bar{\nu}\right) \tilde{K}\left(\boldsymbol{\mu}' + \frac{\boldsymbol{\mu}}{2}, \bar{\nu}\right) {}^o\tilde{\gamma}^{(1,1)}(\boldsymbol{\mu}' - \boldsymbol{\eta})\, d\boldsymbol{\mu}' \right\} . $$

$$. {}^o\tilde{u}^*\left(\boldsymbol{\eta} - \frac{\boldsymbol{\mu}}{2}\right) {}^o\tilde{u}\left(\boldsymbol{\eta} + \frac{\boldsymbol{\mu}}{2}\right) d\boldsymbol{\eta}\,, \tag{9.24} $$

where we have used the substitutions $\boldsymbol{\mu}' + \boldsymbol{\eta} + (\boldsymbol{\mu}/2) \rightarrow \boldsymbol{\mu}'$ and $\boldsymbol{\eta} + (\boldsymbol{\mu}/2) \rightarrow \boldsymbol{\eta}$. This is the well-known relation describing imaging with partially coherent quasi-monochromatic light.

Limiting cases of coherent and incoherent light

By analogy with (4.12) we may write for a fully coherent field

$$\Gamma^{(n,n)}(x_1, \ldots, x_{2n}) = \prod_{j=1}^{n} V^*(x_j)\, V(x_{n+j}) \tag{9.25}$$

for all $x_1, \ldots, x_{2n}$ and all n, where V is a function independent of n. It is obvious that this condition can be fulfilled by a deterministic classical field without fluctuations, as we have mentioned. An analogous relation for stationary fields reads

$$\Gamma^{(n,n)}(\boldsymbol{x}_1, \ldots, \boldsymbol{x}_{2n}; \boldsymbol{\tau}) = \prod_{j=1}^{n} U^*(\boldsymbol{x}_j)\, U(\boldsymbol{x}_{n+j}) \exp\left(i2\pi\nu_0 \sum_{k=2}^{2n} \varepsilon_k \tau_k\right), \tag{9.26}$$

where U is a function of $\boldsymbol{x}$ only (independent of n) and ν_0 is the frequency of the coherent field. A detailed analysis of coherent fields will be given in Sec. 12.2.

Equation (9.26) leads to the conclusion that (4.23) holds again, i.e. the transfer function equals $\tilde{K}(\boldsymbol{\mu}, \nu_0)$ in this case. However, from (9.26) it also follows that

$$\gamma^{(n,n)}(\boldsymbol{x}_1, \ldots, \boldsymbol{x}_{2n}; \boldsymbol{\tau}) = \exp\left\{i \sum_{j=1}^{n} \left[\beta(\boldsymbol{x}_j) - \beta(\boldsymbol{x}_{n+j})\right] + i2\pi\nu_0 \sum_{k=2}^{2n} \varepsilon_k \tau_k\right\}, \tag{9.27a}$$

where β is the phase of U and so

$$\begin{aligned}\gamma^{(n,n)}(\boldsymbol{x}_1, \ldots, \boldsymbol{x}_n, \boldsymbol{x}_n, \ldots, \boldsymbol{x}_1; \boldsymbol{\tau}) &\equiv \gamma^{(n,n)}(\boldsymbol{x}_1, \ldots, \boldsymbol{x}_n; \boldsymbol{\tau}) = \\ &= \exp\left(i2\pi\nu_0 \sum_{k=2}^{2n} \varepsilon_k \tau_k\right).\end{aligned} \tag{9.27b}$$

Thus making use of the Fourier transform we arrive at

$$\tilde{g}_H^{(n,n)}(\boldsymbol{\mu}'_1, \ldots, \boldsymbol{\mu}'_n; \boldsymbol{\nu}) = \prod_{j=1}^{n} \delta(\boldsymbol{\mu}'_j) \prod_{k=2}^{2n} \delta(\nu_k - \nu_0). \tag{9.28}$$

Hence equation (9.16) considered for stationary light $(\nu_1 = -\sum_{j=2}^{2n} \varepsilon_j \nu_j)$ gives us

$$\begin{aligned}&\mathscr{L}_H^{(n,n)}(\boldsymbol{\eta}_1, \ldots, \boldsymbol{\eta}_{2n}; \boldsymbol{\nu}) = \\ &= \prod_{j=1}^{n} \tilde{K}^*(\boldsymbol{\eta}_j, \nu_0)\, \tilde{K}(\boldsymbol{\eta}_{n+j}, \nu_0) \prod_{j=2}^{2n} \delta(\nu_j - \nu_0).\end{aligned} \tag{9.29}$$

For ${}^{i}\tilde{\Gamma}^{(n,n)}$ we obtain

$$ {}^{i}\tilde{\Gamma}^{(n,n)}(\mu_1, \ldots, \mu_{2n}; \tau) = \exp\left(i2\pi\nu_0 \sum_{j=2}^{2n} \varepsilon_j \tau_j\right) \cdot $$

$$ \cdot \prod_{j=1}^{n} \tilde{K}^*(\mu_j, \nu_0)\, \tilde{K}(\mu_{n+j}, \nu_0)\, {}^{o}\tilde{u}^*(\mu_j)\, {}^{o}\tilde{u}(\mu_{n+j}) \,. \tag{9.30} $$

This is in agreement with a spatial Fourier analysis of equation (4.23) performed in the isoplanatic region. However, it can be seen that the relation between the object and its image is nonlinear if the detection is included in the process of imaging even if coherent light is used for imaging.

We shall see in section 17.1 that chaotic light is completely characterized by the second-order correlation function. Thus assuming incoherent light we may restrict ourselves to the second-order correlation function. As $G^{(1,1)}(x_1, x_2; \nu) = G^{(1,1)}(x_1, x_1; \nu)\, \delta(x_1 - x_2)$ (according to (4.2)), equation (9.10) considered for the stationary field reduces to

$$ {}^{i}G^{(1,1)}(x_1', x_2'; \nu) = \int_{\mathscr{A}} K^*(x_1' - x, \nu)\, K(x_2' - x, \nu)\, {}^{o}G^{(1,1)}(x, x; \nu)\, dx \,. \tag{9.31} $$

A spatial Fourier analysis of this equation gives

$$ {}^{i}\tilde{G}^{(1,1)}(\mu_1, \mu_2; \nu) = \tilde{K}^*(-\mu_1, \nu)\, \tilde{K}(\mu_2, \nu)\, {}^{o}\tilde{G}^{(1,1)}(\mu_1 + \mu_2, \nu) \,, \tag{9.32} $$

which shows that the correlation between light beams associated with the spatial frequencies μ_1 and μ_2 in the image is determined by the object function ${}^{o}\tilde{G}^{(1,1)}$ associated with the sum frequency $(\mu_1 + \mu_2)$ in the object.

As $\gamma^{(1,1)}(x_1, x_2; \tau) = \exp(-i2\pi\bar{\nu}\tau)\, \delta(x_1 - x_2)$ $(|\tau| \ll 1/\Delta\nu)$ holds for incoherent quasi-monochromatic light, equation (9.22) gives the standard transfer function for incoherent light in the form of the convolution of the pupil functions $\tilde{K}$,

$$ \mathscr{L}^{(1,1)}(\eta_1, \eta_2) = \int_{-\infty}^{+\infty} \tilde{K}^*(\mu' + \eta_1, \bar{\nu})\, \tilde{K}(\mu' + \eta_2, \bar{\nu})\, d\mu' = $$

$$ = \int_{-\infty}^{+\infty} \tilde{K}^*(\mu', \bar{\nu})\, \tilde{K}(\mu' + \eta_2 - \eta_1, \bar{\nu})\, d\mu' \,. \tag{9.33} $$

Equation (9.31) considered for $x_1' \equiv x_2'$ shows that only the process of imaging with incoherent light is linear if detection is included.

Some further details together with an analysis of weak visibility objects (for which the optical system works as a normal filter) and the matrix formulation of the present results can be found in [363].

9.2 Reconstruction of an object from its image and similarity between object and image

The general formulation of the relation between an object and its image just developed may be used to deduce some consequences about the reconstruction of an object from its image and to study the similarity between an object and its image taking into account coherence effects.

The problem of the reconstruction of the object characteristics from those of the image was considered in [3] for imaging with partially coherent light and in [16] for imaging with coherent or incoherent light. It has been shown, in general, that this problem cannot be solved uniquely. Khurgyn and Yakovlev [16], however, have shown (considering *finite-sized objects* and using the *Paley-Wiener theorem on entire functions*) that a unique solution of the problem exists in the space of square integrable functions L_2 (in the sense of Lebesgue). Consequently the reconstruction of the object is unique.

Reconstruction of the object from its image and analytic continuation

Confining ourselves to quasi-monochromatic light for simplicity (this restriction is not essential for the considerations below) we may start with the following propagation law

$$
{}^{i}\Gamma^{(n,n)}(x_1', \ldots, x_{2n}') = \\
= \int_{\mathscr{A}} \cdots \int \prod_{j=1}^{n} K^*(x_j' - x_j)\, K(x_{n+j}' - x_{n+j})\, {}^{o}\Gamma^{(n,n)}(x_1, \ldots, x_{2n}) \prod_{j=1}^{n} dx_j\, dx_{n+j} \tag{9.34}
$$

which follows from (9.10) by a Fourier transform and by putting $t = 0$ with the assumption that $K(x) \equiv K(x, \bar{\nu})$; only the one-dimensional case will be considered here for simplicity.

We are to determine the object quantity ${}^{o}\Gamma^{(n,n)}$ (i.e. the degree of coherence ${}^{o}\gamma^{(n,n)}$ and the nth-order intensity ${}^{o}I^{(n)}(x) \equiv {}^{o}\Gamma^{(n,n)}(x, \ldots, x)$) from the image specified by ${}^{i}\Gamma^{(n,n)}$. If we use the spatial Fourier analysis of (9.34) we may solve this equation in the form

$$
{}^{o}\Gamma^{(n,n)}(x_1, \ldots, x_n) = \\
= \int_{\Theta} \cdots \int \frac{{}^{i}\tilde{\Gamma}^{(n,n)}(\mu_1, \ldots, \mu_{2n})}{\prod_{j=1}^{n} \tilde{K}^*(-\mu_j)\, \tilde{K}(\mu_{n+j})} \exp\left(-i2\pi \sum_{j=1}^{2n} \mu_j x_j\right) \prod_{j=1}^{2n} d\mu_j + f(x_1, \ldots, x_{2n}), \tag{9.35}
$$

where Θ is a region of spatial frequencies where $\tilde{K}(\mu) \not\equiv 0$ and outside of which $\tilde{K}(\mu) \equiv 0$. The function f is an arbitrary function the Fourier transform of which is zero in Θ. We see indeed that the reconstruction is not in general unique. Assuming that the function ${}^{o}\Gamma^{(n,n)}(x_1, \ldots, x_{2n})$ is non-zero in a region $\Pi = (\Pi')^{2n}$ only, where Π' is a one-dimensional finite interval of a finite-sized object, also assumed to be square integrable in any x_j, then the function

$$ {}^{o}\tilde{\Gamma}^{(n,n)}(\mu_1, \ldots, \mu_{2n}) = \frac{{}^{i}\tilde{\Gamma}^{(n,n)}(\mu_1, \ldots, \mu_{2n})}{\prod_{j=1}^{n} \tilde{K}^*(-\mu_j)\, \tilde{K}(\mu_{n+j})}, \quad \text{in } \Theta \tag{9.36} $$

is an entire function according to the Paley-Wiener theorem [25, 16]. We can define the function ${}^{o}\tilde{\Gamma}^{(n,n)}$ for all μ_j by means of the Taylor series

$$ {}^{o}\tilde{\Gamma}^{(n,n)}(\mu_1, \ldots, \mu_{2n}) = \sum_{r_1, \ldots, r_{2n}} \prod_{j=1}^{2n} \frac{(\mu - \mu_{0j})^{r_j}}{r_j!} \frac{\partial^{r_j}}{\partial \mu_j^{r_j}} {}^{o}\tilde{\Gamma}^{(n,n)}(\mu_1, \ldots, \mu_{2n}) \Bigg|_{\substack{\mu_1 = \mu_{01} \\ \cdots \\ \mu_{2n} = \mu_{02n}}}, \tag{9.37} $$

where $\mu_{01}, \ldots, \mu_{02n} \in \Theta$. Thus the square integrable object function ${}^{o}\Gamma^{(n,n)}$ can be determined uniquely by

$$ {}^{o}\Gamma^{(n,n)}(x_1, \ldots, x_{2n}) = $$

$$ = \int_{-\infty}^{+\infty} \cdots \int \left\{ \sum_{r_1, \ldots, r_{2n}} \prod_{j=1}^{2n} \frac{(\mu - \mu_{0j})^{r_j}}{r_j!} \frac{\partial^{r_j}}{\partial \mu_j^{r_j}} \left[\frac{{}^{i}\tilde{\Gamma}^{(n,n)}(\mu_1, \ldots, \mu_{2n})}{\prod_{k=1}^{n} \tilde{K}^*(-\mu_k)\, \tilde{K}(\mu_{n+k})} \right]_{\substack{\mu_1 = \mu_{01} \\ \cdots \\ \mu_{2n} = \mu_{02n}}} \right\} . $$

$$. \exp\left(-i2\pi \sum_{k=1}^{2n} \mu_k x_k\right) \prod_{k=1}^{2n} d\mu_k \,. \tag{9.38} $$

This is exactly the method mentioned above and used in [368, 374] to obtain super-resolution in connection with the holographic method of deconvolution. In these papers some further possibilities of analytic continuation through orthogonal polynomials were proposed and used in [371] to study the existence of the so-called diagonal representation of the density matrix (section 13.3). Such an analysis assumes the noise to be negligible in the system.

The problem of the correspondence between an object and its image was investigated by Toraldo di Francia [444] from the viewpoint of communication theory using the sampling theorem; this enabled him to introduce the notion of "degrees of freedom". If noise is present in the system, the number of degrees of freedom of the object is infinite while the number of degrees of freedom of the image is finite and consequently some information is lost in the imaging process. This fact expresses the filtering properties of the system and the object cannot be reconstructed uniquely from the image – classes of objects corresponding to a given image exist. A priori knowledge about the object also plays an important role during the reconstruction (e.g. the concept of the resolving power of the optical system). However, if noise may be assumed to be negligible (this question has been discussed in [431]), the number of degrees of freedom of the object as well as of the image is infinite (cf. a discussion in [16], § 6.3). They are contained in the non-filtered part of the spatial spectrum of the image and may be obtained by means of precise measurements or more readily by means of analytic continuation. Thus the reconstruction of the object function from that of the image is unique in this case. Some limitations caused by a given noise level have been discussed in [446, 447].

As a demonstration of the utility of these considerations we mention the concept of the *two-point resolving power* [16]. We consider the one-dimensional case and coherent light for simplicity and assume the object structure to be from two points distance $2b$ apart so that $f(x) = \delta(x - b) + \delta(x + b)$. Choosing the diffraction function in the form

$$K(x) = \frac{\sin 2\pi x}{\pi x}, \tag{9.39}$$

which is the diffraction function of a slit, we obtain

$$\tilde{K}(\mu) = \begin{cases} 1, & |\mu| < 1, \\ 0, & |\mu| > 1. \end{cases} \tag{9.40}$$

Thus this system filters all spatial frequencies $|\mu| > 1$ from the image. The image $g(x')$ can be calculated as

$$g(x') = \int_{-\infty}^{+\infty} K(x' - x) f(x)\, dx = \frac{\sin 2\pi(x' - b)}{\pi(x' - b)} + \frac{\sin 2\pi(x' + b)}{\pi(x' + b)}. \tag{9.41}$$

Using the deconvolution we obtain

$$\tilde{f}(\mu) = \frac{\tilde{g}(\mu)}{\tilde{K}(\mu)} = \begin{cases} 2\cos 2\pi\mu b, & |\mu| < 1, \\ \dfrac{0}{0}, & |\mu| > 1. \end{cases} \tag{9.42}$$

Hence the function $\tilde{f}(\mu)$ is given for various separations $2b$ of the points by a set of cosine functions which end at $\mu = \pm 1$.

The limit of the resolving power of the system may be defined by the minimal root of the equation

$$\left. \frac{d^2 g(x')}{dx'^2} \right|_{x'=0} = 0\,, \tag{9.43a}$$

which gives

$$\tan(2\pi b_0) = \frac{2(2\pi b_0)}{2 - (2\pi b_0)^2}\,, \quad \text{i.e.} \quad 2\pi b_0 \approx 2.1\,. \tag{9.43b}$$

This means that only the curves $b > b_0$ may be determined in the classical sense and the corresponding points may be distinguished only if the *a priori* knowledge that two points are to be considered is used; (even in this case the function $\tilde{f}(\mu)$ must be known for all μ if $f(x)$ is to be determined uniquely). When $b < b_0$ one cannot determine, from the classical point of view, the form of the function $\tilde{f}(\mu) = \tilde{g}(\mu)/\tilde{K}(\mu)$ if it is calculated for $|\mu| < 1$ only. This interval is too short and the function is nearly constant over this interval. The separation $2b$ of the points cannot be determined in this case and the points are not distinguishable. However, taking into account that $f(x)$ is a finite-sized object ($f(x) = 0$, $|x| > b$) then the function $\tilde{f}(\mu)$ is an entire function which can be defined by the Taylor series for all μ. We shall be able to find the form of $\tilde{f}(\mu)$ and we shall arrive at the unique function $f(x) = \delta(x - b) + \delta(x + b)$ without any restrictions on the separation $2b$. In this way we may, in principle, distinguish arbitrarily near points and there is no limit to the resolving power of the system (in the absence of noise).

Of course, an identical analysis can be performed for the correlation function ${}^o\Gamma^{(n,n)}(x_1, \ldots, x_{2n}) = {}^o\gamma^{(n,n)}(x_1, \ldots, x_{2n})\, {}^o u(x_1) \ldots {}^o u(x_{2n})$ with ${}^o u(x) = \delta(x - b) + \delta(x + b)$, which is also a finite-sized function in all variables. Consequently ${}^o\tilde{\Gamma}^{(n,n)}(\mu_1, \ldots, \mu_{2n})$ is an entire function in all variables μ_j.

Similarity between an object and its image — algebraic integral equations

The similarity problem, i.e. finding such an object structure $f(x)$ which is imaged similarly in the sense that there exists a number λ such that the image function $g(x) = \lambda^{-1} f(x)$, was first formulated by Mandelstam [287, 288] for coherent and incoherent light. These results were completed in a certain sense in [16] and this problem was considered for partially coherent light beams in [347, 349, 354, 372, 373]. The similarity problem formulated for coherent light is of basic importance in the theory of optical resonators i.e. for lasers [445, 448]. The role of partial coherence in optical resonators was investigated by Wolf [480, 481] and by Streifer [429].

If we consider the similarity expressed by the condition

$$ {}^{i}\Gamma^{(n,n)}(x_1, \ldots, x_{2n}) = \frac{1}{|\lambda|^{2n}} \, {}^{o}\Gamma^{(n,n)}(x_1, \ldots, x_{2n}) \,, \tag{9.44} $$

we obtain from (9.34) by using a spatial Fourier transform

$$ {}^{o}\tilde{\Gamma}^{(n,n)}(\mu_1, \ldots, \mu_{2n}) = |\lambda|^{2n} \prod_{j=1}^{n} \tilde{K}^*(-\mu_j)\, \tilde{K}(\mu_{n+j})\, {}^{o}\tilde{\Gamma}^{(n,n)}(\mu_1, \ldots, \mu_{2n}) \,. \tag{9.45} $$

As the pupil function $\tilde{K}$ is zero outside a finite region, ${}^{o}\tilde{\Gamma}^{(n,n)}$ must also vanish outside a finite region. Hence, by using the Paley-Wiener theorem, the function ${}^{o}\Gamma^{(n,n)}$ can be continued into the whole complex x-plane by means of the Taylor series and as a consequence of the uniqueness of such a continuation ${}^{o}\Gamma^{(n,n)}$ cannot equal zero in any finite region (if it did it would have to be identically zero). From this result we conclude that for any finite-sized object the similarity problem in the infinite region has no solution, but the similarity problem in a finite region always has a solution.

From (9.45) it follows that one of the following equations

$$ \prod_{j=1}^{n} \tilde{K}^*(-\mu_j)\, \tilde{K}(\mu_{n+j}) = \frac{1}{|\lambda|^{2n}} \,, \tag{9.46} $$

$$ {}^{o}\tilde{\Gamma}^{(n,n)}(\mu_1, \ldots, \mu_{2n}) = 0 \tag{9.47} $$

must be true. Thus, if (9.46) holds in finite regions of the μ-space, then (9.47) must be true outside these regions. If (9.46) holds in the whole space, then K is proportional to the Dirac function and the system is ideal.

There are various possible specifications of object and image [373] and an important case, which we discuss here, is when the image is specified by the intensity only. Confining ourselves to real functions and real similarity coefficients λ for simplicity, we arrive at the similarity equation for the amplitude u

$$ u^{2n}(x) = \lambda^{2n} \int_{\mathscr{A}} \cdots \int L(x, x_1, \ldots, x_{2n}) \prod_{j=1}^{2n} u(x_j)\, dx_j \,, \tag{9.48} $$

where

$$ L(x, x_1, \ldots, x_{2n}) = {}^{o}\gamma^{(n,n)}(x_1, \ldots, x_{2n}) \prod_{j=1}^{2n} K(x, x_j) \,. \tag{9.49} $$

One may assume that $K(x, y) = K(y, x)$ and ${}^{o}\gamma^{(n,n)}(x_1, \ldots, x_{2n})$ are continuous functions and that ${}^{o}\gamma^{(n,n)}$ is symmetric with respect to all pairs of variables $x_1, \ldots, x_{2n}$.

Equation (9.48) represents a special case of the general class of non-linear integral equations which are called *algebraic integral equations*. They were introduced in mathematics by Schmeidler [413] and they were further investigated by Peřinová [376, 377]. The following theorems can be shown to be valid [372, 373].

If $L(x, x_1, \ldots, x_{2n})$ *is a positive continuous function, then there exist a real eigenvalue* $\lambda_0 > 0$ *and a real normalized eigenfunction* $u_0(x) > 0$ *(the norm is defined as* $\|u_0\| = \{\int u_0^{2n}(x)\, dx\}^{1/2n} = 1$*) such that* (9.48) *holds, where*

$$\lambda_0 = \left\{ \int_{\mathscr{A}} \cdots \int L(x, x_1, \ldots, x_{2n}) \prod_{j=1}^{2n} u_0(x_j)\, dx_j\, dx \right\}^{-1/2n}; \tag{9.50}$$

λ_0 *is the smallest eigenvalue and it is simple.*

Besides this theorem on the existence of an eigensolution of (9.48) the following theorem on the countable infinity of eigenvalues can be proved.

Let the following assumptions be valid for every eigenvalue λ_0 *and the corresponding eigenfunction* $u_0(x)$ *of equations* (9.48):

a) $u_0(x)$ *does not equal zero in* $\mathscr{A}$,
b) λ_0^{2n} *is the simple eigenvalue of the kernel*

$$u_0^{-2n+1}(x) \int_{\mathscr{A}} \cdots \int L(x, y, x_2, \ldots, x_{2n}) \prod_{j=2}^{2n} u_0(x_j)\, dx_j\,,$$

c) *the integral* $\int_{\mathscr{A}} \varphi(x)\, \psi^*(x)\, dx$ *is different from zero for the eigenfunction* $\varphi(x) \equiv u_0(x)$ *and the associated eigenfunction* $\psi(x)$ *corresponding to the eigenvalue* λ_0^{2n};

then there exists at most a countably infinite set of eigenvalues.

Because of the non-linearity of equation (9.48) there may in general exist an interval of eigenvalues. As a consequence there may arise *branching* of a structure $u_0(x)$ imaged similarly, corresponding to an eigenvalue λ_0, into several structures imaged similarly for eigenvalues belonging to a neighbourhood of λ_0. This branching effect is caused by *partial coherence* creating *typical non-linearity* in imaging. An outline of the branching theory with application to the optical similarity problem can be found in [373].

For the coherent and incoherent light we obtain as special cases the linear integral equations for the amplitude and the intensity respectively, studied for example by Mandelstam [287, 288].

In this chapter the space-time behaviour of the correlation functions has again been the main subject of investigation and so the results presented here are correct for both the classical and quantum correlation functions.

Chapter 10

FOURTH- AND HIGHER-ORDER COHERENCE PHENOMENA—SEMICLASSICAL TREATMENT

In the preceding chapters of this book, particularly in Chapter 3, we have discussed the possibility of experimental determination of second-order coherence using the interference and diffraction of light. In this chapter we discuss unconventional techniques based on the *correlation photoelectric measurements* and the *photon-counting measurements* through which the statistical properties of optical fields, including effects of arbitrary order, can be determined in principle. In the first case one measures correlation functions of the type $\Gamma^{(n,n)}$ while in the second case one measures the distribution of emitted photoelectrons by a photoelectric detector whose statistics directly reflect the statistics of the absorbed photons. Thus the higher-order moments (correlation functions) may be calculated using the photon-counting distribution. Although the higher-order coherence effects will be investigated by quantum methods in the following chapters we discuss these effects briefly here using the so-called semiclassical treatment which is relatively simple. A typical feature of this semiclassical approach is that the field is treated classically while its interaction with matter in the detection process is described quantum-mechanically.

As we have seen, the second-order coherence effects are characterized by the second-order correlation function which has the physical dimension of intensity (two amplitudes of the field are multiplied and averaged). The fourth- and higher-order correlation functions may be realized as correlations of intensities of the field at various space-time points. Fourth-order measurement of this type was first performed in 1956 by Hanbury Brown and Twiss [92–94], who used fast photoelectric detectors to measure the correlation of intensity fluctuations. A number of experiments of this type were performed both by correlation and photoelectric coincidence techniques (Hanbury Brown and Twiss [95, 96], Twiss, Little and Hanbury Brown [456], Twiss and Little [455], Rebka and Pound [391], Brannen, Ferguson and Wehlau [89] and Harwit [194]). The sixth-order correlation function was measured by Davidson and Mandel [125] (see also [503]). Another means of gaining information about higher-order moments (correlation functions) is based on measuring the photon-counting statistics as we have mentioned. This approach provides higher-order moments more readily; (see Arecchi [41], Arecchi, Berné and Burlamacchi [44], Arecchi, Berné, Sona and Burlamacchi [46], Arecchi, Berné and Sona [45] (for a review see [42]), Freed and Haus [153, 154], Johnson, Jones, McLean and Pike [218], Fray, Johnson, Jones, McLean and Pike [152] (for a review see [386, 387]), Martienssen and Spiller [292, 293], Smith and Armstrong [425, 426] (for a review see [52]) and Chang, Korenman, Alley and Detenbeck [114]).

A new branch of interferometry, the so-called *correlation interferometry*, is based on the fourth- and higher-order correlation measurements. By this technique the angular diameters of stars can be obtained, under better conditions than in measurements using the Michelson stellar interferometer [93, 97, 98]. The correlation measurements also yield information about the state of polarization of light [477, 283, 270] and about the spectral distribution of light beams [148, 149, 162, 164, 170, 171, 478, 482]. A theoretical explanation of the fourth-order correlation effect based on the classical or the semiclassical description of light beams was first given by Purcell [389]. The ideas of this paper have been continued by Wolf [474], Janossy [215, 216], Mandel [263, 264, 269], Kahn [220], Mandel and Wolf [283] and Mandel, Sudarshan and Wolf [281].

The problem was investigated from a quantum-mechanical point of view by Dicke [131, 132], Senitzky [418, 419], Fano [142], Glauber [172–176, 180, 181], Kelley and Kleiner [231], Goldberger and Watson [185, 186], Holliday and Sage [198], Lehmberg [520], Mandel and Meltzer [523], Peřina [357, 364] among others.

A basic relation of the semiclassical theory of photodetection expresses the probability $p(t)\,dt$ of emission (absorption) of a photoelectron (a photon) in the time interval $(t, t + dt)$ by means of the classical intensity $I(t)$ (it is in general stochastic) of the quasi-monochromatic light which is incident on the photocathode

$$p(t)\,dt = \eta\, I(t)\,dt = \eta\, V^*(t)\, V(t)\,dt\,, \tag{10.1}$$

where η represents the photo-efficiency of the photoelectric detector and linearly polarized light is assumed. This relation may be derived with the help of the perturbation theory of quantum mechanics [281] and follows from simple quantum-mechanical consideration (Glauber [173–176] and also Chapter 12). If $V(t)$ is an eigenvalue of the annihilation operator

$\hat{A}^{(+)}(t)$ of a photon in the field in a state $|\rangle$, then the probability of the photodetection process is proportional to the square modulus of the matrix element $\langle f|\hat{A}^{(+)}(t)|\, i\rangle$ connecting the initial state $|_i\rangle$ and the final state $|f\rangle$ of the field in agreement with (10.1).

10.1 Photon-counting distribution

We determine the probability $p(n, T, t)$ of emission of n photoelectrons (i.e. of n photoelectric counts, or alternatively the probability of detection of n photons, see Chapter 14) by the plane photocathode on which light is normally incident for a time interval $(t, t + T)$, where T is fixed. First we assume that the intensity $I(t)$ has no random fluctuations. Dividing the interval T into sub-intervals with the points $t + i\,\Delta T \equiv t_i$, $i = 0, 1, \ldots, T/\Delta T$, we can calculate the probability $p(n, T, t)$ of n counts occurring in the interval $(t, t + T)$. This is the sum over all possible sequences of counts of the product of probabilities of obtaining a count at time t_{r_1}, a count at $t_{r_2}, \ldots,$ a count at t_{r_n}, multiplied by the probabilities of obtaining no counts in the remaining $T/\Delta T - n$ intervals [269],

$$p(n, T, t) =$$

$$= \lim_{\Delta T \to 0} \sum_{r_1=0}^{T/\Delta T} \cdots \sum_{r_n=0}^{T/\Delta T} \frac{1}{n!} \eta^n I(t_{r_1}) \ldots I(t_{r_n}) (\Delta T)^n \frac{\prod_{i=0}^{T/\Delta T} [1 - \eta I(t_i) \Delta T]}{\prod_{j=1}^{n} [1 - \eta I(t_{r_j}) \Delta T]} = \tag{10.2a}$$

$$= \frac{1}{n!} \left[\eta \int_t^{t+T} I(t')\, dt' \right]^n \exp \left[-\eta \int_t^{t+T} I(t')\, dt' \right], \tag{10.2b}$$

which is a Poisson distribution for the number of counts n with the mean value of counts $\langle n\rangle = \eta \int_t^{t+T} I(t')\, dt'$.

If the intensity $I(t)$ is a *stochastic function*, (10.2b) must be averaged over all possible realizations of $I(t)$ and we arrive at

$$p(n, T, t) = \frac{1}{n!} \langle (\eta W)^n \exp(-\eta W) \rangle \equiv$$

$$\equiv \frac{1}{n!} \int_0^\infty (\eta W)^n \exp(-\eta W)\, P(W)\, dW, \tag{10.3a}$$

where

$$W = \int_t^{t+T} I(t')\, dt' \tag{10.3b}$$

and $P(W)$ is the distribution of the integrated intensity W. The *photodetection equation* (10.3a) was first derived by Mandel [263, 264, 269].†

For stationary fields the distribution $p(n, T, t)$ will be independent of t and $p(n, T, t) \equiv p(n, T)$.

In general the distribution of counts will depart from the classical Poisson statistics but an ideal laser source, perfectly stabilized so that the intensity does not fluctuate, leads to the Poisson distribution of counts and will behave like a source of classical particles in this case.

Some properties of the photon-counting distribution

First we calculate the *variance* of the number of counts n. We obtain from (10.3a)

$$\langle n \rangle = \sum_{n=0}^{\infty} p(n, T, t)\, n = \left\langle (\eta W) \frac{d}{d(\eta W)} \left[\sum_{n=0}^{\infty} \frac{(\eta W)^n}{n!} \right] \exp(-\eta W) \right\rangle = \eta \langle W \rangle, \tag{10.4a}$$

$$\langle n^2 \rangle = \sum_{n=0}^{\infty} p(n, T, t)\, n^2 =$$

$$= \left\langle (\eta W) \frac{d}{d(\eta W)} \left\{ (\eta W) \frac{d}{d(\eta W)} \left[\sum_{n=0}^{\infty} \frac{(\eta W)^n}{n!} \right] \right\} \exp(-\eta W) \right\rangle =$$

$$= \eta \langle W \rangle + \eta^2 \langle W^2 \rangle. \tag{10.4b}$$

We can obtain in the same way

$$\langle n^3 \rangle = \eta \langle W \rangle + 3\eta^2 \langle W^2 \rangle + \eta^3 \langle W^3 \rangle, \tag{10.4c}$$

$$\langle n^4 \rangle = \eta \langle W \rangle + 7\eta^2 \langle W^2 \rangle + 6\eta^3 \langle W^3 \rangle + \eta^4 \langle W^4 \rangle, \tag{10.4d}$$

....

The variance of n can be obtained from (10.4a) and (10.4b)

$$\langle (\Delta n)^2 \rangle = \langle n^2 \rangle - \langle n \rangle^2 = \eta \langle W \rangle + \eta^2 \langle (\Delta W)^2 \rangle =$$

$$= \langle n \rangle + \eta^2 \langle (\Delta W)^2 \rangle, \tag{10.5}$$

where

$$\langle (\Delta W)^2 \rangle = \langle W^2 \rangle - \langle W \rangle^2 \tag{10.6}$$

is the variance of the integrated intensity W. The formula (10.5) has a very simple physical interpretation. It shows that the variance of fluctuations

† A quantum derivation of the photodetection equation is given in Chap. 14.

in the number of ejected photoelectrons is the sum of the fluctuations in the number of classical particles obeying the Poisson distribution (the term $\langle n \rangle$, since for the Poisson distribution $\langle n \rangle^n \exp(-\langle n \rangle)/n!$, $\langle (\Delta n)^2 \rangle = \langle n \rangle$) and of the fluctuations in a classical wave field (the wave interference term $\eta^2 \langle (\Delta W)^2 \rangle$). This result generalizes the earlier results by Einstein [137, 138], Bothe [87] and Fürth [155, 156] and is correct for any light beam (thermal or non-thermal and stationary or non-stationary). Although the result refers to the fluctuations of photoelectric counts, it can be regarded as reflecting the fluctuation properties of the light itself. This fact will be pointed out in the quantum description of the statistics of light (Sec. 14.2).

Defining the *characteristic function* as

$$C^{(n)}(ix) = \langle \exp ixn \rangle = \sum_{n=0}^{\infty} p(n, T, t) \exp ixn \tag{10.7}$$

we obtain from (10.3a)

$$C^{(n)}(ix) = \langle \exp [\eta(e^{ix} - 1) W] \rangle = \tag{10.8a}$$

$$= \int_0^{\infty} P(W) \exp [\eta(e^{ix} - 1) W] \, dW = C^{(W)}(\eta(e^{ix} - 1)), \tag{10.8b}$$

where $C^{(W)}(ix)$ is the characteristic function of the probability distribution $P(W)$. The photon-counting distribution $p(n, T, t)$ may be calculated, if the characteristic function $C^{(n)}(ix)$ is given, by means of a Fourier transform

$$p(n, T, t) = \frac{1}{2\pi} \int_0^{2\pi} \exp(-ixn) \, C^{(n)}(ix) \, dx. \tag{10.9}$$

The moments $\langle n^k \rangle$, $k = 0, 1, \ldots$, may be calculated as

$$\langle n^k \rangle = \frac{d^k}{d(ix)^k} C^{(n)}(ix) \Big|_{ix=0}. \tag{10.10}$$

Substituting $\exp(ix) - 1 = is$ in (10.8b) we obtain for the characteristic function $C^{(W)}(i\eta s)$

$$C^{(W)}(i\eta s) = \langle \exp i\eta s W \rangle = \langle (1 + is)^n \rangle, \tag{10.11}$$

and hence $p(n, T, t)$ by means of the derivatives

$$p(n, T, t) = \frac{1}{n!} \frac{d^n}{d(is)^n} C^{(W)}(i\eta s) \Big|_{is=-1}. \tag{10.12}$$

The *factorial moments* are equal to

$$\left\langle \frac{n!}{(n-k)!} \right\rangle = \frac{d^k}{d(is)^k} \langle (1+is)^n \rangle \big|_{is=0} = \eta^k \langle W^k \rangle . \tag{10.13}$$

One can define the *cumulants* $\varkappa_j$ as

$$\varkappa_j^{(n)} = \frac{d^j}{d(ix)^j} \log C^{(n)}(ix) \big|_{ix=0} , \quad j = 1, 2, \ldots , \tag{10.14}$$

so that the characteristic function can be written as

$$C^{(n)}(ix) = \exp\left[\sum_{j=1}^{\infty} \varkappa_j^{(n)} \frac{(ix)^j}{j!} \right]. \tag{10.15}$$

In the same way the cumulants $\varkappa_j^{(W)}$ (called factorial cumulants) are defined by means of $C^{(W)}$. The relation between the cumulants $\varkappa_j^{(n)}$ and $\varkappa_k^{(W)}$ is obtained from (10.8a)

$$\sum_{j=1}^{\infty} \varkappa_j^{(n)} \frac{(ix)^j}{j!} = \sum_{j=1}^{\infty} \varkappa_j^{(W)} \frac{\eta^j (e^{ix} - 1)^j}{j!}. \tag{10.16}$$

Comparing coefficients of the powers x^j in the expansion of (10.16) we arrive at the same relations for the cumulants as for the moments [(10.4)] [264, 269]†

$$\begin{aligned}
\varkappa_1^{(n)} &= \eta \varkappa_1^{(W)} , \\
\varkappa_2^{(n)} &= \eta \varkappa_1^{(W)} + \eta^2 \varkappa_2^{(W)} , \\
\varkappa_3^{(n)} &= \eta \varkappa_1^{(W)} + 3\eta^2 \varkappa_2^{(W)} + \eta^3 \varkappa_3^{(W)} , \\
\varkappa_4^{(n)} &= \eta \varkappa_1^{(W)} + 7\eta^2 \varkappa_2^{(W)} + 6\eta^3 \varkappa_3^{(W)} + \eta^4 \varkappa_4^{(W)} ,
\end{aligned} \tag{10.17}$$

. . . .

The significance of the cumulants is as follows:

$$\varkappa_1^{(n)} = \frac{1}{i\langle \exp ixn \rangle} \frac{d}{dx} \langle \exp ixn \rangle \big|_{ix=0} = \langle n \rangle , \tag{10.18a}$$

† This may be seen by expanding the exponential functions in (10.8); we obtain

$$\sum_{j=1}^{\infty} \langle n^j \rangle \frac{(ix)^j}{j!} = \sum_{j=1}^{\infty} \langle W^j \rangle \frac{\eta^j (e^{ix} - 1)^j}{j!} ,$$

which is the same relation as (10.16).

$$\varkappa_2^{(n)} = \frac{1}{i^2[\langle \exp ixn \rangle]^2} \cdot$$

$$\cdot \left\{ \langle \exp ixn \rangle \frac{d^2}{dx^2} \langle \exp ixn \rangle - \left[\frac{d}{dx} \langle \exp ixn \rangle \right]^2 \right\} \Bigg|_{ix=0} =$$

$$= \langle n^2 \rangle - \langle n \rangle^2 = \langle (\Delta n)^2 \rangle \,, \tag{10.18b}$$

$$\varkappa_3^{(n)} = \langle n^3 \rangle - 3\langle n^2 \rangle \langle n \rangle + 2\langle n \rangle^3 \,, \tag{10.18c}$$

etc. The same relations hold for $\varkappa_j^{(W)}$.

In general we obtain from (10.16)

$$\varkappa_k^{(n)} = \frac{d^k}{d(ix)^k} \sum_{j=1}^{\infty} \varkappa_j^{(W)} \frac{\eta^j}{j!} \sum_{l=0}^{j} \binom{j}{l} (-1)^{j-l} \exp ixl \Big|_{ix=0} =$$

$$= \sum_{j=1}^{k} \varkappa_j^{(W)} \eta^j \sum_{l=0}^{j} \frac{(-1)^{j-l} l^k}{l!\,(j-l)!} \,, \tag{10.19}$$

where the identity $\sum_{l=0}^{j} [(-1)^l \, l^k / l! \, (j-l)!] = 0$ for $j > k \geqq 0$ has been used.

It will be shown in Chapter 14 that the relations (10.4) and (10.17) [(10.19)] are a consequence of the commutation rules for the field operators, which indicates a close formal connection between the classical and quantum descriptions of the statistical properties of light.

When the radiation is very weak, the first term in (10.17) will be dominant and the distribution $p(n, T, t)$ becomes Poissonian ($\varkappa_j^{(n)} = \eta\varkappa_1^{(W)} = \eta\langle W \rangle$ and $C^{(n)}(ix) = \exp[\eta\langle W \rangle (e^{ix} - 1)]$ from (10.15), which is just the generating function for the Poisson distribution, since $\sum_{n=0}^{\infty} [(\eta\langle W \rangle)^n \exp(-\eta\langle W \rangle)/n! \,.\, \exp ixn] = \exp[\eta\langle W \rangle (e^{ix} - 1)]$). Thus, in very weak fields the emitted photoelectrons obey the statistics of classical particles. On the other hand, in strong fields, the last term in (10.17) will be dominant; therefore $\varkappa_j^{(n)} = \eta^j \varkappa_j^{(W)}$ and the distribution of n tends towards the distribution of ηW. The latter distribution is proportional to the distribution of the integrated classical intensity and in this case the output of a photoelectric detector can be regarded as a continuous signal.

10.2 Determination of the integrated intensity distribution from the photon-counting distribution

In practice the fields are usually stationary and the photon-counting distribution $p(n, T, t) \equiv p(n, T)$ is independent of the time t. If we compute the number of emitted photoelectrons within a time interval of length T in many realizations we can determine the photon-counting distribution $p(n, T)$. The measurement may be performed using a capacitor which is charged by the emitted photoelectrons and its voltage is proportional to the number of photoelectrons emitted within the interval T. With knowledge of $p(n, T)$ we can calculate the moments $\langle n^k \rangle$ as well as the factorial moments $\eta^k \langle W^k \rangle = \langle n!/(n-k)! \rangle$ and hence the statistics of the emitted photoelectrons can be determined [41, 44, 46, 45, 42, 153, 154, 218, 152, 386, 387, 292, 293, 425, 426, 52, 345, 260, 114]. The problem now arises of how information about the statistics of the field (i.e. about the distribution $P(W)$) can be obtained from photon-counting statistics $p(n, T) \equiv p(n)$ and this may be solved, from a mathematical point of view, by inverting the photodetection equation (10.3a).

Denoting $n!\, p(n) = M_n$ and $P(W) \exp(-W) = Q(W)$ in (10.3a) (we also put $\eta = 1$ for simplicity) we find that the inversion problem reduces to the well-known *moment problem*, i.e. determining the function $Q(W)$ if the moment sequence

$$\int_0^\infty Q(W)\, W^k\, dW = \langle W^k \rangle = M_k\,, \quad k = 0, 1, \ldots \tag{10.20}$$

is given; the determination of the integrated intensity distribution $P(W)$ from the photon-counting statistics $p(n)$ is such a problem.

One can easily see that a formal solution of (10.3a) ($\eta = 1$) may be written in the form

$$P(W) = \exp(W) \sum_{n=0}^{\infty} (-1)^n\, p(n)\, \delta^{(n)}(W)\,, \tag{10.21}$$

where $\delta^{(n)}(W)$ is the n-th derivative of the Dirac δ-function.

Although such a mathematical quantity having an infinite number of terms may be meaningless sometimes† it is well defined here because of the

† Such a series cannot be considered as defining a normal generalized function either in the space of test functions which are non-zero on a finite interval and continuous in all their derivatives (D-space) or in the space of test functions which decrease like an inverse power of the argument at infinity (S-space): it may be represented as a generalized function in the Z-space of test functions, which is a Fourier transform of the D-space. Such a generalized function is sometimes called ultradistribution Z'. We shall discuss these questions in connection with the existence of the diagonal representation of the density matrix in section 13.3.

analyticity of the characteristic function

$$C^{(W)}(ix) = \int_0^{\infty} P(W) \exp ixW \, dW ; \tag{10.22}$$

this is analytic in the upper half of the complex x-plane, since $P(W) = 0$ for $W < 0$. Expressing the $\delta^{(n)}$-function in (10.21) by means of the Fourier integral,

$$\delta^{(n)}(W) = \frac{1}{2\pi} \int_{-\infty}^{+\infty} (-ix)^n \exp(-ixW) \, dx , \tag{10.23}$$

we obtain from (10.21) [485]

$$P(W) = \frac{1}{2\pi} \exp(W) \int_{-\infty}^{+\infty} C(ix) \exp(-ixW) \, dx , \tag{10.24}$$

where

$$C(ix) = \sum_{n=0}^{\infty} (ix)^n p(n) , \tag{10.25}$$

which is the characteristic function of $P(W) \exp(-W)$. From (10.22) and (10.24) $C(ix) = C^{(W)}(ix - 1)$. The analyticity of $C(ix)$ makes it possible to obtain the values of $C(ix)$ for all x by analytic continuation if the series (10.25) has a finite radius of convergence. Thus $P(W)$, determined in this way, is unique.

For example, with the Bose-Einstein distribution $p(n) = \langle n \rangle^n / (1 + \langle n \rangle)^{1+n}$ we obtain $C(ix) = (1 + \langle n \rangle - ix\langle n \rangle)^{-1}$ and $P(W) = \langle n \rangle^{-1} \exp(-W/\langle n \rangle)$, which is an exponential distribution. For the Poisson distribution $p(n) = \langle n \rangle^n \exp(-\langle n \rangle)/n!$ we obtain $C(ix) = \exp[-\langle n \rangle (1 - ix)]$ and $P(W) = \delta(W - \langle n \rangle)$, where δ is the Dirac function. Under certain assumptions, the first case describes *chaotic light* while the second case describes *light of the ideal laser* (Sec. 10.3).

As every term of (10.21) represents a generalized function with one-point support (roughly speaking only at this point is the function non-zero) we must know the infinity of $p(n)$ to be able to reconstruct $P(W)$ as an ordinary function. This cannot be reached in experiments and so we give a method based on a decomposition of $P(W)$ in terms of Laguerre polynomials [22, 58, 388, 17], in which every term has as its support the whole complex plane. Such a prescription for constructing $P(W)$ also provides an approximate means of determining $P(W)$ giving a finite number of $p(n)$ from experiment. This method will be used in Sec. 13.3 in a generalized form to study the existence of the diagonal representation of the density matrix [369, 371].

If $P(W) \in L_2$, we may look for a solution of the inverse problem for the photodetection equation in the form

$$P(W) = \exp\left[-(\zeta - 1)\, W\right] \sum_{j=0}^{\infty} c_j\, L_j^0(\zeta W)\,, \tag{10.26}$$

where L_j^0 are the Laguerre polynomials defined by ([22], Chapter 6)

$$L_j^{\mu}(x) = [\Gamma(j + \mu + 1)]^2 \sum_{s=0}^{j} \frac{(-x)^s}{s!\,(j - s)!\,\Gamma(s + \mu + 1)}\,, \tag{10.27}$$

obeying the orthogonality condition

$$\int_0^{\infty} x^{\mu} \exp(-x)\, L_j^{\mu}(x)\, L_k^{\mu}(x)\, dx = \delta_{jk} \frac{[\Gamma(j + \mu + 1)]^3}{\Gamma(j + 1)}\,; \tag{10.28}$$

$\zeta \geqq 1$ is a real number and $\Gamma(\mu) = \int_0^{\infty} x^{\mu-1} \exp(-x)\, dx$. Multiplying (10.26) by $L_k^0(\zeta W) \exp(-W)$, integrating over ζW and using (10.28) with $\mu = 0$ we obtain

$$c_k = \frac{\zeta}{(k!)^2} \int_0^{\infty} P(W) \exp(-W)\, L_k^0(\zeta W)\, dW = \tag{10.29a}$$

$$= \zeta \sum_{s=0}^{k} p(s) \frac{(-\zeta)^s}{(k - s)!\, s!}\,, \tag{10.29b}$$

where (10.27) and (10.3a) have also been used. The accuracy of an approximation in which $p(n)$ is obtained for the first N terms $(n = 0, 1, \ldots, N)$ from measurements is given by (we put $\zeta = 1$)

$$\int_0^{\infty} [P(W) - P^{(N)}(W)]^2 \exp(-W)\, dW = \sum_{n=N+1}^{\infty} (n!)^2\, c_n^2 < \varepsilon\,, \tag{10.30}$$

where $P^{(N)}$ denotes a function constructed as the N-th partial sum of (10.26) and ε is an arbitrarily small number. As $|L_k^0(x)| \leqq \exp(x/2)$ we have $|L_k^0(\zeta W)\,.\, \exp[-(\zeta - 1)\, W]| \leqq \exp(\zeta W/2) \exp[-(\zeta - 1)\, W] = \exp(-\zeta W/2 + W) \leqq 1$ for $\zeta \geqq 2$. In this case the series (10.26) will be uniformly convergent if $\sum_{j=0}^{\infty} |c_j| < \infty$.

Sometimes it is more suitable for calculations to use a slightly different decomposition showing the explicit dependence on the number of degrees of freedom (modes) M; this may be written in the form

$$P(W) = W^{M-1} \sum_{j=0}^{\infty} c_j\, L_j^{M-1}(W)\,, \tag{10.31}$$

where

$$c_j = \frac{j!}{\Gamma(j + M)} \sum_{s=0}^{j} \frac{(-1)^s}{(j - s)!\, \Gamma(s + M)}\, p(s)\,. \tag{10.32}$$

If we wish to determine a function $Q(W)$ from its moments $\langle W^k \rangle = \int_0^\infty Q(W)\, W^k\, dW = M_k$, it is sufficient to substitute $P(W) \exp(-W) = Q(W)$ and $p(s)\, s! = M_s$ in (10.31) and (10.32).

The quality of the integrated intensity distribution

We can use some theorems from the moment theory of mathematical statistics [1,4] to make clear the conditions under which the function (10.26) represents either an ordinary non-negative function or a generalized function with the support composed of a finite number of points [371].

We define the moment sequence

$$M_n = \int_0^\infty W^n\, dF(W)\,, \quad n = 0, 1, \ldots, \tag{10.33}$$

where $dF(W) = F'(W)\, dW = P(W)\, dW$; (in the same way one can consider the sequence $\{p(n)\}_{n=0}^{\infty}$ denoting $M_n = p(n)\, n!$ and $F'(W) = \exp(-W)\,.\, P(W)$). We introduce the following system of quadratic forms [1]

$$q_m = \sum_{i,k}^{m} M_{i+k} u_i u_k = \int_0^\infty \left(\sum_{i=0}^{m} W^i u_i\right)^2 dF(W)\,, \quad m = 0, 1, \ldots, \tag{10.34}$$

where $\{u_i\}$ is an arbitrary non-trivial real vector. A necessary and sufficient condition for the measure $dF(W)$ to be non-negative is

$$q_m \geqq 0 \tag{10.35}$$

for all m and an arbitrary non-trivial vector $\{u_i\}$. The positiveness of the quadratic forms q_m is also necessary and sufficient for the existence of a non-decreasing function $F(W)$ $(0 \leqq W < \infty)$ fulfilling (10.33) and having an infinite number of points at which the function increases [1]. Thus if $q_m > 0$ for every m and for an arbitrary non-trivial vector $\{u_i\}$, it is possible to construct the distribution $F'(W) = P(W) \geqq 0$ from the given moment sequence $\{M_n\}_{n=0}^{\infty}$ in a unique way. Another case occurs if for some m and non-trivial vector $\{u_i\}$ the quadratic forms (10.34) equal zero. It can be shown [4] in this case that there exists a unique distribution $P(W)$ the support of which is composed of a finite number of points equal to the minimal number of the m's for which $q_m = 0$.

These conclusions are illustrated by the following examples. Let us consider the system of determinants

$$\mathscr{D}_m = \begin{vmatrix} M_0 & M_1 & \dots & M_m \\ M_1 & M_2 & \dots & M_{m+1} \\ \dots & \dots & \dots & \dots \\ M_m & M_{m+1} & \dots & M_{2m} \end{vmatrix}, \quad m = 0, 1, \dots . \tag{10.36}$$

If $\mathscr{D}_m > 0$, then $q_m > 0$ $(m = 0, 1, \dots)$. These conditions can easily be shown to be valid for chaotic light, for which $M_n = n! \langle n \rangle^n$, as we shall see later (equation (10.40)); here $\langle n \rangle$ is the mean number of photons in the field. Thus the function $P(W)$ must exist as a unique ordinary function; indeed, we shall see that $P(W) = \langle n \rangle^{-1} \exp(-W/\langle n \rangle)$ in this case (equation (10.39)). For coherent light $M_n = \langle n \rangle^n$ and so $\mathscr{D}_m = 0$ for $m \geqq 1$ and $\mathscr{D}_0 = = 1$. Thus $q_0 > 0$ and $q_m = 0$ for $m \geqq 1$. Using the above theorem we arrive at the conclusion that the corresponding function $P(W)$ is a generalized function having a one-point support and in fact this distribution is equal to $\delta(W - \langle n \rangle)$ (see (10.61)).

Thus it is possible, giving the moments M_n from an experiment, to decide using these results and the determinants (10.36) if the distribution $P(W)$ exists as an ordinary function which is non-negative or as a generalized function with a support composed of a finite number of points.

10.3 Short-time measurements

We begin with a description of the chaotic field which is the most usual state of the field in nature. Such a field is generated by a thermal source composed of many independent atomic radiators and consists of superpositions of waves of many different frequencies lying within some continuous range. These elementary waves can be regarded as independent waves of indeterminate phase. Using the central limit theorem of mathematical statistics, we can conclude that the field represents a Gaussian random process for the amplitude with zero mean value; this was verified by van Cittert [457, 458] and Janossy [215, 216] by direct calculations and a quantum-mechanical treatment leads to the same result (Sec. 17.1).

A different situation occurs for laser sources (where stimulated emission is dominant) and for Cerenkov radiation; neither can be regarded as Gaussian processes.

Consider first a linearly polarized chaotic field. Since $V^{(r)}(t)$ is a Gaussian variable with zero mean value then, according to (2.7) and (2.31b), $V^{(i)}(t)$ is also a Gaussian variable with zero mean value, uncorrelated with $V^{(r)}(t)$.

Further, as the variance of $V^{(r)}$ and the variance of $V^{(i)}$ are equal, say $2\sigma^2$ (compare (2.31a)), it follows that the joint probability distribution of $V^{(r)}$ and $V^{(i)}$ is given by

$$P(V^{(r)}, V^{(i)})\, dV^{(r)}\, dV^{(i)} = \frac{1}{2\pi\sigma^2} \exp\left[-\frac{V^{(r)2} + V^{(i)2}}{2\sigma^2}\right] dV^{(r)}\, dV^{(i)}\,. \tag{10.37}$$

However $(V^{(r)2} + V^{(i)2})$ is equal to the intensity I of the field and (10.37) can be rewritten in the form

$$P(\sqrt{I}, \arg V)\, d(\sqrt{I})\, d(\arg V) = \frac{1}{2\pi\sigma^2} \exp\left(-\frac{I}{2\sigma^2}\right) \sqrt{(I)}\, d(\sqrt{I})\, d(\arg V)\,. \tag{10.38}$$

This probability density is independent of the phase of V, i.e. all phase angles in the range $0 \leqq \arg V \leqq 2\pi$ are equally probable. So, integrating over the phase and taking into account that $\langle I \rangle = 2\sigma^2$, we finally arrive at

$$P(I)\, dI = \frac{1}{\langle I \rangle} \exp\left(-\frac{I}{\langle I \rangle}\right) dI\,, \tag{10.39}$$

which is an exponential distribution in the intensity I. The k-th moment is

$$\langle I^k \rangle = k!\, \langle I \rangle^k \tag{10.40}$$

and the variance

$$\langle (\Delta I)^2 \rangle = \langle I^2 \rangle - \langle I \rangle^2 = \langle I \rangle^2\,. \tag{10.41}$$

Now consider a resolving time T of the detector much smaller than the coherence time. The integrated intensity W is equal to IT since the intensity $I(t)$ may be regarded as practically constant over such a time interval and the photodetection equation becomes

$$p(n, T) \equiv p(n, T, t) = \int_0^\infty \frac{(\eta I T)^n}{n!} \exp(-\eta I T)\, P(I)\, dI\,. \tag{10.42}$$

Such short-time measurements provide a deeper insight into the physical problem since the intensity has a simple physical meaning whereas the integrated intensity is a rather more complicated quantity. The distribution $p(n, T)$ of photoelectric counts can be obtained by substituting (10.39) into (10.42) and

$$p(n, T) = \frac{\langle n \rangle^n}{(1 + \langle n \rangle)^{1+n}}\,, \tag{10.43}$$

where $\langle n\rangle = \eta\langle I\rangle\, T$. This is the well-known Bose-Einstein distribution for n identical particles in one quantum state. (In one cell of a phase space, this can be understood as follows: in the direction of the beam photons cannot be distinguished in the linear distance $cT(T \approx 1/\Delta\nu)$ as a consequence of the uncertainty principle and so they occupy one cell of the phase space; that is they behave like Bose-Einstein particles.) The variance (10.5) can be calculated as

$$\begin{aligned}\langle(\Delta n)^2\rangle &= \eta\langle I\rangle\, T + \eta^2\langle(\Delta I)^2\rangle\, T^2 = \\ &= \eta\langle I\rangle\, T + \eta^2\langle I\rangle^2\, T^2 = \langle n\rangle\,(1 + \langle n\rangle)\,,\end{aligned} \tag{10.44}$$

where (10.41) has been used. This variance exceeds $\langle n\rangle$, the value of the variance for the Poisson distribution.

The characteristic function $C^{(W)}(i\eta s)$ is

$$C^{(W)}(i\eta s) = \int_0^\infty \frac{\exp\left(is\eta IT - \dfrac{I}{\langle I\rangle}\right)}{\langle I\rangle}\, dI = \frac{1}{1 - is\langle n\rangle}\,. \tag{10.45}$$

The results just obtained may be generalized to M degrees of freedom (e.g. $M = T/\tau_c$ if $T \gg \tau_c$, Sec. 10.5) and to partially polarized light.

Note that if $W = \sum_j W_j$ is a stochastic quantity, where the W_j (having the probability distributions $P_j(W_j)$) are statistically independent, then the probability distribution $P(W)$ of W is equal to the convolution of $P_j(W_j)$, namely

$$P(W) = \int_0^\infty \ldots \int \delta(W - \sum_j W_j) \prod_j P_j(W_j)\, dW_j\,. \tag{10.46}$$

Consequently the corresponding characteristic function of $P(W)$ equals the product of the characteristic functions of $P_j(W_j)$ and we have

$$\langle \exp isW\rangle = \prod_j \langle \exp isW_j\rangle \tag{10.47a}$$

or

$$C^{(W)}(is) = \prod_j C^{(W_j)}(is)\,. \tag{10.47b}$$

Applying this result to a system with M independent degrees of freedom with the same mean values $\langle n\rangle/M$ per degree of freedom we obtain for the resulting characteristic function

$$C^{(W)}(i\eta s) = \frac{1}{\left(1 - is\dfrac{\langle n\rangle}{M}\right)^M}\,. \tag{10.48}$$

This gives [264, 449]

$$P(W) = \frac{1}{2\pi}\int_{-\infty}^{+\infty} \frac{1}{\left(1 - is\dfrac{\langle n\rangle}{\eta M}\right)^M} \exp(-isW)\, ds =$$

$$= \left(\frac{\eta M}{\langle n\rangle}\right)^M \frac{W^{M-1}}{(M-1)!} \exp\left(-\frac{\eta W M}{\langle n\rangle}\right), \qquad (10.49)$$

$$p(n, T) = \frac{1}{n!}\frac{d^n}{d(is)^n} \left.\frac{1}{\left(1 - is\dfrac{\langle n\rangle}{M}\right)^M}\right|_{is=-1} =$$

$$= \frac{(n+M-1)!}{n!\,(M-1)!}\left(1 + \frac{M}{\langle n\rangle}\right)^{-n}\left(1 + \frac{\langle n\rangle}{M}\right)^{-M}, \qquad (10.50)$$

$$\langle W^k\rangle = \frac{1}{\eta^k}\frac{d^k}{d(is)^k} \left.\frac{1}{\left(1 - is\dfrac{\langle n\rangle}{M}\right)^M}\right|_{is=0} =$$

$$= \left(\frac{\langle n\rangle}{\eta}\right)^k \frac{(k+M-1)!}{(M-1)!\,M^k}, \qquad (10.51)$$

and

$$\langle(\Delta n)^2\rangle = \langle n\rangle\left(1 + \frac{\langle n\rangle}{M}\right). \qquad (10.52)$$

The number $\langle n\rangle/M$ represents the degeneracy parameter.†

Consider next partially polarized light: if I_1 and I_2 are the intensities of two linearly independent modes of polarization of Gaussian light, we obtain for the characteristic function

$$C^{(W)}(i\eta s) = \frac{1}{1 - is\langle n_1\rangle}\,\frac{1}{1 - is\langle n_2\rangle}, \qquad (10.53)$$

where

$$\langle n_j\rangle = \eta\langle I_j\rangle T, \quad j = 1, 2. \qquad (10.54)$$

Using (6.56), we can write

$$\langle n_1\rangle = \tfrac{1}{2}(1 + P)\,\langle n\rangle, \qquad (10.55a)$$

$$\langle n_2\rangle = \tfrac{1}{2}(1 - P)\,\langle n\rangle \qquad (10.55b)$$

† If M is not integer one must use the Γ-functions in (10.49), (10.50) and (10.51) instead of the factorials.

(assuming $\langle I_1 \rangle \geqq \langle I_2 \rangle$ without loss of generality), where $\langle n \rangle = \langle n_1 \rangle + \langle n_2 \rangle$. Writing (10.53) in the form

$$C^{(W)}(i\eta s) = \frac{1}{\langle n_1 \rangle - \langle n_2 \rangle} \left[\frac{\langle n_1 \rangle}{1 - is\langle n_1 \rangle} - \frac{\langle n_2 \rangle}{1 - is\langle n_2 \rangle} \right] \tag{10.56}$$

we obtain [270]

$$P(I) = \frac{1}{P\langle I \rangle} \left[\exp\left(- \frac{2I}{(1 + P)\langle I \rangle} \right) - \exp\left(- \frac{2I}{(1 - P)\langle I \rangle} \right) \right] \tag{10.57}$$

and

$$p(n, T) = \frac{1}{P\langle n \rangle} \left[\left(1 + \frac{2}{1 + P} \langle n \rangle \right)^{-n-1} - \left(1 + \frac{2}{1 - P} \langle n \rangle \right)^{-n-1} \right]. \tag{10.58}$$

The variance of n is equal to

$$\langle (\Delta n)^2 \rangle = \langle n^2 \rangle - \langle n \rangle^2 = \langle n \rangle \left[1 + \tfrac{1}{2}(1 + P^2) \langle n \rangle \right], \tag{10.59}$$

so that $M = 2/(1 + P^2)$. If the light is completely polarized $P = 1$ and we obtain (10.44); if it is unpolarized $P = 0$ and we have

$$\langle (\Delta n)^2 \rangle = \langle n \rangle \left(1 + \tfrac{1}{2} \langle n \rangle \right) \tag{10.60}$$

and $M = 2$.

Another important kind of light is that from an ideal laser whose intensity can be regarded as perfectly stabilized (at least approximately) and the intensity probability distribution is a δ-function distribution,

$$P(I) = \delta(I - \langle I \rangle). \tag{10.61}$$

The photodetection equation (10.42) gives in this case

$$p(n, T) = \frac{\langle n \rangle^n}{n!} \exp\left(- \langle n \rangle \right), \tag{10.62}$$

which is the Poisson distribution. This result can also be obtained using a configuration space of wave functions [272] and it follows simply in the quantum description (Sec. 13.1, also [30, 253]). The variance, $\langle (\Delta n)^2 \rangle = \langle n \rangle$, is in agreement with (10.5) since $\langle (\Delta W)^2 \rangle = 0$ in this case.

Experimental verification of the applicability of the distributions (10.43) and (10.62) to chaotic and laser light respectively was carried out by Arecchi [41] (also [44, 46]) and others (further details can be found in Chapter 17).

A more realistic description of laser light above the threshold of oscillations uses the model of the superposition of coherent and chaotic light, which will be discussed using quantum methods in Sec. 17.3. Chaotic light in Arecchi's experiments was obtained by sending the light of an amplitude-stabilized single-mode He—Ne laser onto a moving ground-glass disk. Such light, called *pseudothermal light*, was first introduced by Martienssen and Spiller [291]. The photon-counting distribution $p(n, T)$ for the chaotic field with M degrees of freedom given by (10.50), and for polarized and unpolarized light by (10.58), was experimentally verified by Martienssen and Spiller [292] obtaining very good agreement with theory. Experimental verification of the validity of the Bose-Einstein distribution for the laser operating below threshold was performed by Freed and Haus [153].

10.4 Bunching effect of photons

We have noted that the photon-counting distribution will depart, in general, from Poisson statistics and that the variance of the number of counts is usually in excess of that given by the Poisson distribution. This must imply that the photoelectrons (and also photons) do not arrive at random but have *bunching properties.*

In order to describe the bunching effect of photons, as well as the fourth-order correlation effect first observed by Hanbury Brown and Twiss, we need to use the fourth-order correlation function which, for linearly polarized light, is

$$\Gamma^{(2,2)}(\boldsymbol{x}_1, \boldsymbol{x}_2, \boldsymbol{x}_1, \boldsymbol{x}_2; 0, -\tau, 0, -\tau) = \langle I_1(t + \tau)\, I_2(t)\rangle =$$

$$= \langle V_1^*(t + \tau)\, V_1(t + \tau)\, V_2^*(t)\, V_2(t)\rangle\,; \qquad (10.63)$$

this expresses the *intensity correlation.* For the *chaotic field* we must calculate the fourth-order correlation function for the Gaussian distribution function: it was shown by Wang and Uhlenbeck [461] and Reed [392] using the method of stochastic processes and by Glauber [174, 176] using the method of quantum theory (cf. section 17.1) that the $(m + n)$th-order correlation function for the chaotic field has the form

$$\Gamma^{(m,n)}(x_1, \ldots, x_{m+n}) = \delta_{mn} \sum_{\pi} \Gamma^{(1,1)}(x_1, x_{n+1}) \ldots \Gamma^{(1,1)}(x_n, x_{2n})\,, \qquad (10.64a)$$

where $\sum_{\pi}$ stands for the sum of $n!$ possible permutations of the indices $n + 1, \ldots, 2n$ (or $1, \ldots, n$). Equation (10.64a) shows that the $2n$th-order correlation function for the chaotic field is determined completely by the second-order correlation function. Putting $x_1 \equiv x_2 \equiv \ldots \equiv x_{2n}$ we obtain

$$\Gamma^{(n,n)}(x, \ldots, x) = \langle I^n(x) \rangle = n! \, \langle I(x) \rangle^n \,, \tag{10.64b}$$

which agrees with (10.40).

Applying (10.64a) to (10.63) we obtain [474]

$$\langle I_1(t + \tau) \, I_2(t) \rangle = \Gamma_{11}(0) \, \Gamma_{22}(0) + |\Gamma_{12}(\tau)|^2 =$$
$$= \langle I_1 \rangle \langle I_2 \rangle \left[1 + |\gamma_{12}(\tau)|^2\right] \tag{10.65a}$$

and the correlation of fluctuations $\Delta I_1 = I_1 - \langle I_1 \rangle$ and $\Delta I_2 = I_2 - \langle I_2 \rangle$ is

$$\langle \Delta I_1(t + \tau) \, \Delta I_2(t) \rangle = \langle I_1 I_2 \rangle - \langle I_1 \rangle \langle I_2 \rangle = \langle I_1 \rangle \langle I_2 \rangle \, |\gamma_{12}(\tau)|^2 \,. \tag{10.65b}$$

This shows that the modulus of the degree of coherence may be determined by measuring the correlation of intensity fluctuations and this result is the basis of correlation interferometry.

If partially polarized light is assumed, we can obtain (in analogy to (10.59))

$$\langle \Delta I_1(t + \tau) \, \Delta I_2(t) \rangle = \tfrac{1}{2} \langle I_1 \rangle \langle I_2 \rangle \, (1 + P^2) \, |\gamma_{12}(\tau)|^2 \,, \tag{10.66}$$

which shows that measurements of the correlation of intensity fluctuations may also provide information about the degree of polarization of beams.

The conditional probability $p_c(t \mid t + \tau) \, \Delta\tau$ is the probability that a photoelectric count will be registered in the time interval $(t + \tau, t + \tau + \Delta\tau)$ if a count has been registered at a time t [269]. The probability of observing a count at both times t and $t + \tau$ within dt and $d\tau$ is obviously $\eta^2 \, I(t) \, I(t + \tau) \, dt \, d\tau$ and the probability of finding two counts separated by the interval τ is $\langle \eta^2 \, I(t) \, I(t + \tau) \rangle \, dt \, d\tau$. If this is divided by the ensemble average of the probability $\eta \, I(t) \, dt$ of finding one count at t within dt, we arrive at the conditional probability $p_c(\tau)$ of obtaining a second count τ seconds after the first one,

$$p_c(\tau) \, d\tau = \frac{\eta^2 \langle I(t + \tau) \, I(t) \rangle \, dt \, d\tau}{\eta \langle I(t) \rangle \, dt} =$$
$$= \eta \langle I(t) \rangle \left[1 + \tfrac{1}{2}(1 + P^2) \, |\gamma(\tau)|^2\right] d\tau \,, \tag{10.67a}$$

where (10.66) has been used. Assuming completely polarized light we have, putting $P = 1$,

$$p_c(\tau) = \eta \langle I(t) \rangle \left[1 + |\gamma(\tau)|^2\right] . \tag{10.67b}$$

Since $|\gamma(\tau)| \approx 1$ for $\tau \ll \tau_c = 1/\Delta\nu$ and $|\gamma(\tau)| \approx 0$ for $\tau \gg \tau_c$, $p_c(\tau)$ starts at the value $2\eta \langle I(t) \rangle$ for small τ and tends to $\eta \langle I(t) \rangle$ for large τ. This illustrates the bunching properties of a photon beam. The bunching effect was observed by Arecchi, Gatti and Sona [48] and Morgan and Mandel [326]. Their results are given in Figs. 10.1 and 10.2 and the experimental results are in very good agreement with the theoretical predictions.

In these measurements, light from a quasi-monochromatic chaotic source is incident on a photomultiplier with a fast response. The output

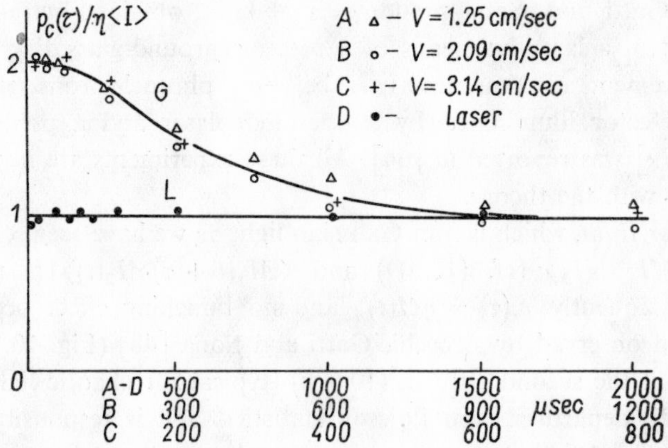

Fig. 10.1. The conditional probability $p_c(\tau)$ of a second count occurring at a time τ after a first has occurred at time $\tau = 0$. Experimental results apply to a laser (*L*) and to an artificially synthesized chaotic source (*G*) created by passing the laser radiation through ground glass which is rotated at speeds $V = 1{\cdot}25$ cm/sec, $2{\cdot}09$ cm/sec, and $3{\cdot}14$ cm/sec. [After F. T. Arecchi, E. Gatti and A. Sona, *Phys. Lett.* **20**, 27 (1966); reprinted with permission.]

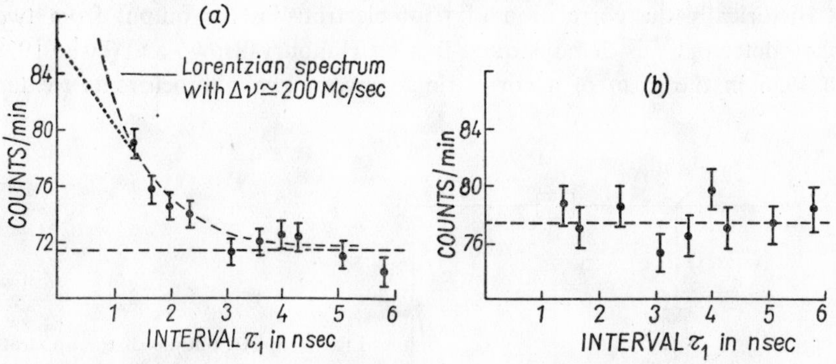

Fig. 10.2. Counting rates illustrating the phenomena of photon bunching with light from a) a ^{198}Hg source, b) the tungsten lamp as light source. The ordinate represents essentially a quantity which is proportional to the integral $\int_{\tau_1}^{\tau_2} p_c(\tau)\,d\tau$, where τ_2 is a constant. In part (b) the wide frequency spectrum of the conventional tungsten lamp leads to intensity correlations in an immeasurably short time interval. [After B. L. Morgan and L. Mandel, *Phys. Rev. Lett.* **16**, 1012 (1966); reprinted with permission.]

of the photomultiplier reaches a coincidence counter via two different paths, one of which contains a time delay line so that only pulses in the output of the photomultiplier which are separated by the time delay τ can produce an output from the coincidence counter. The conditional probability $p_c(\tau)$ may then be determined from an analysis of the distribution of coincidences as a function of τ. While in the experiment of Magyar and Mandel light from a low pressure ^{198}Hg light source was used, in the experiment of Arecchi, Gatti and Sona pseudothermal light, obtained by randomizing the output of a laser by means of a rotating ground-glass disk, was used. A measurement of time intervals between photoelectrons emitted by a photodetector illuminated by a one-mode laser and a pseudothermal source [179] was reported in [64]. All these experiments are in very good agreement with the theory.

For *laser light*, which is non-Gaussian light as we have seen, $\langle I_1(t + \tau) . . I_2(t)\rangle = \langle I_1\rangle \langle I_2\rangle$ (cf. (10.61)) and $\langle \Delta I_1(t + \tau) \Delta I_2(t)\rangle$ is practically zero. Consequently $p_c(\tau) = \eta\langle I(t)\rangle$ and no bunching effect occurs. This was indeed observed by Arecchi, Gatti and Sona [48] (Fig. 10.1). It may be said that the second term in (10.67b) (typical for chaotic light) is connected with departure from Poisson statistics, and is responsible for the bunching effect of photons.

10.5 Hanbury Brown - Twiss effect — Correlation interferometry of the fourth order

Historically the correlation of photoelectrons in the output from two photodetectors was demonstrated first by Hanbury Brown and Twiss [92] in 1956 in the form of a correlation between two photocurrents treated

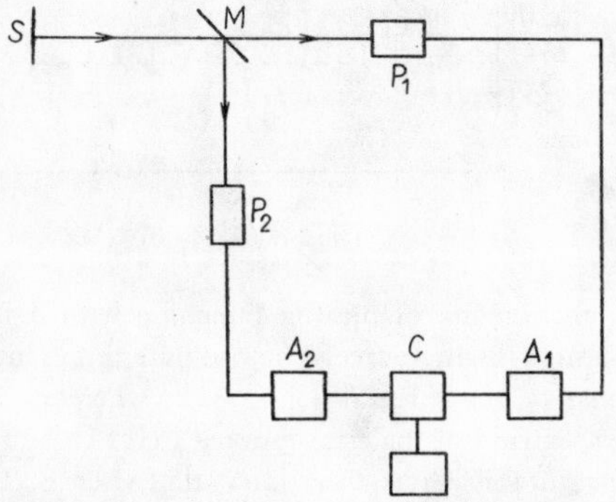

Fig. 10.3. An outline of the apparatus for demonstrating the correlation between the intensity fluctuations; S is a thermal source, M is a half-silvered mirror, P_1 and P_2 are photomultipliers, A_1 and A_2 are amplifiers, C is a correlater and Mo is an integrating motor.

as continuous signals. The scheme of their apparatus is shown in Fig. 10.3. Light from a thermal source (mercury lamp) S was divided into two beams by a half-silvered mirror M and fell on two photocells P_1, P_2 whose outputs were sent through bandlimited amplifiers A_1, A_2 to a correlater C (one of the photomultipliers is movable to ensure changes of the time delay τ). The averaged value of the product of intensity fluctuations was recorded by an integrating motor Mo. The experiment has also been repeated by Martienssen and Spiller [291] with a pseudothermal source.

To explain this experiment we must use a deeper physical model working with the field intensity rather than with the integrated intensity.

The variance of the number of photoelectrons in one photodetector (assuming linearly polarized light) is, using the photodetection equation (10.3a),

$$\langle n^2 \rangle = \langle n \rangle + \eta^2 \iint_t^{t+T} \langle I(t')\, I(t'') \rangle \, dt' \, dt'' =$$

$$= \langle n \rangle + \eta^2 \iint_0^T \langle I(t + t' - t'')\, I(t) \rangle \, dt' \, dt'' , \qquad (10.68)$$

where $\langle n \rangle = \eta \langle I \rangle T$ and the stationary condition is assumed. Denoting

$$T\xi(T) = \iint_0^T |\gamma(t' - t'')|^2 \, dt' \, dt'' = 2 \int_0^T (T - t')\, |\gamma(t')|^2 \, dt' \quad (10.69)$$

[263, 264, 269] we obtain from (10.68)

$$\langle n^2 \rangle = \langle n \rangle + \eta^2 \langle I \rangle^2 T^2 + \eta^2 \langle I \rangle^2 T^2 \frac{\xi(T)}{T} =$$

$$= \langle n \rangle^2 + \langle n \rangle \left(1 + \langle n \rangle \frac{\xi(T)}{T} \right), \qquad (10.70)$$

where (10.65a) has also been used. Since $|\gamma(t' - t'')| \leqq 1$ then $\xi(T) \leqq T$ and if $T \ll 1/\Delta\nu$, then $\gamma(\tau)$ is practically equal to one and $\xi(T) \approx T$. Thus the variance following from (10.70) agrees with (10.44). However, if $T \gg 1/\Delta\nu$, then since $\gamma(\tau)$ is non-vanishing only on an interval of lengt h a few times

$$\xi(T) \approx \xi(\infty) = \int_{-\infty}^{+\infty} |\gamma(t)|^2 \, dt \, ; \qquad (10.71)$$

$1/\Delta\nu$, this can be adopted as the coherence time τ_c (cf. (2.53)) [263, 264, 269].

From (10.70) we obtain

$$\langle (\Delta n)^2 \rangle = \langle n \rangle \left(1 + \langle n \rangle \frac{\tau_c}{T}\right), \tag{10.72}$$

which is the variance of the number of bosons divided into T/τ_c cells of phase space – the length of the cell is $c\tau_c$. The photon-counting distribution (10.50) becomes for $M = T/\tau_c$

$$p(n, T) = \frac{\Gamma(n + T/\tau_c)}{n!\,\Gamma(T/\tau_c)} \left(1 + \frac{T}{\tau_c \langle n \rangle}\right)^{-n} \left(1 + \frac{\langle n \rangle \tau_c}{T}\right)^{-T/\tau_c}, \tag{10.73}$$

where we have written the Γ-function instead of the factorials since T/τ_c need not be integer. It can be shown [61] that this formula is valid not only for $T \gg \tau_c$ but represents a very good approximation for all T. Similar results have been obtained for the superposition of coherent and chaotic light in [509, 510, 528, 532].

Departures from classical Poissonian counting statistics, for which $\langle (\Delta n)^2 \rangle = \langle n \rangle$ (valid for ideal laser light), depend on the degeneracy parameter

$$\delta = \frac{\langle n \rangle \tau_c}{T} = \eta \langle I \rangle \tau_c, \tag{10.74}$$

which gives the mean number of counts per cell of phase space: $\delta \ll 1$ for thermal light and $\delta \gg 1$ for laser light as we have noted. The fact that $\delta \ll 1$ for thermal light implies that a direct measurement of the excess photon noise is difficult. The position is much better for pseudothermal light.

The correlation between counts of two photodetectors may be obtained in the same manner

$$\langle n_1 n_2 \rangle = \eta_1 \eta_2 \iint_t^{t+T} \langle I_1(t')\, I_2(t'') \rangle \, dt'\, dt'' =$$

$$= \eta_1 \eta_2 \iint_0^T \langle I_1(t + t' - t'')\, I_2(t) \rangle \, dt'\, dt''. \tag{10.75}$$

If we assume the cross-spectral purity condition $\gamma_{12}(\tau) = \gamma_{12}(0)\,\gamma_{11}(\tau)$ to be valid, we obtain

$$\langle n_1 n_2 \rangle = \langle n_1 \rangle \langle n_2 \rangle \left(1 + \frac{\xi(T)}{T} |\gamma_{12}(0)|^2\right) \tag{10.76a}$$

and

$$\langle \Delta n_1 \, \Delta n_2 \rangle = \frac{\xi(T)}{T} |\gamma_{12}(0)|^2 \langle n_1 \rangle \langle n_2 \rangle . \tag{10.76b}$$

The normalized correlation is

$$\frac{\langle \Delta n_1 \, \Delta n_2 \rangle}{[\langle (\Delta n_1)^2 \rangle \langle (\Delta n_2)^2 \rangle]^{1/2}} = \frac{\delta}{1 + \delta} |\gamma_{12}(0)|^2 , \tag{10.77}$$

where the degeneracy parameter δ is given by (10.74) ($\langle n_1 \rangle = \langle n_2 \rangle$). As $\delta \approx 10^{-3}$ for a thermal source, the effect is very difficult to observe. For pseudothermal light the conditions for observation are much better, since $\delta \approx 10^{12}$.

If partially polarized light is assumed one obtains, using (10.66)

$$\langle (\Delta n)^2 \rangle = \langle n \rangle \left[1 + \tfrac{1}{2}(1 + P^2) \langle n \rangle \frac{\xi(T)}{T} \right] , \tag{10.78a}$$

$$\langle \Delta n_1 \, \Delta n_2 \rangle = \tfrac{1}{2}(1 + P^2) \langle n_1 \rangle \langle n_2 \rangle |\gamma_{12}(0)|^2 \frac{\xi(T)}{T} . \tag{10.78b}$$

We have mentioned that the measurements of Hanbury Brown and Twiss were performed in terms of classical continuous outputs of photodetectors rather than in terms of particles – photoelectrons. The currents of photoelectrons were high enough to make n_1 and n_2 proportional to the continuous currents $S_1(t)$ and $S_2(t)$ in the outputs of two photodetectors; T played the role of a "resolving time". The correlation of fluctuations of photocurrents can be obtained in the form

$$\langle \Delta S_1 \, \Delta S_2 \rangle = \tfrac{1}{2}(1 + P^2) \langle S_1 \rangle \langle S_2 \rangle |\gamma_{12}(0)|^2 \frac{\xi(T)}{T} \tag{10.79}$$

and the variance $\langle (\Delta S)^2 \rangle$ is equal to $\langle \Delta S_1 \, \Delta S_2 \rangle|_{x_1 \equiv x}$. This exactly corresponds to (10.78b). The term corresponding to $\langle n \rangle$ in (10.78a), which is a typical quantum term, is lost here. This can be regarded as a reflection of the fact that in the strong-field approximation the last terms in (10.17) are dominant so that the first term in (10.78a) can be neglected.

The cumulants defined in section 10.1 for chaotic light are, [269],

$$\varkappa_1^{(W)} = \langle I \rangle \, T ,$$

$$\varkappa_j^{(W)} = (j - 1)! \, \langle I \rangle^j \int_0^T \ldots \int \gamma(t_1 - t_2) \, \gamma(t_2 - t_3) \ldots \gamma(t_j - t_1) \, dt_1 \ldots dt_j , \quad j \geqq 2 \tag{10.80}$$

and they are derived in section 17.1.

Experiments using the coincidence technique instead of the correlation technique have been carried out by Twiss, Little and Hanbury Brown [456], Twiss and Little [455], Rebka and Pound [391], Brannen, Ferguson and Wehlau [89], and Martienssen and Spiller [291] among others. These experiments employ a coincidence circuit instead of a correlator.

If *non-Gaussian light* is assumed one can write for the conditional probability $p_c(\tau)$

$$p_c(\tau) = \eta^2[\langle I_1\rangle \langle I_2\rangle + \langle \Delta I_1(t+\tau)\, \Delta I_2(t)\rangle] = \\ = \eta^2\langle I_1\rangle \langle I_2\rangle\,(1 + |\zeta_{12}(\tau)|^2)\,, \qquad (10.81)$$

where

$$|\zeta_{12}(\tau)|^2 = \frac{\langle \Delta I_1(t+\tau)\, \Delta I_2(t)\rangle}{\langle I_1\rangle \langle I_2\rangle}\,. \qquad (10.82)$$

For thermal light $|\zeta_{12}(\tau)| = |\gamma_{12}(\tau)|$. The correlation of counts is given as

$$\langle n_1 n_2\rangle = \langle n_1\rangle \langle n_2\rangle + \eta_1\eta_2 \iint_0^T \langle \Delta I_1(t')\, \Delta I_2(t'')\rangle\, dt'\, dt''\,. \qquad (10.83)$$

The correlation of fluctuations is equal to

$$\langle \Delta n_1\, \Delta n_2\rangle = \langle n_1\rangle \langle n_2\rangle \frac{\xi(T)}{T}\,, \qquad (10.84)$$

where

$$\xi(T) = \frac{1}{T\langle I_1\rangle \langle I_2\rangle} \iint_0^T \langle \Delta I_1(t')\, \Delta I_2(t'')\rangle\, dt'\, dt'' = \\ = \frac{2}{T\langle I_1\rangle \langle I_2\rangle} \int_0^T (T-\tau)\, \langle \Delta I_1(\tau)\, \Delta I_2(0)\rangle\, d\tau \qquad (10.85)$$

if the stationary condition is assumed. As we have pointed out no Hanbury Brown-Twiss effect will occur for light from an ideal laser since $\Delta I \approx 0$ and consequently $\langle \Delta n_1\, \Delta n_2\rangle \approx 0$. There is no departure from classical Poisson statistics in this case.

The correlation technique has been applied to measurements of angular diameters of stars by Hanbury Brown and Twiss [93, 97, 98] and forms the basis of correlation interferometry. A scheme of the correlation interferometer is shown in Fig. 10.4. According to (10.79) such correlation measurements allow $|\gamma_{12}(0)|$ to be determined when the separation d of the mirrors is changed (the results obtained will follow the curve shown in Fig. 3.5). The angular radius of a star can be determined in the way described in connection with the Michelson stellar interferometer (Chapter 3). More recently a large stellar interferometer was built at Narrabri (Australia) [90] (see also [483]). Further applications of intensity interferometry in physics and astronomy have been reviewed in [454].

There are some advantages in the correlation technique for measuring diameters of stars compared with the use of the Michelson stellar interferometer. Correlation measurements involve the intensity $I = |V|^2$ which varies slowly in comparison with the rapidly varying amplitude V and the

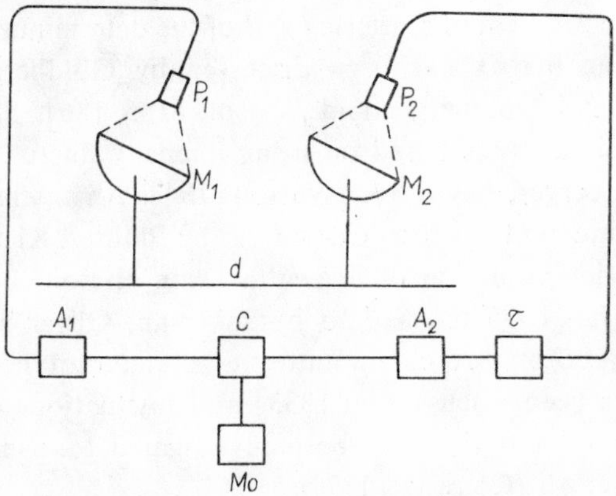

Fig. 10.4. An outline of a stellar correlation interferometer; M_1 and M_2 are mirrors, P_1 and P_2 are photodetectors, A_1 and A_2 are amplifiers, τ is a delay line, C is a correlater and Mo is an integrating motor.

phase of V is not present (the degree of coherence appears in the equations only as the modulus $|\gamma_{12}(0)|$); consequently phase distortions due to turbulence in the air, changes in the index of refraction of atmosphere, etc., do not affect the measurements, whereas they may be intolerable for the Michelson stellar interferometer. Also much larger separations d can be used in the correlation interferometer than with the Michelson stellar interferometer, making it possible to observe stars with very small angular dimensions.

On the other hand, the phase information about the degree of coherence is lost in the correlation measurements, even if such measurements (as well as photon bunching measurements) provide information about the degree of polarization and the modulus of the degree of coherence. The problem of recovering the phase of the degree of coherence was discussed in Sec. 4.4. Some information about the spectrum can also be gained from the spectrum

$h(\nu)$ of $|\gamma(\tau)|^2$ using measurements of $p_c(\tau)$ [268, 269, 279, 423, 482] since

$$h(\nu) = \int_{-\infty}^{+\infty} |\gamma(\tau)|^2 \exp i2\pi\nu\tau \, d\tau = \int_0^{\infty} g(\nu') \, g(\nu' + \nu) \, d\nu' , \quad (10.86)$$

where g is the normalized power spectrum of light. An experimental recovery of the phase of γ employing the sixth-order correlation function has been proposed by Gamo [163, 164] (see also [269]). Some other experimental possibilities of determining the phase of γ were discussed in Sec. 4.4.

Finally let us note that an interesting use of the intensity correlation technique in the problem of scattering and of the determination of the phase of the scattering matrix has been discussed by Goldberger, Lewis and Watson [183] and Goldberger and Watson [185, 186]. Some techniques for spectroscopic analysis based on using intensity fluctuations have been studied by Goldberger, Lewis and Watson [184]. An intensity correlation linewidth measurement has been carried out by Phillips, Kleiman and Davis [381], while a determination of linewidth using photon-counting statistics measurements has been referred to by Jakeman, Oliver and Pike [211]. The role of the photon field commutator in the quantum theory of intensity correlations has been discussed in [333] and fluctuations of the intensity of beams in space and time has been investigated following the analysis of Goldberger et al. (Conway [117]).

10.6 Higher-order coherence effects

We have shown that light from a chaotic source can be described by a Gaussian probability distribution function, which is fully determined by the second-order correlation function (the first-order moment is zero). For a full description of the statistics of non-Gaussian light (for example laser light) the moments – correlation functions – of all orders are needed. The correlation functions of higher order can in principle be measured by means of multiple coincidence measurements.

Consider a system of N photodetectors illuminated by beams of partially coherent linearly polarized light. If n_j is the number of counts registered by the jth detector in a time interval $(t_j, t_j + T_j)$, then, according to (10.3a),

$$p(n_1, \ldots, n_N, T_1, \ldots, T_N, t_1, \ldots, t_N) = \\ = \left\langle \prod_{j=1}^{N} \frac{[\eta_j W_j(T_j, t_j)]^{n_j}}{n_j!} \exp\left[-\eta_j W_j(T_j, t_j)\right] \right\rangle , \quad (10.87)$$

where η_j is the photoefficiency of the jth photodetector and

$$W_j(T_j, t_j) = \int_{t_j}^{t_j+T_j} I_j(x_j, t_j')\, dt_j'\,. \tag{10.88}$$

The brackets in (10.87) stand for the average over the N-fold probability distribution $P(W_1, \ldots, W_N)$. As with our previous results, we can introduce the characteristic function

$$C^{(\{n_j\})}(\{ix_j\}) = \langle \prod_{j=1}^{N} \exp ix_j n_j \rangle = \sum_{\{n_j\}} p(\{n_j\}, \{T_j\}, \{t_j\}) \prod_{j=1}^{N} \exp ix_j n_j\,, \tag{10.89}$$

where $\{n_j\} \equiv (n_1, \ldots, n_N)$, etc., and we obtain from (10.87)

$$C^{(\{n_j\})}(\{ix_j\}) = C^{(\{W_j\})}(\{\eta_j(e^{ix_j} - 1)\})\,, \tag{10.90}$$

where

$$C^{(\{W_j\})}(\{i\eta_j x_j\}) = \langle \prod_{j=1}^{N} \exp i\eta_j x_j W_j \rangle = \sum_{\{n_j\}} p(\{n_j\}, \{T_j\}, \{t_j\}) \prod_{j=1}^{N} (1 + ix_j)^{n_j}\,. \tag{10.91}$$

The correlation of the number of counts is

$$\langle n_1 \ldots n_N \rangle = \prod_{j=1}^{N} \frac{\partial}{\partial(ix_j)} \langle \prod_{k=1}^{N} \exp ix_k n_k \rangle \Big|_{\{ix_j\}=0} =$$

$$= \eta_1 \ldots \eta_N \int_{t_1}^{t_1+T_1} \ldots \int_{t_N}^{t_N+T_N} \Gamma^{(N,N)}(\boldsymbol{x}_1, \ldots, \boldsymbol{x}_N, \boldsymbol{x}_N, \ldots, \boldsymbol{x}_1; t_1', \ldots, t_N', t_N', \ldots, t_1')\,.$$
$$.\, dt_1' \ldots dt_N'\,, \tag{10.92}$$

where $\Gamma^{(N,N)}$ is the $2N$th-order correlation function. Thus the correlation of photoelectric counts is fully determined by the $2N$th-order correlation function and if the field is assumed to be stationary, then $\int_{t_j}^{t_j+T_j}$ can be replaced by $\int_0^{T_j}$ and the correlation function is given by

$$\Gamma^{(N,N)}(\boldsymbol{x}_1, \ldots, \boldsymbol{x}_N, \boldsymbol{x}_N, \ldots, \boldsymbol{x}_1; T_1, \ldots, T_N, T_N, \ldots, T_1) =$$

$$= \frac{\partial^N}{\partial T_1 \ldots \partial T_N} \frac{\langle n_1 \ldots n_N \rangle}{\eta_1 \ldots \eta_N}\,. \tag{10.93}$$

Writing (10.92) in the form

$$\langle n_1 \ldots n_N \rangle = \eta_1 \ldots \eta_N \langle W_1 \ldots W_N \rangle \tag{10.94}$$

we can show that

$$\langle \Delta n_1 \dots \Delta n_N \rangle = \eta_1 \dots \eta_N \langle \Delta W_1 \dots \Delta W_N \rangle . \tag{10.95}$$

This correlation of fluctuations will be small for laser light since $\Delta W \approx 0$.

Some applications of the present N-fold formalism to the detection of Gaussian light have been studied by Bédard [60], who has shown that information about the spectral profile of the light can be obtained from the three-fold joint photon-counting distribution. The two-fold photon-counting distribution for Gaussian light has been experimentally investigated by Arecchi, Berné and Sona [45]. Measurements of higher-order quantities have been made by Davidson and Mandel [125] (see also [503]), who directly measured the sixth-order correlation function, Chang, Detenbeck, Korenman, Alley and Hochuli [113], Chang, Korenman and Detenbeck [115] and Chang, Korenman, Alley and Detenbeck [114], who measured the cumulants $\varkappa_2^{(W)}$, $\varkappa_3^{(W)}$ and $\varkappa_4^{(W)}$ of the second, third and fourth order for laser light at the threshold of oscillations; the third reduced factorial moment $H_3 = \langle n(n-1)(n-2) \rangle / \langle n \rangle^3 - 1$ was given by Arecchi [42].

An exact approach to the study of the coherence properties of optical fields including higher-order correlation effects, based on the theory of stochastic processes and particularly a study of experiments described by the fourth-order correlation function was given by Picinbono and Boileau [385] and Boileau and Picinbono [84]. Statistical properties of photoelectrons, including some special models of the field, have been studied in a similar way by Rousseau [411].

We have now completed the classical and semiclassical treatments of the coherence properties of optical fields. The next chapters will be devoted to the quantum description of the statistical properties of fields (employing quantum electrodynamics) as well as to the relation between the classical and the quantum descriptions.

Chapter 11

BASIC IDEAS OF THE QUANTUM THEORY OF THE ELECTROMAGNETIC FIELD

With the improvement of measuring techniques and the development of new techniques (e.g. coincidence circuits) new possibilities of detecting single photons in the visible region occurred allowing a measurement of, for example, the statistics of photons; this was in contrast to earlier investigations where the detecting devices were able to respond only to intensive currents of photons. Thus the particle character of the electromagnetic field may be studied directly. Under such circumstances it is obviously advantageous to describe the coherence phenomena in the field by the methods of quantum electrodynamics, where a field operator corresponds to the classical field and the quanta of the field correspond to the classical wave field. As is well-known this correspondence is realized by the second quantization of the field.

The standard treatment of quantum electrodynamics is formulated on the basis of perturbation theory and such a formulation lends itself to the discussion of processes involving a few particles. The treatment of many-particle processes is mathematically complex and consequently the classical limit of quantum electrodynamics has never been fully developed. In contrast a discussion of phenomena such as coherence involves a considerable number of photons and their mutual relationships in space and time. (It is only in this way that the operation of a laser source – a typical non-linear device functioning only at sufficiently high field intensities – can be explained.) Consequently it is necessary to develop quantum electrodynamics in a manner which exhibits clearly the classical limit. To this end we will not use the Fock states, the basis of the usual formulation, but instead use *coherent states*; these are linear combinations of Fock states and have the advantage that they retain information about the phase state of the field – information lost when Fock states are used. With these coherent states it is possible to obtain a close *formal* correspondence between the quantum and classical descriptions; of course this says nothing about the physical identity of these descriptions. In general, one can say that the classical description fails for any measurements which are sensitive to fluctuations of the physical vacuum. Such measurements are described by functions of the field operators in orderings other than the normal ordering. In the normal order all creation operators stand to the left of all the annihilation operators and the vacuum expectation value is zero; in other orderings the vacuum expectation value may be non zero. The principles of the quantum theory of optical coherence employing coherence states were given by Glauber [173–176, 180, 181] in 1963.

In the following we shall not pay attention to the mechanism of photon emission and to the role of coherence in this process but instead study photon propagation in space and time (the sources are not present in the Maxwell equations for the field operators of the electric and magnetic strength $\boldsymbol{E}$ and $\boldsymbol{H}$, respectively) and coherence problems of free fields. A discussion of coherence in processes involving interactions of the field with matter lies outside the framework of this book as we have pointed out in the Introduction.

11.1 Quantum description of the field

We give in this chapter some basic ideas and results of the second quantization of the electromagnetic field *in vacuo*. For a fuller treatment we refer the reader to the texts [2, 7, 9, 27, 21].

In the quantum theory of the electromagnetic field the electric and magnetic strengths, $\hat{\boldsymbol{E}}$ and $\hat{\boldsymbol{H}}$, are regarded as operators in the space of states which describe the field. These quantities satisfying Maxwell's equations can be derived from the vector potential operator $\hat{\boldsymbol{A}}$ according to the relations

$$\hat{\boldsymbol{E}} = -\frac{1}{c}\frac{\partial \hat{\boldsymbol{A}}}{\partial t}, \tag{11.1a}$$

$$\hat{\boldsymbol{H}} = \nabla \times \hat{\boldsymbol{A}}. \tag{11.1b}$$

Considering a field in a normalization volume L^3 with periodic boundary conditions we can decompose the vector potential operator $\hat{\boldsymbol{A}}$ in the form

$$\hat{\boldsymbol{A}}(x) = \frac{(2\pi\hbar c)^{1/2}}{L^{3/2}} \sum_{\boldsymbol{k}} \sum_{s=1}^{2} \frac{1}{\sqrt{k}} \{\boldsymbol{e}^{(s)}(\boldsymbol{k})\, \hat{a}_{\boldsymbol{k}s} \exp[i(\boldsymbol{k}\,.\,\boldsymbol{x} - ckt)] + \\ + \boldsymbol{e}^{(s)*}(\boldsymbol{k})\, \hat{a}_{\boldsymbol{k}s}^{+} \exp[-i(\boldsymbol{k}\,.\,\boldsymbol{x} - ckt)]\}, \tag{11.2}$$

where $\hat{a}_{\boldsymbol{k}s}$, is the *annihilation operator* for a photon of momentum $\hbar\boldsymbol{k}$ and polarization s and, $\hat{a}_{\boldsymbol{k}s}^{+}$ (the Hermitian conjugate to $\hat{a}_{\boldsymbol{k}s}$) is a *creation operator*. These operators obey the following commutation rules

$$[\hat{a}_\lambda, \hat{a}^+_{\lambda'}] = \delta_{\lambda\lambda'}, \quad [\hat{a}_\lambda, \hat{a}_{\lambda'}] = [\hat{a}^+_\lambda, \hat{a}^+_{\lambda'}] = 0; \tag{11.3}$$

$\lambda \equiv (\boldsymbol{k}, s)$ is the mode index, $[\hat{A}, \hat{B}] = \hat{A}\hat{B} - \hat{B}\hat{A}$ stands for the commutator of operators $\hat{A}$, $\hat{B}$ and $\delta_{\lambda\lambda'} = \delta_{kk'}\delta_{ss'}$. The unit polarization vector $\boldsymbol{e}^{(s)}(\boldsymbol{k})$ satisfies

$$\boldsymbol{e}^{(s)*}(\boldsymbol{k}) \, . \, \boldsymbol{e}^{(s')}(\boldsymbol{k}) = \delta_{ss'}, \tag{11.4a}$$

$$\boldsymbol{k} \, . \, \boldsymbol{e}^{(s)}(\boldsymbol{k}) = 0, \tag{11.4b}$$

that is the vectors $\boldsymbol{e}^{(1)}$, $\boldsymbol{e}^{(2)}$ and $\boldsymbol{k}/k$ $(k = |\boldsymbol{k}|)$ form an orthogonal system

$$\sum_{s=1}^{2} e^{(s)*}_\mu(\boldsymbol{k})\, e^{(s)}_\nu(\boldsymbol{k}) + \frac{k_\mu k_\nu}{k^2} = \delta_{\mu\nu}, \quad \mu, \nu = 1, 2, 3. \tag{11.5}$$

In (11.2) only transverse polarizations are present $(s = 1, 2)$ since longitudinal and scalar photons are eliminated by the Lorentz condition $\sum_{\mu=1}^{4} \partial \hat{A}_\mu / \partial x_\mu = 0$ $(x_4 = ict)$ [2].

We define the *number operator* $\hat{n} = \hat{a}^+\hat{a}$ (omitting the mode index λ for simplicity), for which, with the use of the commutation rules (11.3),

$$[\hat{a}, \hat{n}] = \hat{a}, \tag{11.6a}$$

$$[\hat{n}, \hat{a}^+] = \hat{a}^+. \tag{11.6b}$$

Repeated use of these relations gives

$$\hat{n}\hat{a}^m = \hat{a}^m (\hat{n} - m), \tag{11.7a}$$

$$\hat{n}\hat{a}^{+m} = \hat{a}^{+m}(\hat{n} + m) \tag{11.7b}$$

for $m = 0, 1, 2, \ldots$ and also

$$\hat{a}^{+m}\hat{a}^m = \hat{n}(\hat{n} - 1) \ldots (\hat{n} - m + 1), \tag{11.8a}$$

$$\hat{a}^m\hat{a}^{+m} = (\hat{n} + 1)(\hat{n} + 2) \ldots (\hat{n} + m). \tag{11.8b}$$

Assume that there exists an eigenstate $|n\rangle$ of the number operator $\hat{n}$ such that

$$\hat{n}|n\rangle = n|n\rangle, \tag{11.9}$$

where n is a real number. From this equation $\langle n| \hat{n} |n\rangle = n\langle n | n\rangle$, where $\langle n|$ is Hermitian conjugate to $|n\rangle$. Thus $n = \langle n| \hat{a}^+\hat{a} |n\rangle / \langle n | n\rangle \geqq 0$. Further, if $|n\rangle$ is an eigenstate of $\hat{n}$, then $\hat{a}^+\hat{a}(\hat{a}^k|n\rangle) = (n - k)(\hat{a}^k|n\rangle)$ from (11.7a), i.e. the state $(\hat{a}^k|n\rangle)$ is also an eigenstate of $\hat{n}$ with eigenvalues $(n - k)$, $k = 0, 1, 2, \ldots$. As $n \geqq 0$, n must be a positive integer or zero to terminate this sequence and it can be seen from (11.7b) that all integers are eigenvalues of $\hat{n}$ since $\hat{n}(\hat{a}^{+k}|n\rangle) = (n + k)(\hat{a}^{+k}|n\rangle)$, $k = 0, 1, 2, \ldots$. The *ground state* (*vacuum*) can be defined as the state with $n = 0$, that is $\hat{n}(\hat{a}^{+k}|0\rangle) = k(\hat{a}^{+k}|0\rangle)$ (making use of (11.7b)), and so the state $\hat{a}^{+k}|0\rangle$ must be proportional to the state $|k\rangle$. From (11.8b) $\langle 0| \hat{a}^k\hat{a}^{+k}|0\rangle = k!$ and we can define the normalized states

$$|n\rangle = \frac{1}{\sqrt{n!}}\, \hat{a}^{+n}|0\rangle\,; \tag{11.10}$$

these are called the *Fock states* or *occupation number states*. It is obvious that these states are orthogonal, $\langle n \mid m\rangle = \delta_{nm}$, and

$$\hat{a}^+|n\rangle = \sqrt{(n+1)}\;|n+1\rangle\,, \tag{11.11a}$$

$$\hat{a}|n\rangle = \sqrt{n}\;|n-1\rangle\,. \tag{11.11b}$$

The vacuum stability condition demands that $\hat{a}|0\rangle = 0$. Equations (11.11) show that the operators $\hat{a}^+$ and $\hat{a}$ can indeed be called the creation and annihilation operators of a photon in the field.

If we write the Hamiltonian of the field in the form

$$\hat{H}_F = \frac{1}{8\pi}\int_{L^3} (\hat{\boldsymbol{E}}^2 + \hat{\boldsymbol{H}}^2)\, d^3\boldsymbol{x} \tag{11.12}$$

we obtain by using (11.2) and (11.1)

$$\hat{H}_F = \sum_\lambda (\hat{n}_\lambda + \tfrac{1}{2})\, \hbar\omega_\lambda\,, \tag{11.13}$$

where $\omega_\lambda = kc$. This operator is, apart from the additive constant $\sum_\lambda(\hbar\omega_\lambda/2)$ (the energy of the physical vacuum), the Hamiltonian of a set of dynamically independent harmonic oscillators with energies $n_\lambda\hbar\omega_\lambda$. Introducing the canonical variables $\hat{p}$ and $\hat{q}$ by the relations

$$\hat{a} = \frac{1}{\sqrt{(2\hbar\omega)}}(\omega\hat{q} + i\hat{p})\,,$$

$$\hat{a}^+ = \frac{1}{\sqrt{(2\hbar\omega)}}(\omega\hat{q} - i\hat{p})\,, \tag{11.14}$$

so that

$$\hat{q} = \sqrt{\left(\frac{\hbar}{2\omega}\right)}(\hat{a} + \hat{a}^+)\,,$$

$$\hat{p} = -i\sqrt{\left(\frac{\hbar\omega}{2}\right)}(\hat{a} - \hat{a}^+)\,, \tag{11.15}$$

the commutation rule becomes

$$[\hat{q}, \hat{p}] = i\hbar \tag{11.16}$$

and (11.13) gives us

$$\hat{H}_F = \sum_\lambda \tfrac{1}{2}(\hat{p}_\lambda^2 + \omega_\lambda^2 \hat{q}_\lambda^2), \tag{11.17}$$

which shows clearly that the free electromagnetic field is equivalent to an infinite set of harmonic oscillators. With this result Dirac began the development of quantum electrodynamics.

The states $|n\rangle$ for all n form a complete orthonormal system, which is expressed by the *completeness condition* (condition of resolution of unity)

$$\sum_{n=0}^{\infty} |n\rangle \langle n| = \hat{1}, \tag{11.18}$$

where $\hat{1}$ is the identity operator. Thus every state $|\,\rangle$ may be decomposed into states $|n\rangle$ in the form

$$|\,\rangle = \sum_{k=0}^{\infty} c_k |k\rangle, \tag{11.19}$$

where

$$c_k = \langle k|\,\rangle. \tag{11.20}$$

The space they span is the familiar Fock space of quantum field theory and nearly all treatments of quantum electrodynamics have been formulated using these states. As we have mentioned such formulations do not retain any information about the phase of the field. In early applications this was unimportant since such information played a minimal role but in discussing coherence the information is vital and to retain it we must use coherent states. These states also allow one to consider the classical limit of quantum electrodynamics when the fields are strong and n is large and uncertain.

The vector potential $\hat{\boldsymbol{A}}$ is Hermitian ($\hat{\boldsymbol{A}}^+ = \hat{\boldsymbol{A}}$) since the classical field is real and we see from (11.2) that $\hat{\boldsymbol{A}}$ (and also $\hat{\boldsymbol{E}}$ and $\hat{\boldsymbol{H}}$) can be decomposed into the sum of a positive frequency part $\hat{\boldsymbol{A}}^{(+)}$ and a negative frequency part $\hat{\boldsymbol{A}}^{(-)}$ as follows

$$\hat{\boldsymbol{A}} = \hat{\boldsymbol{A}}^{(+)} + \hat{\boldsymbol{A}}^{(-)}, \tag{11.21}$$

where

$$\hat{\boldsymbol{A}}^{(+)}(x) = \frac{\sqrt{(2\pi\hbar c)}}{L^{3/2}} \sum_{\boldsymbol{k},s} \frac{1}{\sqrt{k}}\, \boldsymbol{e}^{(s)}(\boldsymbol{k})\, \hat{a}_{\boldsymbol{k}s} \exp\left[i(\boldsymbol{k}\,.\,\boldsymbol{x} - ckt)\right] \tag{11.22}$$

and $\hat{\boldsymbol{A}}^{(-)} = [\hat{\boldsymbol{A}}^{(+)}]^+$. Hence the operators $\hat{\boldsymbol{A}}^{(+)}$ and $\hat{\boldsymbol{A}}^{(-)}$ also represent the annihilation and creation operators respectively. They correspond to the classical fields V and V^* and play an important role in the process of detection of fields. For this reason the decomposition of the field operator $\hat{A}$ into two complex quantities has a basic significance in quantum theory, whereas in the classical theory it is largely a matter of mathematical convenience. In the classical theory, where energy quantities with values comparable with the field quantum $\hbar\omega$ are neglected, one cannot distinguish between absorption and emission; only the real part of the complex field $\hat{A}^{(+)}$ can be measured. On the other hand in the quantum theory, atoms considered as detectors measure the annihilation operator $\hat{A}^{(+)}$ if they are in the normal unexcited states and the creation operator $\hat{A}^{(-)}$ if they are in excited states.

To study the detection of the electromagnetic field we also introduce the so-called *detection operator* ([27], § 10, [275])

$$\hat{\mathbf{A}}(x) = L^{-3/2} \sum_{k,s} e^{(s)}(\boldsymbol{k})\, \hat{a}_{\boldsymbol{k}s} \exp\left[i(\boldsymbol{k}\,.\,\boldsymbol{x} - ckt)\right], \tag{11.23}$$

which is closely related to the positive frequency part of the vector potential operator. The number operator $\hat{n}$ of photons localized in a finite volume V (the linear dimensions of which are large compared with the wavelength) is

$$\hat{n}_{Vt} = \int_V \hat{\mathbf{A}}^+(x)\,.\,\hat{\mathbf{A}}(x)\, d^3\boldsymbol{x}, \tag{11.24a}$$

which gives for $V \equiv L^3$

$$\hat{n} = \sum_\lambda \hat{a}_\lambda^+ \hat{a}_\lambda\,. \tag{11.24b}$$

For quasi-monochromatic fields, the most usual case in practice, all the operators $\hat{E}^{(+)}$, $\hat{H}^{(+)}$, $\hat{A}^{(+)}$ and $\hat{\mathbf{A}}$ are proportional to one another.

11.2 Statistical states

From the quantum-mechanical point of view all information about the statistics of a system is contained in the *density matrix* $\hat{\varrho}$ (it is Hermitian, $\hat{\varrho}^+ = \hat{\varrho}$, see e.g. [18]). It can be decomposed in terms of the Fock states into

$$\hat{\varrho} = \sum_{n,m} \varrho(n, m)\, |n\rangle \langle m|, \tag{11.25}$$

where $\varrho(n, m) = \langle n|\, \hat{\varrho}\, |m\rangle$ since $\langle n \mid m\rangle = \delta_{nm}$ and the normalization condition is

$$\mathrm{Tr}\,\hat{\varrho} = 1\,, \quad \text{or} \quad \sum_n \varrho(n, n) = 1\,. \tag{11.26}$$

A pure state of the field is described by $\hat{\varrho} = |n\rangle\langle n|$.

As an example we can discuss thermal radiation in thermal equilibrium at temperature T. Denoting $\Theta = \hbar\omega/KT$, where K is Boltzmann constant, we have

$$\hat{\varrho} = \sum_{n=0}^{\infty} \varrho(n) \quad |n\rangle\langle n|\,, \tag{11.27}$$

where

$$\varrho(n) \equiv \varrho(n, n) = [1 - \exp(-\Theta)] \exp(-n\Theta)\,, \tag{11.28}$$

and so

$$\hat{\varrho} = \sum_{n=0}^{\infty} [1 - \exp(-\Theta)] \exp(-\Theta\hat{a}^+\hat{a})\,|n\rangle\langle n| =$$

$$= [1 - \exp(-\Theta)] \exp\left(-\frac{\Theta}{\hbar\omega}\hat{H}\right) = \frac{\exp\left(-\dfrac{\Theta}{\hbar\omega}\hat{H}\right)}{\mathrm{Tr}\left\{\exp\left(-\dfrac{\Theta}{\hbar\omega}\hat{H}\right)\right\}}\,, \tag{11.29}$$

where (11.18) has been used and $\hat{H}$ is the renormalized Hamiltonian. The expectation value of $\hat{n}$ is equal to

$$\langle\hat{n}\rangle = \mathrm{Tr}\,\{\hat{\varrho}\hat{a}^+\hat{a}\} = \sum_{n=0}^{\infty} n \exp(-n\Theta)\,[1 - \exp(-\Theta)] =$$

$$= -[1 - \exp(-\Theta)] \frac{d}{d\Theta} \frac{1}{1 - \exp(-\Theta)} = \frac{1}{\exp\Theta - 1} = \langle n\rangle\,, \tag{11.30}$$

so that the mean energy $\langle U\rangle$ is equal to

$$\langle U\rangle = \hbar\omega\langle\hat{n}\rangle = \frac{\hbar\omega}{\exp\Theta - 1}\,. \tag{11.31}$$

From (11.28) one has for the matrix elements of the density matrix

$$\varrho(n) = \frac{\langle n\rangle^n}{(1 + \langle n\rangle)^{1+n}}\,, \tag{11.32}$$

which is the well-known Bose-Einstein distribution. The quantum characteristic function becomes

$$\operatorname{Tr}\{\hat{\varrho} \exp ix\hat{n}\} = \sum_{n=0}^{\infty} \frac{1}{1+\langle n\rangle}\left[\frac{\exp(ix)\langle n\rangle}{1+\langle n\rangle}\right]^n =$$

$$= \frac{1}{1+\langle n\rangle} \frac{1}{1-\dfrac{\langle n\rangle}{1+\langle n\rangle}\exp ix} = \frac{1}{1-\langle n\rangle(\exp ix - 1)}. \quad (11.33)$$

Similarly we find for the Poisson distribution,

$$\varrho(n) = \frac{1}{n!}\langle n\rangle^n \exp(-\langle n\rangle), \quad (11.34)$$

that

$$\operatorname{Tr}\{\hat{\varrho}\hat{n}\} = \langle n\rangle \quad (11.35a)$$

and

$$\operatorname{Tr}\{\hat{\varrho} \exp ix\hat{n}\} = \exp[(e^{ix}-1)\langle n\rangle]. \quad (11.35b)$$

11.3 Multimode description

To describe multimode fields we introduce the Fock states

$$|\{n_\lambda\}\rangle \equiv \prod_\lambda |n_\lambda\rangle \quad (11.36)$$

for which the completeness condition reads

$$\sum_{\{n_\lambda\}} |\{n_\lambda\}\rangle\langle\{n_\lambda\}| = \hat{1}. \quad (11.37)$$

The density matrix has the form

$$\hat{\varrho} = \sum_{\{n_\lambda\}}\sum_{\{m_\lambda\}} \varrho(\{n_\lambda\},\{m_\lambda\})\,|\{n_\lambda\}\rangle\langle\{m_\lambda\}|, \quad (11.38)$$

where $\varrho(\{n_\lambda\},\{m_\lambda\}) = \langle\{n_\lambda\}|\,\hat{\varrho}\,|\{m_\lambda\}\rangle$. A pure state is described by the density matrix $\hat{\varrho} = |\{n_\lambda\}\rangle\langle\{n_\lambda\}|$.

11.4 Calculation of commutators of the field operators

The correct correspondence between classical fields and the field operators is given by the commutation rules (11.3). However, the commutation rules for space-time field operators, such as $\hat{\boldsymbol{A}}, \hat{\boldsymbol{E}}, \hat{\boldsymbol{H}}$, etc., are sometimes needed and therefore we derive some of them here, for use in some later calculations. It will be sufficient for us to derive the commutation rules for $\hat{A}$ and $\hat{\mathbf{A}}$ since the commutation rules for $\hat{E}$ and $\hat{H}$ follow using (11.1).

Making use of (11.2) and (11.3) we obtain

$$[\hat{A}_\mu(x), \hat{A}_\nu(x')] = \frac{2\pi\hbar c}{L^3} \sum_{k,s} \frac{1}{k}\, e_\mu^{(s)}(k)\, e_\nu^{(s)*}(k)\,.$$

$$\{\exp[ik(x - x')] - \exp[-ik(x - x')]\} =$$

$$= \frac{2\pi\hbar c}{L^3} \sum_k \frac{1}{k} \left(\delta_{\mu\nu} - \frac{k_\mu k_\nu}{k^2}\right) \{\exp[ik(x - x')] - \exp[-ik(x - x')]\} \tag{11.39}$$

(where (11.5) has also been used), $kx = \boldsymbol{k} \cdot \boldsymbol{x} - kct$ and

$$[\hat{A}_\mu^{(+)}(x), \hat{A}_\nu^{(-)}(x')] = \frac{2\pi\hbar c}{L^3} \sum_{k,s} \frac{1}{k}\, e_\mu^{(s)}(k)\, e_\nu^{(s)*}(k) \exp[ik(x - x')] =$$

$$= \frac{2\pi\hbar c}{L^3} \sum_k \frac{1}{k} \left(\delta_{\mu\nu} - \frac{k_\mu k_\nu}{k^2}\right) \exp[ik(x - x')]\,. \tag{11.40}$$

The commutation rule for the detection operator (11.23) becomes

$$[\hat{\mathbf{A}}_\mu(x), \hat{\mathbf{A}}_\nu^+(x')] = \frac{1}{L^3} \sum_k \left(\delta_{\mu\nu} - \frac{k_\mu k_\nu}{k^2}\right) \exp[ik(x - x')] \tag{11.41}$$

and for $t = t'$ it reduces to

$$[\hat{\mathbf{A}}_\mu(\boldsymbol{x}, t), \hat{\mathbf{A}}_\nu^+(\boldsymbol{x}', t)] = \delta_{\mu\nu}\, \delta(\boldsymbol{x} - \boldsymbol{x}') - \frac{1}{(2\pi)^3} \int \frac{k_\mu k_\nu}{k^2} \exp[i\boldsymbol{k} \cdot (\boldsymbol{x} - \boldsymbol{x}')]\, d^3\boldsymbol{k}\,. \tag{11.42}$$

Here we have replaced the sum $\sum_k$ by the integral $(L/2\pi)^3 \int \ldots d^3\boldsymbol{k}$. It is easily seen that

$$[\hat{A}_\mu^{(+)}(x), \hat{A}_\nu^{(+)}(x')] = [\hat{A}_\mu^{(-)}(x), \hat{A}_\nu^{(-)}(x')] =$$

$$= [\hat{\mathbf{A}}_\mu(x), \hat{\mathbf{A}}_\nu(x')] = [\hat{\mathbf{A}}_\mu^+(x), \hat{\mathbf{A}}_\nu^+(x')] = 0\,. \tag{11.43}$$

The functions on the right-hand sides of (11.39)–(11.42) are singular functions which are zero outside the light cone, that is, when $(\boldsymbol{x} - \boldsymbol{x}')^2 > c^2(t - t')^2$. This is a reflection of the fact that measurements of fields at such pairs of points can be performed with arbitrary accuracy and they cannot influence one another (we remember that if two operators $\hat{A}$, $\hat{B}$ satisfy $[\hat{A}, \hat{B}] = C$ then $\langle(\Delta\hat{A})^2\rangle \langle(\Delta\hat{B})^2\rangle \geqq |C|^2/4$, cf. (13.24)). This expresses the principle of causality in quantum electrodynamics (signals cannot propagate more rapidly than with the velocity of light *c in vacuo*) [409].

In analogy with the commutation rules (11.6) one can derive the following commutation rules [277]

$$[\hat{\mathbf{A}}(\boldsymbol{x}, t), \hat{n}_{Vt'}] = \begin{cases} \hat{\mathbf{A}}(\boldsymbol{x}, t), & \text{if } (\boldsymbol{x}, t) \text{ is } \textit{conjoint} \text{ with } (V, t') \\ 0, & \text{if } (\boldsymbol{x}, t) \text{ is } \textit{disjoint} \text{ with } (V, t') \end{cases} \tag{11.44a}$$

and

$$[\hat{n}_{Vt'}, \hat{\mathbf{A}}^+(\boldsymbol{x}, t)] = \begin{cases} \hat{\mathbf{A}}^+(\boldsymbol{x}, t), & \text{if } (\boldsymbol{x}, t) \text{ is } \textit{conjoint} \text{ with } (V, t'), \\ 0, & \text{if } (\boldsymbol{x}, t) \text{ is } \textit{disjoint} \text{ with } (V, t'). \end{cases} \tag{11.44b}$$

The term conjoint and disjoint is defined in the following way [277]. We define the function $U(\boldsymbol{x}, V) = 1$ if $\boldsymbol{x} \in V$ and $U = 0$ if $\boldsymbol{x} \notin V$; then $(\boldsymbol{x}, t)$ is conjoint with (V, t') if $U(\boldsymbol{x}', V) = 1$ and disjoint if $U(\boldsymbol{x}', V) = 0$, where $\boldsymbol{x}' = \boldsymbol{x} - c\boldsymbol{k}(t - t')/|\boldsymbol{k}|$.

11.5 Time development in quantum electrodynamics

It is well known that in quantum electrodynamics a state of the field is described by a vector in the Hilbert space of states. The time development of a state can be interpreted either as the time development of the vector in the space if the system of coordinates is fixed (the Schrödinger picture) or as the time development of the system of coordinates if the state vector is fixed (the Heisenberg picture).

In the Schrödinger picture the Schrödinger equation holds for the state vector $|\psi(t)\rangle$

$$i\hbar \frac{\partial}{\partial t} |\psi(t)\rangle = \hat{H}|\psi(t)\rangle, \tag{11.45a}$$

where $\hat{H}$ is the Hamiltonian, independent of time, and for an operator $\hat{F}$

$$-i\hbar \frac{\partial \hat{F}}{\partial t} = 0. \tag{11.45b}$$

Performing the canonical transformations

$$\hat{F}(t) = \exp\left(i\frac{\hat{H}t}{\hbar}\right)\hat{F}\exp\left(-i\frac{\hat{H}t}{\hbar}\right), \tag{11.46a}$$

$$|\psi(t)\rangle = \exp\left(-i\frac{\hat{H}t}{\hbar}\right)|\psi\rangle, \tag{11.46b}$$

which provide the correspondence between quantities in the Schrödinger and Heisenberg pictures, we obtain in the Heisenberg picture,

$$-i\hbar \frac{\partial}{\partial t} \hat{F}(t) = [\hat{H}, \hat{F}(t)] \,, \tag{11.47a}$$

$$i\hbar \frac{\partial|\psi\rangle}{\partial t} = 0 \,. \tag{11.47b}$$

For the density matrix $\hat{\varrho} = \{|\,\rangle\langle\,|\}_{\text{average}}$ we obtain

$$i\hbar \frac{\partial\hat{\varrho}(t)}{\partial t} = [\hat{H}, \hat{\varrho}(t)] \tag{11.48a}$$

in the Schrödinger picture whereas

$$i\hbar \frac{\partial\hat{\varrho}}{\partial t} = 0 \tag{11.48b}$$

in the Heisenberg picture and

$$\hat{\varrho}(t) = \exp\left(-i\frac{\hat{H}t}{\hbar}\right)\hat{\varrho}(0)\exp\left(i\frac{\hat{H}t}{\hbar}\right). \tag{11.49}$$

In calculations of the expectation values of an operator $\hat{F}(t)$ it may be convenient to use the following rule

$$\operatorname{Tr}\{\hat{\varrho}\,\hat{F}(t)\} = \operatorname{Tr}\left\{\hat{\varrho}\exp\left(i\frac{\hat{H}t}{\hbar}\right)\hat{F}\exp\left(-i\frac{\hat{H}t}{\hbar}\right)\right\} =$$

$$= \operatorname{Tr}\left\{\exp\left(-i\frac{\hat{H}t}{\hbar}\right)\hat{\varrho}\exp\left(i\frac{\hat{H}t}{\hbar}\right)\hat{F}\right\} = \operatorname{Tr}\{\hat{\varrho}(t)\,\hat{F}\} \,. \tag{11.50}$$

Sometimes it is convenient to introduce the interaction picture. If $\hat{H} = \hat{H}_0 + \hat{H}_i$, where $\hat{H}_0$ and $\hat{H}_i$ are the free and interaction Hamiltonians, it follows that

$$i\hbar \frac{\partial|\phi(t)\rangle}{\partial t} = \mathscr{H}_i|\phi(t)\rangle \,, \tag{11.51a}$$

$$-i\hbar \frac{\partial\mathscr{F}(t)}{\partial t} = [\hat{H}_0, \mathscr{F}(t)] \,, \tag{11.51b}$$

where

$$|\phi(t)\rangle = \exp\left(i\frac{\hat{H}_0 t}{\hbar}\right)|\psi(t)\rangle \,, \tag{11.52a}$$

$$\mathscr{H}_i = \exp\left(i\frac{\hat{H}_0 t}{\hbar}\right)\hat{H}_i \exp\left(-i\frac{\hat{H}_0 t}{\hbar}\right), \tag{11.52b}$$

$$\mathscr{F}(t) = \exp\left(i\frac{\hat{H}_0 t}{\hbar}\right)\hat{F} \exp\left(-i\frac{\hat{H}_0 t}{\hbar}\right). \tag{11.52c}$$

The Schrödinger equation (11.45a) (with $\hat{H}$ generally dependent on t) can be solved in the form

$$|\psi(t)\rangle = \hat{S}(t)\,|\psi(0)\rangle\,, \tag{11.53}$$

where $\hat{S}(t)$ is the time development (unitary) operator obeying the Schrödinger equation

$$i\hbar\frac{\partial \hat{S}(t)}{\partial t} = \hat{H}(t)\,\hat{S}(t)\,. \tag{11.54}$$

Solving this equation by the usual perturbation method with the condition $\hat{S}(0) = \hat{1}$ (see e.g. [2]) we obtain

$$\hat{S}(t) = \hat{T}\exp\left[-\frac{i}{\hbar}\int_0^t \hat{H}(t')\,dt'\right] =$$

$$= \sum_{n=0}^{\infty}\frac{(-i/\hbar)^n}{n!}\int_0^t \ldots \int \hat{T}\{\hat{H}(t_1)\ldots\hat{H}(t_n)\}\,dt_1\ldots dt_n\,, \tag{11.55}$$

where the operator $\hat{T}$ is the time-ordering operator. This operator formally reorders the terms so that all operator products are taken with the factors in an increasing time-order sequence (from the right to the left).

Chapter 12

OPTICAL CORRELATION PHENOMENA

Having prepared the basic quantum formalism we can start a quantum-mechanical treatment of coherence phenomena in the electromagnetic field and show what kind of experiments are described by the quantum correlation functions. We consider the detection of a field using the interaction of this radiation field with matter. The detection device, composed of atoms, will be assumed to be in its ground state so that photons are registered by *absorption*. Such detectors of light are photodetectors, photographic plates, etc.

12.1 Quantum correlation functions

Consider an ideal detector which is quite small in size and has a sensitivity independent of the photon frequency. One atom may be considered as such a detector. Suppose the initial state of the field is characterized by the state vector $|i\rangle$ and the atom is initially in its ground state. If $|f\rangle$ is a final state of the field after the absorption of a photon described by the annihilation operator $\hat{A}^{(+)}(x)$ (a linearly polarized field may be assumed again), then the transition amplitude of the process is equal to $\langle f|\,\hat{A}^{(+)}(x)\,|i\rangle$. The transition probability per unit time from the state $|i\rangle$ to the state $|f\rangle$ of the radiation field due to the absorption of a photon at a space-time point (x, t) is then proportional to $|\langle f|\,\hat{A}^{(+)}(x)\,|i\rangle|^2$. The final state $|f\rangle$ of the field usually remains unobserved so that we must sum over all possible final states $|f\rangle$ which gives

$$\sum_f \langle i|\,\hat{A}^{(-)}(x)\,|f\rangle\,\langle f|\,\hat{A}^{(+)}(x)\,|i\rangle = \langle i|\,\hat{A}^{(-)}(x)\,\hat{A}^{(+)}(x)\,|i\rangle \qquad (12.1)$$

because of the completeness condition $\sum_f |f\rangle\,\langle f| = \hat{1}$. The initial states

$|i\rangle$ of the field depend on some random or unknown parameters and consequently all field measurements must be carried out as averages over the ensemble of ways in which the field can be prepared. As we know the information about the statistics of the field as a statistical dynamic system is contained in the density matrix $\hat{\varrho} = \{|i\rangle\langle i|\}_{\text{average}}$ and we arrive at

$$\{\langle i|\,\hat{A}^{(-)}(x)\,\hat{A}^{(+)}(x)\,|i\rangle\}_{\text{average}} = \mathrm{Tr}\,\{\hat{\varrho}\,\hat{A}^{(-)}(x)\,\hat{A}^{(+)}(x)\}\,, \tag{12.2}$$

which represents the mean intensity at x. This intensity is a particular value (at $x' = x$) of the correlation function

$$\Gamma_{\mathcal{N}}^{(1,1)}(x, x') = \mathrm{Tr}\,\{\hat{\varrho}\,\hat{A}^{(-)}(x)\,\hat{A}^{(+)}(x')\} \tag{12.3}$$

which corresponds to the classical correlation function $\Gamma^{(1,1)}(x, x')$. If the vectorial properties of the field are taken into account vectorial indices have to be added to $\hat{A}^{(+)}$ and $\hat{A}^{(-)}$ and we obtain the second-rank tensor $\Gamma_{\mathcal{N},\mu\nu}^{(1,1)}$ as earlier. For simplicity we use x to denote $(\boldsymbol{x}, t, \mu)$. Sometimes one considers fields obtained using polarizing filters which select photons of polarization $\boldsymbol{e}$; then the use of only the scalar field operator $\hat{A} = \boldsymbol{e} \,.\, \hat{\boldsymbol{A}}$ is sufficient.

More generally, we define the transition probability per unit (time)^n that a photon of a polarization μ_1 is absorbed at $(\boldsymbol{x}_1, t_1) \equiv x_1$, a photon of a polarization μ_2 at $(\boldsymbol{x}_2, t_2) \equiv x_2$, etc. and a photon of a polarization μ_n at $(\boldsymbol{x}_n, t_n) \equiv x_n$. We obtain in the same way that this probability is proportional to

$$\begin{aligned} &\Gamma_{\mathcal{N}}^{(n,n)}(x_1, \ldots, x_n, x_n, \ldots, x_1) = \\ &= \{\sum_f |\langle f|\,\hat{A}^{(+)}(x_1) \ldots \hat{A}^{(+)}(x_n)\,|i\rangle|^2\} = \\ &=_{\text{average}} = \mathrm{Tr}\,\{\hat{\varrho}\,\hat{A}^{(-)}(x_1) \ldots \hat{A}^{(-)}(x_n)\,\hat{A}^{(+)}(x_n) \ldots \hat{A}^{(+)}(x_1)\}\,. \end{aligned} \tag{12.4}$$

This is a particular value of the general correlation function (tensor) $\Gamma_{\mathcal{N}}^{(m,n)}(x_1, \ldots, x_{m+n})$ defined by

$$\Gamma_{\mathcal{N}}^{(m,n)}(x_1, \ldots, x_{m+n}) = \mathrm{Tr}\,\{\hat{\varrho}\,\hat{A}^{(-)}(x_1) \ldots \hat{A}^{(-)}(x_m)\,\hat{A}^{(+)}(x_{m+1}) \ldots \hat{A}^{(+)}(x_{m+n})\} \tag{12.5}$$

if $m = n$ and $x_j = x_{n+j}$, $j = 1, 2, \ldots, n$. Note that all the $\hat{A}^{(-)}$ as well as the $\hat{A}^{(+)}$ can be interchanged since the commutation rules (11.43) hold. These correlation functions agree exactly with the classical correlation functions (8.4) via the correspondence $\hat{A}^{(+)} \leftrightarrow V$, $\hat{A}^{(-)} \leftrightarrow V^*$, $\hat{\varrho} \leftrightarrow P$. Of course the operators $\hat{A}^{(+)}$ and $\hat{A}^{(-)}$ do not commute in general (cf. (11.40)) so that this correspondence (of the classical and quantum correlation functions) is not unique. The Hanbury Brown-Twiss effect discussed in Chapter 10 is described by the quantum correlation function $\Gamma_{\mathcal{N}}^{(2,2)}(x_1, x_2, x_2, x_1)$.

An important property of the correlation functions (12.5) is that the operators $\hat{A}^{(+)}$ and $\hat{A}^{(-)}$ are in their *normal form*, i.e. all annihilation operators $\hat{A}^{(+)}$ stand to the right of all creation operators $\hat{A}^{(-)}$. A consequence of this, if the vacuum stability condition $\hat{A}^{(+)}|0\rangle = 0$ is used, is that the vacuum expectation value of these normally ordered products is zero,

$$\Gamma_{\mathcal{N},\mathrm{vac}}^{(n,n)}(x_1, \ldots, x_n, x_n, \ldots, x_1) = \langle 0| \prod_{j=1}^{n} \hat{A}^{(-)}(x_j) \prod_{k=1}^{n} \hat{A}^{(+)}(x_k) |0\rangle = 0\,, \tag{12.6}$$

and the physical vacuum gives no contribution to the quantities characterizing these photoelectric correlation measurements; it does give contribution to measurements which are related to other orderings of field operators. As an example we shall discuss in section 16.4 some devices (quantum counters) whose operation is connected with the *antinormally* ordered products of field operators ($\hat{A}^{(+)}$ and $\hat{A}^{(-)}$ are interchanged). These devices detect the field by means of stimulated emission rather than by absorption. In this case $\langle 0| \hat{A}^{(+)}(x) \hat{A}^{(-)}(x') |0\rangle = [\hat{A}^{(+)}(x), \hat{A}^{(-)}(x')]$, where the commutator is given in (11.40) and the physical vacuum contributes. Just these circumstances are responsible for the fact that the probability distributions of absorbed photons and emitted photoelectrons have a similar form whereas the distribution of photons and the distribution of counts of a quantum counter differ from one another (as a consequence of the contribution of the physical vacuum).

From the correlation functions defined in terms of $\hat{A}$ we can obtain using (11.1) the correlation functions (tensors) defined in terms of $\hat{E}$,

$$\mathrm{Tr}\,\{\hat{\varrho} \prod_{j=1}^{m} \hat{E}^{(-)}(x_j) \prod_{k=m+1}^{m+n} \hat{E}^{(+)}(x_k)\}\,, \tag{12.7}$$

(or in terms of $\hat{H}$) as well as various mixed forms. The correlation functions (12.7) were used in Glauber's treatment of quantum coherence [173–176, 180, 181]. In practice quasi-monochromatic light is mostly used and then all these quantities are proportional to one another. Further kinds of correlation functions (tensors), based on the electromagnetic antisymmetric tensor $\hat{F}_{\mu\nu}(x) = \partial \hat{A}_\nu / \partial x_\mu - \partial \hat{A}_\mu / \partial x_\nu$, may also be introduced [173] and some of their properties have been investigated [239, 128].

A more precise treatment of the detection of an electromagnetic field must use (time-dependent) perturbation theory. This was done making the electric-dipole approximation in [176]; the interaction Hamiltonian

$$\hat{H}_i = -e \sum_{\gamma} \hat{\boldsymbol{q}}_{\gamma} \cdot \hat{\boldsymbol{E}}(\boldsymbol{x}, t), \tag{12.8a}$$

where $-e$ is the charge of an electron, $\hat{\boldsymbol{q}}_{\gamma}$ is the spatial coordinate operator of the γth electron of the atom relative to its nucleus, assumed to be located at $\boldsymbol{x}$. More generally, the interaction Hamiltonian can be written as

$$\hat{H}_i = -e \sum_{\gamma} \hat{\boldsymbol{p}}_{\gamma} \cdot \hat{\boldsymbol{A}}(\boldsymbol{x}, t), \tag{12.8b}$$

where $\hat{\boldsymbol{p}}_{\gamma}$ is the momentum operator and $\hat{\boldsymbol{A}}$ is the vector potential operator.

Consider N atoms as detectors placed at points $\boldsymbol{x}_1, \ldots, \boldsymbol{x}_N$ and apply the formalism for the time development contained in section 11.5. We can solve the Schrödinger equation (11.45a) with $\hat{H} = \hat{H}_i(t)$ (the interaction picture) in the form (11.53), where the time development operator $\hat{S}$ is given by (11.55). Using the Nth-order approximation of perturbation theory (Nth term in (11.55)), we obtain for the probability that the first atom has undergone a transition in the interval $(0, T_1)$, the second one in $(0, T_2)$, etc. and Nth one in $(0, T_N)$ [176]

$$p^{(N)}(T_1, \ldots, T_N) = \\ = \int_0^{T_1} \ldots \int_0^{T_N} \Gamma_{\mathcal{N}}^{(N,N)}(\boldsymbol{x}_1, \ldots, \boldsymbol{x}_N, \boldsymbol{x}_N, \ldots, \boldsymbol{x}_1; t'_1, \ldots, t'_N, t''_N, \ldots, t''_1) \, . \\ \cdot \prod_{j=1}^{N} \mathscr{S}(t''_j - t'_j) \, dt'_j \, dt''_j \, , \tag{12.9}$$

where $\mathscr{S}$ is a weight function (a "response" function of the detectors).

For detectors assumed to be broadband (they do not distinguish between frequencies)

$$\mathscr{S}(t'' - t') = \eta \, \delta(t'' - t'), \tag{12.10}$$

where η is the sensitivity factor and we have, from (12.9),

$$p^{(N)}(T_1, \ldots, T_N) = \\ = \eta^N \int_0^{T_1} \ldots \int_{0]}^{T_N} \Gamma_{\mathcal{N}}^{(N,N)}(\boldsymbol{x}_1, \ldots, \boldsymbol{x}_N, \boldsymbol{x}_N, \ldots, \boldsymbol{x}_1; t'_1, \ldots, t'_N, t'_N, \ldots, t'_1) \prod_{j=1}^{N} dt'_j \, . \tag{12.11}$$

This probability must equal the expectation value $\langle \hat{n}_1 \ldots \hat{n}_N \rangle$ which is simply the probability that all the detectors $1, 2, \ldots, N$ have contributed counts in $(0, T_1), (0, T_2), \ldots, (0, T_N)$ respectively; this is a similar result to that obtained in the semiclassical description (equation (10.92)).

The joint counting rate corresponding to times $T_1, \ldots, T_N$ will be equal to

$$w(T_1, \ldots, T_N) = \frac{\partial^N}{\partial T_1 \ldots \partial T_N} p^{(N)}(T_1, \ldots, T_N) =$$

$$= \eta^N \Gamma_{\mathcal{N}}^{(N,N)}(\boldsymbol{x}_1, \ldots, \boldsymbol{x}_N, \boldsymbol{x}_N, \ldots, \boldsymbol{x}_1, T_1, \ldots, T_N, T_N, \ldots, T_1) \quad (12.12)$$

in analogy to (10.93).

Equation (12.9) can be interpreted as applicable to point detectors which are not sensitive to the instantaneous value of the field but are sensitive rather to an average of the field over a time interval. More generally we can consider detectors which are not sensitive even to the field at point $\boldsymbol{x}$ but to the averaged field over spatial regions $V_1, \ldots, V_N$ and we obtain

$$p^{(N)} = \int_0^{T_1} \ldots \int_0^{T_N} \int_{V_1} \ldots \int_{V_N} \Gamma_{\mathcal{N}}^{(N,N)}(x'_1, \ldots, x'_N, x''_N, \ldots, x''_1) \,.$$

$$\cdot \prod_{j=1}^{N} \mathcal{S}(\boldsymbol{x}''_j - \boldsymbol{x}'_j, t''_j - t'_j)\, d^4x'_j d^4x''_j \,. \quad (12.13)$$

Equation (12.9) is then obtained for $\mathcal{S}(\boldsymbol{x}'' - \boldsymbol{x}', t'' - t') = \delta(\boldsymbol{x}'' - \boldsymbol{x}') \,.$ $. \,\delta(\boldsymbol{x}' - \boldsymbol{x})\, \mathcal{S}(t'' - t')$. In (12.13) x denotes $(\boldsymbol{x}, t, \mu)$ and the integrals over x imply in addition a summation over the polarization indices μ.

Some properties of the correlation functions were summarized in Chapter 2 and here we briefly outline a derivation of some of them and give some further properties.

Equation (2.39) follows from the inequality

$$\mathrm{Tr}\,\{\hat{\varrho}\hat{B}^+\hat{B}\} \geqq 0 \quad (12.14)$$

(valid for an arbitrary operator $\hat{B}$) putting $\hat{B} = \hat{A}^{(+)}(x_1) \ldots \hat{A}^{(+)}(x_n)$. If

$$\hat{B} = \sum_{j=1}^{n} c_j\, \hat{A}^{(+)}(x_j)\,, \quad (12.15)$$

we obtain the quantum analog use of (4.7a)

$$\sum_{j,k}^{n} c_j^* c_k\, \Gamma_{\mathcal{N}}^{(1,1)}(x_j, x_k) \geqq 0\,. \quad (12.16)$$

This condition leads to

$$\mathrm{Det}\,\{\Gamma_{\mathcal{N}}^{(1,1)}(x_j, x_k)\} \geqq 0\,; \quad (12.17)$$

for $n = 1$ we have $\Gamma_{\mathcal{N}}^{(1,1)}(x, x) \geqq 0$ and for $n = 2$

$$\Gamma_{\mathcal{N}}^{(1,1)}(x_1, x_1)\, \Gamma_{\mathcal{N}}^{(1,1)}(x_2, x_2) \geqq \left|\Gamma_{\mathcal{N}}^{(1,1)}(x_1, x_2)\right|^2 . \tag{12.18}$$

If

$$\hat{B} = c_1\, \hat{A}^{(+)}(x_1) \ldots \hat{A}^{(+)}(x_n) + c_2\, \hat{A}^{(+)}(x_{n+1}) \ldots \hat{A}^{(+)}(x_{2n}) , \tag{12.19}$$

(2.41) follows at once.

Another interesting property of $\Gamma_{\mathcal{N}}^{(n,n)}$ is that this function is zero for states with a finite number of photons, say N, in the field when $n > N$, since $\hat{A}^{(+)}$ is the annihilation operator.

As we have already mentioned the arguments $x_1, \ldots, x_m$ as well as $x_{m+1}, \ldots, x_{m+n}$ can be interchanged without changing $\Gamma_{\mathcal{N}}^{(m,n)}$ because the operators $\hat{A}^{(+)}$ as well as $\hat{A}^{(-)}$ mutually commute.

12.2 Quantum coherence

Second-order phenomena

Second-order coherence has been discussed in Chapter 3 in connection with the Young experiment and the form of the mutual coherence function for the coherent field was derived in section 4.2. Equation (9.25) has been adopted to define the coherent field in higher orders and we shall discuss this point more fully in this section.

The description of the Young experiment in terms of quantum correlation functions proceeds in an identical way to the classical treatment leading to the results discussed in section 3.1;

$$\begin{aligned} I(x) &= \operatorname{Tr}\{\hat{\varrho}[a_1^*\, \hat{A}^{(-)}(x_1) + a_2^*\, \hat{A}^{(-)}(x_2)]\, [a_1\, \hat{A}^{(+)}(x_1) + a_2\, \hat{A}^{(+)}(x_2)]\} = \\ &= |a_1|^2\, \Gamma_{\mathcal{N}}^{(1,1)}(x_1, x_1) + |a_2|^2\, \Gamma_{\mathcal{N}}^{(1,1)}(x_2, x_2) + \\ &\quad + |a_1 a_2|\, \Gamma_{\mathcal{N}}^{(1,1)}(x_1, x_2) + |a_1 a_2|\, \Gamma_{\mathcal{N}}^{(1,1)}(x_2, x_1) . \end{aligned} \tag{12.20}$$

The interference terms $\Gamma_{\mathcal{N}}^{(1,1)}(x_1, x_2)$ and $\Gamma_{\mathcal{N}}^{(1,1)}(x_2, x_1)$ in (12.20) express the fact that it is not possible to distinguish from which pinhole a photon came. We have seen that full coherence gives in second order the maximum visibility of interference fringes and

$$\left|\Gamma_{\mathcal{N}}^{(1,1)}(x_1, x_2)\right|^2 = \Gamma_{\mathcal{N}}^{(1,1)}(x_1, x_1)\, \Gamma_{\mathcal{N}}^{(1,1)}(x_2, x_2) . \tag{12.21}$$

This is the boundary value of (12.18) corresponding, according to the definition of the degree of coherence [(2.42a, b, c)], to

$$\left|\gamma_{\mathcal{N}}^{(1,1)}(x_1, x_2)\right| = 1 \tag{12.22}$$

valid for all x_1 and x_2; in general

$$0 \leqq |\gamma_{\mathcal{N}}^{(1,1)}(x_1, x_2)| \leqq 1 \,. \tag{12.23}$$

Writing (12.21) in the form

$$\Gamma_{\mathcal{N}}^{(1,1)}(x_1, x_2) = A(x_1)\, B(x_2)\,, \tag{12.24}$$

we have from the cross-symmetry condition

$$A(x_1)\, B(x_2) = A^*(x_2)\, B^*(x_1) \tag{12.25}$$

and so

$$\frac{A(x_1)}{B^*(x_1)} = \frac{A^*(x_2)}{B(x_2)} = k\,, \tag{12.26}$$

where k is a real constant. Therefore $A(x) = k\, B^*(x)$ and defining the function $V(x) = \sqrt{(k)}\, B(x)$ we can rewrite (12.24) in the form

$$\Gamma_{\mathcal{N}}^{(1,1)}(x_1, x_2) = V^*(x_1)\, V(x_2)\,, \tag{12.27}$$

which is just (4.12).

We can now ask in what manner the higher-order correlation functions are restricted by the factorization condition (12.27) [441]. If (12.27) holds, then (12.21) is also fulfilled and this implies that

$$\mathrm{Tr}\,\{\hat{\varrho}\hat{B}^+\hat{B}\} = 0\,, \tag{12.28}$$

where

$$\hat{B} = \hat{A}^{(+)}(x) - \frac{\Gamma_{\mathcal{N}}^{(1,1)}(x_0, x)}{\Gamma_{\mathcal{N}}^{(1,1)}(x_0, x_0)}\, \hat{A}^{(+)}(x_0)\,; \tag{12.29}$$

here x_0 is an arbitrary space-time point. Therefore $\hat{\varrho}\hat{B}^+ = \hat{B}\hat{\varrho} = 0$ and we obtain

$$\hat{A}^{(+)}(x)\, \hat{\varrho} = \frac{\Gamma_{\mathcal{N}}^{(1,1)}(x_0, x)}{\Gamma_{\mathcal{N}}^{(1,1)}(x_0, x_0)}\, \hat{A}^{(+)}(x_0)\, \hat{\varrho}\,, \tag{12.30a}$$

$$\hat{\varrho}\hat{A}^{(-)}(x) = \frac{\Gamma_{\mathcal{N}}^{(1,1)}(x, x_0)}{\Gamma_{\mathcal{N}}^{(1,1)}(x_0, x_0)}\, \hat{\varrho}\hat{A}^{(-)}(x_0)\,. \tag{12.30b}$$

Considering identities (12.30a) and (12.30b) at x_2 and x_1 respectively we obtain

$$\mathrm{Tr}\,\{\hat{\varrho}\hat{A}^{(-)}(x_1)\, \hat{A}^{(+)}(x_2)\} =$$

$$= \frac{\Gamma_{\mathcal{N}}^{(1,1)}(x_1, x_0)\, \Gamma_{\mathcal{N}}^{(1,1)}(x_0, x_2)}{[\Gamma_{\mathcal{N}}^{(1,1)}(x_0, x_0)]^2}\, \mathrm{Tr}\,\{\hat{\varrho}\hat{A}^{(-)}(x_0)\, \hat{A}^{(+)}(x_0)\}\,, \tag{12.31a}$$

or

$$\Gamma_{\mathcal{N}}^{(1,1)}(x_1, x_2) = \frac{\Gamma_{\mathcal{N}}^{(1,1)}(x_1, x_0)\, \Gamma_{\mathcal{N}}^{(1,1)}(x_0, x_2)}{\Gamma_{\mathcal{N}}^{(1,1)}(x_0, x_0)} = V^*(x_1)\, V(x_2) \quad (12.31b)$$

in agreement with (12.27), where $V(x) = \Gamma_{\mathcal{N}}^{(1,1)}(x_0, x)/[\Gamma_{\mathcal{N}}^{(1,1)}(x_0, x_0)]^{1/2}$.

Higher-order phenomena

For the $(m + n)$th-order correlation function we obtain in an identical manner:

$$\Gamma_{\mathcal{N}}^{(m,n)}(x_1, \ldots, x_{m+n}) = \gamma^{(m,n)} \prod_{j=1}^{m} V^*(x_j) \prod_{k=m+1}^{m+n} V(x_k)\,, \qquad (12.32)$$

where

$$\gamma^{(m,n)} = \frac{\Gamma_{\mathcal{N}}^{(m,n)}(x_0, \ldots, x_0)}{[\Gamma_{\mathcal{N}}^{(1,1)}(x_0, x_0)]^{(m+n)/2}}\,. \qquad (12.33)$$

However, x_0 is an arbitrary point so that (12.33) must be independent of x_0. Consequently the factorization condition (12.27) expressing the second-order coherence leads to the factorization of all correlation functions into the form (12.32).

By analogy with the factorization condition (12.27) for second-order coherence one may construct the following set of factorization conditions,

$$\Gamma_{\mathcal{N}}^{(m,n)}(x_1, \ldots, x_{m+n}) = \prod_{j=1}^{m} V^*(x_j) \prod_{k=m+1}^{m+n} V(x_k)\,, \qquad (12.34)$$

in which V is independent of m and n. Thus we can speak of $2N$th-order coherence if (12.34) holds for $m, n \leqq N$. If (12.34) holds for all m, n (in practice for all $m = n$), then the field possesses full coherence. It is clear from a classical point of view that fully coherent fields are noiseless fields whose distribution function equals the Dirac function. This fact illustrates the close relation between the noiselessness of fields and coherence. In the quantum theory there also exist quantum states of fields for which (12.34) holds. These are called the coherent states [174] and their properties will be investigated in the next chapter.

Comparing (12.34) with (12.32) we see that these two equations differ from one another by the factor $\gamma^{(m,n)}$ and second-order coherence does not lead, in general, to higher-order coherence, for which

$$\gamma^{(m,n)} = 1\,. \qquad (12.35)$$

If a finite number of photons, say M, are present in the field, then $\gamma^{(n,n)} = 0$ for $n > M$ and such a field cannot possess coherence to all orders. On the other hand, for the chaotic field we obtain from (10.64b)

$$\gamma^{(m,n)} = \delta_{mn} n! \,, \tag{12.36}$$

and we can conclude that chaotic radiation cannot possess coherence of order higher than the second ($m = n = 1$), if the definition (2.42b) of the degree of coherence is used. Expression (12.36) as well as some further examples can also be obtained using the results of Chapter 11. Considering one mode only we have from (11.8a)

$$\gamma^{(k,k)} = \frac{\langle \hat{a}^{+k} \hat{a}^k \rangle}{\langle \hat{a}^+ \hat{a} \rangle^k} = \frac{\langle \hat{n}(\hat{n} - 1) \dots (\hat{n} - k + 1) \rangle}{\langle \hat{n} \rangle^k} \,. \tag{12.37}$$

For a Fock state $|n\rangle$ the density matrix $\hat{\varrho} = |n\rangle\langle n|$ and so

$$\gamma^{(k,k)} = \frac{n!}{(n-k)!\, n^k} \,. \tag{12.38}$$

An interesting property of the Fock state is that the normal variance is negative: $\langle (\Delta \hat{n})^2 \rangle_{\mathcal{N}} = \langle \hat{a}^{+2} \hat{a}^2 \rangle - \langle \hat{a}^+ \hat{a} \rangle^2 = n(n-1) - n^2 = -n < 0$ and so the variance of $\hat{n}$ is $\langle (\Delta \hat{n})^2 \rangle = \langle \hat{n} \rangle + \langle (\Delta \hat{n})^2 \rangle_{\mathcal{N}} = 0$. This is an anticorrelation effect for these states and the Fock states have no classical analogue [315, 81]. (Note that for the coherent states (Chapter 13) $\langle \hat{a}^{+2} \hat{a}^2 \rangle - \langle \hat{a}^+ \hat{a} \rangle^2 = 0$.)

For the thermal radiation $\hat{\varrho}$ is given by (11.29) and

$$\gamma^{(k,k)} = [\operatorname{Tr} \{\exp(-\Theta \hat{n})\}]^{k-1} \frac{\operatorname{Tr} \left\{ \exp(-\Theta \hat{n}) \dfrac{\hat{n}!}{(\hat{n} - k)!} \right\}}{[\operatorname{Tr} \{\exp(-\Theta \hat{n})\, \hat{n}\}]^k} \,. \tag{12.39}$$

Further

$$\begin{aligned} \operatorname{Tr} \left\{ \exp(-\Theta \hat{n}) \frac{\hat{n}!}{(\hat{n}-k)!} \right\} &= \sum_{n=0}^{\infty} \exp(-\Theta n) \frac{n!}{(n-k)!} = \\ &= \exp(-\Theta k) \frac{d^k}{d[\exp(-\Theta)]^k} \sum_{n=0}^{\infty} \exp(-\Theta n) = \\ &= \exp(-\Theta k) \frac{d^k}{d[\exp(-\Theta)]^k} \frac{1}{1 - \exp(-\Theta)} = \\ &= \exp(-\Theta k) \frac{k!}{[1 - \exp(-\Theta)]^{k+1}} \,. \end{aligned} \tag{12.40}$$

Substituting this into (12.39) we obtain

$$\gamma^{(k,k)} = \frac{1}{[1-\exp(-\Theta)]^{k-1}} \frac{\exp(-\Theta k)\, k!}{[1-\exp(-\Theta)]^{k+1}} \frac{[1-\exp(-\Theta)]^{2k}}{\exp(-\Theta k)} = k! \tag{12.41}$$

in agreement with (12.36).

The conditions (12.35) and (12.36) (i.e. the validity of factorization conditions) for laser and chaotic light were experimentally verified by Jakeman, Oliver and Pike [212]. Very good agreement was found up to twelfth order ($m = n$).

Taking into account that for the fully coherent field

$$\mathrm{Tr}\,\{\hat{\varrho}\,\hat{A}^{(-)}(x)\,\hat{A}^{(+)}(x)\} - |\mathrm{Tr}\,\{\hat{\varrho}\hat{A}^{(+)}(x)\}|^2 = 0 \tag{12.42}$$

we arrive at

$$V(x) = \mathrm{Tr}\,\{\hat{\varrho}\hat{A}^{(+)}(x)\} = \Gamma_{\mathcal{N}}^{(0,1)}(x)\,. \tag{12.43}$$

We have seen that the idea of second-order coherence can be derived from the Young experiment and that it has a clear physical significance – the interference fringes have the maximum visibility. We can also ask for the physical significance of higher-order coherence conditions, but in this case the situation is not so clear. A natural experiment to investigate this is the coincidence photon-counting experiment.

Equation (12.34) implies, if the definition (2.42b) of the degree of coherence is used, that $|{}^{(G)}\gamma_{\mathcal{N}}^{(n,n)}(x_1, \ldots, x_{2n})| \equiv 1$, i.e. ${}^{(G)}\gamma_{\mathcal{N}}^{(n,n)}(x_1, \ldots, x_n, x_n, \ldots$ $\ldots, x_1) \equiv 1$ and consequently

$$\Gamma_{\mathcal{N}}^{(n,n)}(x_1, \ldots, x_n, x_n, \ldots, x_1) = \prod_{j=1}^{n} \Gamma_{\mathcal{N}}^{(1,1)}(x_j, x_j)\,. \tag{12.44}$$

This means, according to our earlier results, that the N-fold joint counting rate defined by (12.12) is equal to the product of the counting rates which would be measured by each of the N counters in the absence of all others; the responses of the counters are statistically independent of one another. If the average intensity of the field is independent of time (the field is stationary), the counters detect no tendency toward any sort of correlation in the arrival times of photons. This is in agreement with the semiclassical analysis contained in Chapter 10 of the bunching effect and the Hanbury Brown-Twiss effect for laser light and in this case no Hanbury Brown-Twiss effect will occur. Fields for which this effect occurs cannot be coherent in the fourth-or higher - order. A typical example is the chaotic field, for which there is a tendency for photons to arrive in pairs (cf. Figs. 10.1 and 10.2).

The full coherence conditions are equivalent to the fact that the photon statistics are Poissonian (cf. Chapter 10). Since $\langle \hat{a}^{+n}\hat{a}^{n}\rangle = \langle \hat{a}^{+}\hat{a}\rangle^{n}$ the normally ordered characteristic function $\langle \exp is\hat{n}\rangle_{\mathcal{N}}$ is given by

$$\langle \exp is\hat{n}\rangle_{\mathcal{N}} = \sum_{n=0}^{\infty} \frac{(is)^n}{n!} \langle \hat{a}^{+n}\hat{a}^{n}\rangle = \sum_{n=0}^{\infty} \frac{(is)^n}{n!} \langle \hat{a}^{+}\hat{a}\rangle^{n} = \exp is\langle \hat{n}\rangle . \tag{12.45}$$

From (10.11) with $\eta = 1$

$$\sum_{n=0}^{\infty} \varrho(n, n)\,(1 + is)^n = \langle \exp is\hat{n}\rangle_{\mathcal{N}} \tag{12.46}$$

and by analogy with (10.12), we obtain

$$\varrho(n, n) = \frac{1}{n!}\,\frac{d^n}{d(is)^n} \langle \exp is\hat{n}\rangle_{\mathcal{N}}\big|_{is=-1} = \frac{\langle n\rangle^n}{n!} \exp\left(-\langle n\rangle\right), \tag{12.47}$$

which is the Poisson distribution; this agrees with (10.62).

While the degrees of coherence $\gamma_{\mathcal{N}}$, ${}^{(G)}\gamma_{\mathcal{N}}$, and ${}^{(S)}\gamma_{\mathcal{N}}$ defined by (2.42a, b, c) are equivalent if the factorization condition (12.34) is fulfilled and $|\gamma_{\mathcal{N}}| = |{}^{(G)}\gamma_{\mathcal{N}}| = |{}^{(S)}\gamma_{\mathcal{N}}| = 1$, they are in general different (they are identical for $n = m = 1$). As we have seen the degree of coherence ${}^{(G)}\gamma_{\mathcal{N}}$ is suitable for analyzing the photon correlation experiments but it cannot be used as a measure of the partial coherence of higher order since it has no bound. For this purpose the degrees of coherence $\gamma_{\mathcal{N}}$ and ${}^{(S)}\gamma_{\mathcal{N}}$ can be used, since the inequalities (2.43a, b) hold under the assumption that $\gamma_{\mathcal{N}}$ and ${}^{(S)}\gamma_{\mathcal{N}}$ exist; (2.43a) holds for fields having a classical analogue [441] for which the weight function of the diagonal representation of the density matrix (Chapter 13) is positive-definite. Adopting for example the definition (2.42a) we can speak of partially coherent fields in higher order if $0 < |\gamma_{\mathcal{N}}^{(n,n)}| < 1$ (in analogy with the case $n = 1$). From the definition of the degree of coherence in quantum terms it follows that this quantity can be defined if the number of photons in the field is larger than n. To define all the sequence an infinite number of photons must be present in the field. A deeper physical meaning of the various definitions of the degree of coherence is not yet very clear and we can only note that the definition (2.42a) is useful in connection with imaging problems [372, 363] (cf. Chapter 9), while (2.42c) serves to characterize the visibility in the two-slit experiment if the resulting measured quantity is the n-fold intensity [17]. It may be said that the various definitions of the degree of coherence correspond to various types of experiments.

An interesting characterization of coherent fields, in terms of classical stochastic functions having a close analogy to the results of quantum theory given above, was proposed by Picinbono [383]. If $\gamma^{(1,1)} = 1$ for the quantity (12.33) he called the field *weakly coherent* while if $\gamma^{(n,n)} = 1$ for every n it is *strongly coherent*. It is clear that if for a weakly coherent field $|\gamma_{\mathcal{N}}^{(1,1)}(x_1, x_2)| \equiv 1$, then $|\gamma_{\mathcal{N}}^{(n,n)}(x_1, \ldots, x_{2n})| \equiv 1$ in the sense of (2.42a) and vice versa and the same is true in the sense of (2.42c); this can easily be verified by using (12.32). This means that relation (12.22) completely determines a weakly coherent field. Note for strongly coherent fields, considered as classical fields or as quantum fields (with non-negative weight function of the diagonal representation of the density matrix – Chapter 13), a sufficient specification is $\gamma^{(1,1)} = \gamma^{(2,2)} = 1$ [441, 383].

12.3 Quantum characteristic functional

In analogy to (8.5) we can define the characteristic functional

$$C^{(\mathcal{N})}\{y(x)\} = \operatorname{Tr}\left\{\hat{\varrho} \exp\left[\int y(x)\, \hat{A}^{(-)}(x)\, d^4x\right] \exp\left[-\int y^*(x)\, \hat{A}^{(+)}(x)\, d^4x\right]\right\}, \tag{12.48}$$

from which the complete set of normally ordered correlation functions can be derived.

$$\operatorname{Tr}\{\hat{\varrho}\, \hat{A}^{(-)}(x_1) \ldots \hat{A}^{(-)}(x_m)\, \hat{A}^{(+)}(x_{m+1}) \ldots \hat{A}^{(+)}(x_{m+n})\} =$$

$$= \prod_{j=1}^{m} \frac{\delta}{\delta y(x_j)} \prod_{k=m+1}^{m+n} \frac{\delta}{\delta(-y^*(x_k))} C^{(\mathcal{N})}\{y(x)\}\big|_{y(x)\equiv 0}\,; \tag{12.49}$$

here $x \equiv (\mathbf{x}, t, \mu)$ again so that the integrals over x include the summation over the polarization indices μ.

12.4 Measurements for mixed-order correlation functions

Quadratic detectors only allow the even-order correlation functions of the type (n, n) to be measured, but some further experiments are possible permitting, at least in principle, the measurement of mixed-order correlation functions of the type (m, n) for $m \neq n$. Suggestions have been given by Glauber [181].

Considering a stationary field and returning to the frequency stationary conditions (2.21), $\sum_{j=1}^{m+n} \varepsilon_j \nu_j = 0$, which are necessary if the $\Gamma_{\mathcal{N}}^{(m,n)}$ are to be non-vanishing, we see that they are easily fulfilled if $m = n$ and $\nu_j = \nu_{n+j}$. In this case one time dependence cancels the other and it is easy to have a non-vanishing function. A more complicated situation will occur if $m \neq n$, because in this case a very special set of frequencies must be occupied in the field. For example for a non-vanishing $\Gamma_{\mathcal{N}}^{(1,2)}$ it is necessary that $\nu_1 = \nu_2 + \nu_3$, a condition valid in the parametric amplifier [322, 323, 318, 187] or in the Raman effect. This condition is necessary in order that the correlation function is non-vanishing but a statistical dependence of modes must also exist for if modes are independent, then the density matrix factorizes into the product of the mode density matrices and $\Gamma_{\mathcal{N}}^{(1,2)} = 0$. In the parametric amplifier and the Raman effect this dependence exists and $\Gamma_{\mathcal{N}}^{(1,2)} \neq 0$. For the second harmonic generation in a non-linear medium we also have $2\nu = \nu + \nu$ [5], since in this case pairs of red photons are joined together to form blue photons. In the outgoing field there are two components; one of them is precisely twice the frequency of the other, that is (see e.g. [69])

$$\hat{A}^{(+)}(t) = \hat{B}^{(+)}(t) + b\,\hat{B}^{(+)2}(t) = \hat{U}^{(+)}(t) \exp i\omega t + b\,\hat{U}^{(+)2}(t) \exp 2i\omega t\,, \tag{12.50}$$

where $\hat{B}^{(+)}(t) = \hat{U}^{(+)}(t) \exp i\omega t$ is the incident field, $\omega = 2\pi\nu$ and b is a constant; there are also phase relations between these two fields and so $\Gamma_{\mathcal{N}}{}^{(1,2)} \neq 0$.

Measuring the correlation functions with $m \neq n$ implies a measurement of the high frequency field against the low frequency field. Assume we have a photodetector (an atom) with a threshold for photon detection which lies higher in frequency than the red frequency used. A single red photon will not give a photoelectron while two red photons will and so two red photons can be observed simultaneously. The photoelectric effect will be proportional to $\hat{A}_R^{(+)^2}$. Of course, a blue photon will also make a photoelectric transition and these two processes interfere with one another giving the correlation function

$$\Gamma_{\mathcal{N}}^{(1,2)} = \mathrm{Tr}\,\{\hat{\varrho}\,\hat{A}_B^{(-)}\,\hat{A}_R^{(+)}\,\hat{A}_R^{(+)}\}\,. \tag{12.51}$$

Another possibility of measuring $\Gamma_{\mathcal{N}}^{(1,2)}$ (and $\Gamma_{\mathcal{N}}^{(2,1)}$) is to use two converters C_1 and C_2 (Fig. 12.1) formed from non-linear dielectrics. The first converter C_1 is illuminated by red photons, some of which are converted to blue photons and some of which remain red. Some of the remaining red photons are converted to blue ones by the second converter C_2 and the

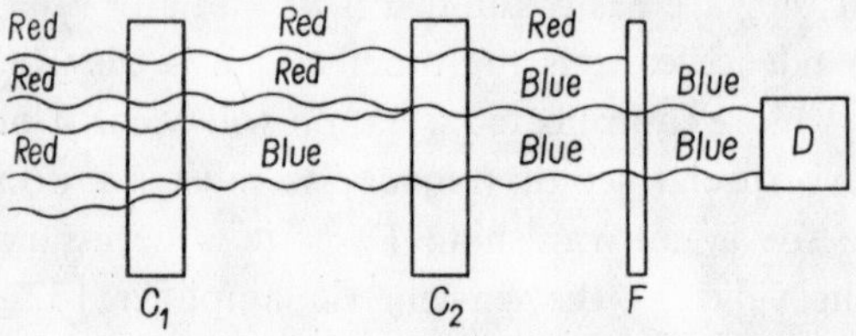

Fig. 12.1 A scheme for measuring $\Gamma_{\mathcal{N}}^{(1,2)}$.

remaining red photons are filtered by a filter F. The detector D detects the function $\Gamma_{\mathcal{N}}^{(1,1)}$. Denoting the detected field by $\hat{A}_B^{(+)}$ we have ($b = 1$)

$$\hat{A}_B^{(+)} = \hat{B}_B^{(+)} + \hat{B}_R^{(+)2} . \tag{12.52}$$

The detector measures the following correlation function

$$\begin{aligned} \mathrm{Tr}\,\{\hat{\varrho}\,\hat{A}_B^{(-)}\,\hat{A}_B^{(+)}\} = \\ &= \mathrm{Tr}\,\{\hat{\varrho}(\hat{B}_B^{(-)} + \hat{B}_R^{(-)2})\,(\hat{B}_B^{(+)} + \hat{B}_R^{(+)2})\} = \mathrm{Tr}\,\{\hat{\varrho}\,\hat{B}_B^{(-)}\,\hat{B}_B^{(+)}\} + \\ &+ \mathrm{Tr}\,\{\hat{\varrho}\,\hat{B}_R^{(-)2}\,\hat{B}_B^{(+)}\} + \mathrm{Tr}\,\{\hat{\varrho}\,\hat{B}_B^{(-)}\,\hat{B}_R^{(+)2}\} + \mathrm{Tr}\,\{\hat{\varrho}\,\hat{B}_R^{(-)2}\,\hat{B}_R^{(+)2}\}\,, \end{aligned} \tag{12.53}$$

which also includes the correlation functions $\Gamma_{\mathcal{N}}^{(1,2)}$ and $\Gamma_{\mathcal{N}}^{(2,1)}$.

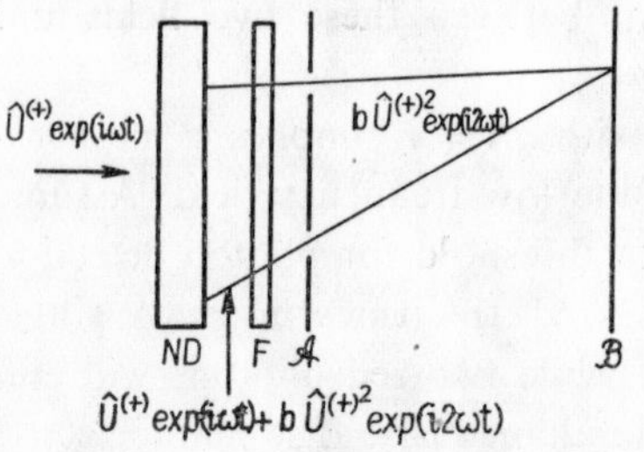

Fig. 12.2. Combination of the non-linear dielectric (ND) with Young's arrangement for measuring $\Gamma_{\mathcal{N}}^{(2,2)}$.

However, in usual experiments with stationary fields the correlation functions with $m \neq n$ are practically zero [303] and only in the peculiar circumstances of non-linear optics is $\Gamma_{\mathcal{N}}^{(m,n)} \neq 0$ for $m \neq n$.

An analogue of the Hanbury Brown - Twiss correlation experiment measuring the correlation function $\Gamma_{\mathcal{N}}^{(2,2)}$ can also be performed by using non-linear dielectrics (instead of two quadratic detectors with the outputs correlated in a correlater). This was suggested by Beran, De Velis and Parrent [69]. In Fig. 12.2 we see a scheme of the experimental arrangement, where a non-linear dielectric ND is combined with the Young two-slit arrangement. If the incident field has the complex amplitude $\hat{U}^{(+)}(t)\,.$ $\exp i\omega t$, then, according to (12.50), the outgoing field has the complex amplitude $\hat{U}^{(+)}(t) \exp i\omega t + b\, \hat{U}^{(+)2}(t) \exp 2i\omega t$. If the first component is filtered by the filter F, then one can observe interference fringes, described by the fourth-order correlation function $\Gamma_{\mathcal{N}}^{(2,2)} = \mathrm{Tr}\,\{\varrho\, \hat{U}^{(-)^2}\, \hat{U}^{(+)^2}\}$, on the screen $\mathscr{B}$.

Chapter 13

COHERENT-STATE DESCRIPTION OF THE ELECTROMAGNETIC FIELD

In the following we show that in the quantum theory of the electromagnetic field there exist quantum states called coherent states, in which the quantum correlation functions are factorized in the form (12.34), i.e. a field in such a state fulfils the full coherence conditions. These states form an overcomplete set and they can be used as a basis for the decomposition of vectors and operators in quantum mechanics. Furthermore, these states make it possible to formulate the quantum theory of optical coherence in a form very close to the classical description developed in earlier chapters. Although the coherent states were first introduced by Schrödinger [414] in 1927 and were studied by others (e.g. [234]), they were fully utilized in connection with quantum optics by Glauber [173–176], who used them to study the quantum coherence of optical fields.

13.1 Coherent states of the electromagnetic field

Definitions

Let us introduce the displaced vacuum states

$$|\alpha\rangle = \hat{D}(\alpha)\,|0\rangle\,, \tag{13.1}$$

where

$$\hat{D}(\alpha) = \exp\left(\alpha\hat{a}^{+} - \alpha^{*}\hat{a}\right) \tag{13.2}$$

is a displacement operator. The states $|\alpha\rangle$ are called the *coherent states.* Here $|0\rangle$ is the vacuum state, α is a complex number and $\hat{a}$ and $\hat{a}^{+}$ are the annihilation and creation operators of a photon respectively. Using the Baker-Hausdorff identity [19]

$$\exp\left(\hat{A} + \hat{B}\right) = \exp\hat{A}\exp\hat{B}\exp\left(-\tfrac{1}{2}[\hat{A}, \hat{B}]\right), \tag{13.3}$$

where $\hat{A}$ and $\hat{B}$ are operators for which the commutator $[\hat{A}, \hat{B}]$ is a c-number, i.e. $[[\hat{A}, \hat{B}], \hat{A}] = [[\hat{A}, \hat{B}], \hat{B}] = 0$, we can rewrite (13.1) in the form

$$|\alpha\rangle = \exp(-\tfrac{1}{2}|\alpha|^2) \exp(\alpha\hat{a}^+) \exp(-\alpha^*\hat{a}) |0\rangle = \exp(-\tfrac{1}{2}|\alpha|^2) \sum_{n=0}^{\infty} \frac{\alpha^n}{\sqrt{n!}} |n\rangle , \tag{13.4}$$

where we have expanded both $\exp(\alpha\hat{a}^+)$ and $\exp(-\alpha^*\hat{a})$ in a Taylor series and used the vacuum stability condition $\hat{a}|0\rangle = 0$ together with definition (11.10) of the Fock state $|n\rangle$. Making use of the relation (11.11b) we can easily verify that

$$\hat{a}|\alpha\rangle = \alpha|\alpha\rangle , \tag{13.5a}$$

$$\langle\alpha| \hat{a}^+ = \langle\alpha| \alpha^* , \tag{13.5b}$$

i.e. the coherent state $|\alpha\rangle$ is an eigenstate of the annihilation operator $\hat{a}$ with eigenvalue α, which is in general a complex number since $\hat{a}$ is not a Hermitian operator. Equation (13.5a) shows that $|\alpha\rangle$ must contain an infinite number of photons since the annihilation operator $\hat{a}$ does not change this state.

One can prove [314] that it is not possible to construct states $||\alpha\rangle\rangle$, with α finite such that

$$\hat{a}^+||\alpha\rangle\rangle = \alpha||\alpha\rangle\rangle , \tag{13.6a}$$

$$\langle\langle\alpha|| \hat{a} = \langle\langle\alpha|| \alpha^* , \tag{13.6b}$$

i.e. no eigenstate of the creation operator $\hat{a}^+$ exists with finite eigenvalue. Indeed assuming that

$$||\alpha\rangle\rangle = \sum_{n=0}^{\infty} |n\rangle \langle n||\alpha\rangle\rangle , \tag{13.7}$$

we obtain the recursion relation, for the coefficient $\langle n||\alpha\rangle\rangle$ using (13.6a) and Hermitian adjoint to (11.11b))

$$\alpha\langle n||\alpha\rangle\rangle = \sqrt{(n)} \langle n-1||\alpha\rangle\rangle , \tag{13.8}$$

which, by repeated use, gives

$$\langle n||\alpha\rangle\rangle = \frac{\sqrt{n!}}{\alpha^n} \langle 0||\alpha\rangle\rangle . \tag{13.9}$$

Substituting (13.9) into (13.7) we have

$$||\alpha\rangle\rangle = \langle 0||\alpha\rangle\rangle \sum_{n=0}^{\infty} \frac{\sqrt{n!}}{\alpha^n} |n\rangle \tag{13.10}$$

and consequently we have for the squared norm of $||\alpha\rangle\rangle$

$$\langle\langle\alpha||\alpha\rangle\rangle = |\langle 0||\alpha\rangle\rangle|^2 \sum_{n=0}^{\infty} \frac{n!}{|\alpha|^{2n}}, \tag{13.11}$$

(we recall that $\langle n \mid m\rangle = \delta_{nm}$). This is divergent for all finite complex amplitudes α and the states $||\alpha\rangle\rangle$ cannot be regarded as physically admissible states of the radiation field. The states $|n\rangle$ and $|\alpha\rangle$ seem to be the only physically useful bases for representations of the radiation field.

From (13.4) it is seen that the states $|\alpha\rangle$ are normalized, i.e. $\langle\alpha \mid \alpha\rangle = 1$ since $\langle n \mid m\rangle = \delta_{nm}$ and also

$$\langle n \mid \alpha\rangle = \exp\left(-\tfrac{1}{2}|\alpha|^2\right) \frac{\alpha^n}{\sqrt{n!}}, \tag{13.12a}$$

which gives

$$|\langle n \mid \alpha\rangle|^2 = \exp\left(-|\alpha|^2\right) \frac{|\alpha|^{2n}}{n!}. \tag{13.12b}$$

This is a Poisson distribution with $\langle n\rangle = |\alpha|^2$ and so the probability that a mode of the field contains n quanta is given by the Poisson distribution in agreement with (12.47) and (10.62).

The scalar product of two coherent states is

$$\langle\alpha \mid \beta\rangle = \exp\left(-\tfrac{1}{2}|\alpha|^2 - \tfrac{1}{2}|\beta|^2\right) \sum_{n,m} \frac{\alpha^{*n}\beta^m}{\sqrt{(n!\, m!)}} \langle n \mid m\rangle =$$

$$= \exp\left(-\tfrac{1}{2}|\alpha|^2 - \tfrac{1}{2}|\beta|^2 + \alpha^*\beta\right) \tag{13.13a}$$

and so

$$|\langle\alpha \mid \beta\rangle|^2 = \exp\left(-|\alpha - \beta|^2\right), \tag{13.13b}$$

which shows that the coherent states are not orthogonal; they can be regarded as approximately orthogonal if $|\alpha - \beta| \gg 1$.

Expansions in terms of coherent states

Even if the coherent states are not orthogonal they do form an overcomplete system of states, enabling us to expand an arbitrary vector or operator (particularly the density operator) in terms of the coherent states.

We first note that the coherent states are mutually dependent, which is the most characteristic property of an overcomplete set of vectors. On multiplying (13.4) by α^k and integrating over the whole complex α-plane we conclude that

$$\int \alpha^k |\alpha\rangle \, d^2\alpha = 0 \,, \quad k = 1, 2, \ldots,$$

where $\alpha = r \exp i\varphi$, r and φ are real and $d^2\alpha = d(\mathrm{Re}\,\alpha)\, d(\mathrm{Im}\,\alpha) = r\, dr\, d\varphi$. The Fock state $|n\rangle$ may be expanded in terms of $|\alpha\rangle$ by multiplying (13.4) by $\pi^{-1}(n!)^{-1/2}\, \alpha^{*n} \exp(-|\alpha|^2/2)$ and integrating over all α giving

$$|n\rangle = \frac{1}{\pi} \int \exp(-\tfrac{1}{2}|\alpha|^2) \frac{\alpha^{*n}}{\sqrt{n!}} |\alpha\rangle \, d^2\alpha \,. \tag{13.14}$$

Substituting (13.14) into (11.18) we obtain

$$\frac{1}{\pi^2} \iint \exp\left(-\frac{|\alpha|^2}{2} - \frac{|\beta|^2}{2} + \alpha\beta^*\right) |\beta\rangle \langle\alpha| \, d^2\alpha \, d^2\beta = 1 \,. \tag{13.15}$$

The integral over β is

$$\frac{1}{\pi} \int |\beta\rangle \exp \frac{|\beta|^2}{2} \exp(-|\beta|^2 + \alpha\beta^*) \, d^2\beta =$$

$$= \frac{1}{\pi} \sum_{n=0}^{\infty} \frac{|n\rangle}{\sqrt{n!}} \sum_{m=0}^{\infty} \frac{\alpha^m}{m!} \int \beta^n \beta^{*m} \exp(-|\beta|^2) \, d^2\beta = \exp\left(\frac{|\alpha|^2}{2}\right) |\alpha\rangle \,, \tag{13.16}$$

where we have expressed $|\beta\rangle \exp(|\beta|^2/2)$ using (13.4), we have written $\exp \alpha\beta^*$ in the form of series and we have used the integral

$$\int \beta^n \beta^{*m} \exp(-s|\beta|^2) \, d^2\beta = \frac{1}{2} \int_0^\infty |\beta|^{n+m} \exp(-s|\beta|^2) \, d|\beta|^2 \,.$$

$$\cdot \int_0^{2\pi} \exp[i(n - m) \arg \beta] \, d(\arg \beta) = \pi \int_0^\infty |\beta|^{n+m} \exp(-s|\beta|^2) \, d|\beta|^2 \, \delta_{nm} =$$

$$= \frac{\pi n!}{s^{n+1}} \delta_{nm} \,, \quad \mathrm{Re}\, s > 0 \tag{13.17}$$

for $s = 1$. Thus we arrive at the "resolution of unity" in terms of the coherent states

$$\frac{1}{\pi} \int |\alpha\rangle \langle\alpha| \, d^2\alpha = \hat{1} \,. \tag{13.18}$$

It can easily be shown that for an entire function $f(\alpha)$, in analogy to (13.16),

$$\frac{1}{\pi}\int f(\beta)\exp\left(-|\beta|^2+\alpha\beta^*\right)d^2\beta = f(\alpha)\,. \tag{13.19}$$

By using the identity (13.18) we may also express an arbitrary normalized *vector* $|\,\rangle$ as

$$|\,\rangle = \frac{1}{\pi}\int|\alpha\rangle\,\langle\alpha\,|\,\rangle\,d^2\alpha \tag{13.20}$$

and an arbitrary *operator* $\hat{F}$ as

$$\hat{F} = \frac{1}{\pi^2}\iint|\alpha\rangle\,\langle\alpha|\,\hat{F}|\beta\rangle\,\langle\beta|\,d^2\alpha\,d^2\beta\,. \tag{13.21}$$

The coefficient $\langle\alpha|\,\rangle$ of the decomposition (13.20) can be written, expanding the state in Fock states, as

$$\langle\alpha|\sum_n f_n|n\rangle = \exp\left(-\tfrac{1}{2}|\alpha|^2\right)\sum_n\frac{\alpha^{*n}}{\sqrt{n!}}f_n = \exp\left(-\tfrac{1}{2}|\alpha|^2\right)f(\alpha^*)\,. \tag{13.22}$$

As the squared norm of $|\,\rangle$ is $\langle\,|\,\rangle = \sum_n|f_n|^2 = 1$ the series $f(\alpha^*) = \sum_n\alpha^{*n}f_n/(n!)^{1/2}$ converges for all values of α in the finite plane and therefore $f(\alpha^*)$ is an entire function.

Equation (13.16) can serve as an example of the decomposition (13.20) when $|\,\rangle \equiv |\alpha\rangle$.

Minimum-uncertainty wave packets

Defining $\Delta\hat{p}$ and $\Delta\hat{q}$ by

$$\Delta\hat{p} = \hat{p} - \langle\hat{p}\rangle\,,\quad \Delta\hat{q} = \hat{q} - \langle\hat{q}\rangle \tag{13.23}$$

we have from the commutation rule (11.16)

$$\tfrac{1}{2}\hbar = \tfrac{1}{2}|\langle[\hat{q},\hat{p}]\rangle| = \tfrac{1}{2}|\langle[\Delta\hat{q},\Delta\hat{p}]\rangle| \leqq |\langle\Delta\hat{q}\,\Delta\hat{p}\rangle| \leqq \{\langle(\Delta\hat{q})^2\rangle\,\langle(\Delta\hat{p})^2\rangle\}^{1/2}\,. \tag{13.24}$$

If we use eigen-properties (13.5) of the coherent states we obtain from (11.15)

$$\langle\hat{q}\rangle = \sqrt{\left(\frac{\hbar}{2\omega}\right)}(\alpha+\alpha^*)\,,\quad \langle\hat{p}\rangle = -i\sqrt{\left(\frac{\hbar\omega}{2}\right)}(\alpha-\alpha^*)\,, \tag{13.25a}$$

$$\langle(\Delta\hat{q})^2\rangle = \langle\hat{q}^2\rangle - \langle\hat{q}\rangle^2 = \frac{\hbar}{2\omega}\,,\quad \langle(\Delta\hat{p})^2\rangle = \frac{\hbar\omega}{2} \tag{13.25b}$$

and for the coherent states the inequality (13.24) reduces to the equality

$$\langle(\Delta\hat{q})^2\rangle\,\langle(\Delta\hat{p})^2\rangle = \frac{\hbar^2}{4}\,. \tag{13.26}$$

Thus the coherent states are in a sense as nearly classical as quantum theory allows.

The role of the coherent states in a model of the interaction of radiation and matter applied to the laser was investigated by Picard and Willis [382]. Using the arithmetic mean-geometric mean inequality and $\langle(\Delta\hat{p})^2\rangle = \omega^2\langle(\Delta\hat{q})^2\rangle$ we have

$$\tfrac{1}{2}[\langle(\Delta\hat{p})^2\rangle + \omega^2\langle(\Delta\hat{q})^2\rangle] \geqq \omega[\langle(\Delta\hat{q})^2\rangle\,\langle(\Delta\hat{p})^2\rangle]^{1/2} = \frac{\hbar\omega}{2}\,. \tag{13.27}$$

It can easily be seen that the inequality (13.27) reduces to the equality for the coherent states. Thus the quantity

$$\tfrac{1}{2}[\langle(\Delta\hat{p})^2\rangle + \omega^2\langle(\Delta\hat{q})^2\rangle] = \tfrac{1}{2}[\langle\hat{p}^2\rangle + \omega^2\langle\hat{q}^2\rangle] - \tfrac{1}{2}[\langle\hat{p}\rangle^2 + \omega^2\langle\hat{q}\rangle^2] \tag{13.28}$$

represents the difference between the total energy of the field and the coherent energy of the field, i.e. it is the incoherent energy in the field. Therefore it may be said that in the coherent state the incoherent energy takes on its minimum value $\hbar\omega/2$ contributing only the energy of the zero-point fluctuations.

Some properties of displacement operator $\hat{D}(\alpha)$

From the definition of the displacement operator (13.2) it follows that

$$\hat{D}^+(\alpha) = \hat{D}^{-1}(\alpha) = \hat{D}(-\alpha)\,. \tag{13.29}$$

Further

$$\hat{a}\,\hat{D}(\alpha) = \exp\left(-\tfrac{1}{2}|\alpha|^2\right)\hat{a}\exp\left(\alpha\hat{a}^+\right)\exp\left(-\alpha^*\hat{a}\right) =$$

$$= \exp\left(-\tfrac{1}{2}|\alpha|^2\right)\sum_{n=0}^{\infty}\frac{\alpha^n}{n!}\,\hat{a}\hat{a}^{+n}\exp\left(-\alpha^*\hat{a}\right), \tag{13.30}$$

where we have used (13.3). The commutation rule gives $\hat{a}\hat{a}^{+n} = \hat{a}^{+n}\hat{a} + n\hat{a}^{+n-1}$ and so

$$\hat{a}\,\hat{D}(\alpha) = \hat{D}(\alpha)\,\hat{a} + \alpha\,\hat{D}(\alpha)\,. \tag{13.31}$$

Therefore the operator $\hat{D}(\alpha)$ represents a displacement operator in the sense that

$$\hat{D}^{-1}(\alpha)\,\hat{a}\;\;\hat{D}(\alpha) = \hat{a}\;\; + \alpha\,, \qquad (13.32)$$
$$\hat{D}^{-1}(\alpha)\,\hat{a}^{+}\,\hat{D}(\alpha) = \hat{a}^{+} + \alpha^{*}\,.$$

More generally for an operator function $\hat{F}(\hat{a}^{+}, \hat{a})$

$$\hat{D}^{-1}(\alpha)\,\hat{F}(\hat{a}^{+}, \hat{a})\,\hat{D}(\alpha) = \hat{F}(\hat{a}^{+} + \alpha^{*}, \hat{a} + \alpha)\,. \qquad (13.33)$$

The product of the two displacement operators is equal to

$$\hat{D}(\alpha)\,\hat{D}(\beta) = \exp(\alpha\hat{a}^{+} - \alpha^{*}\hat{a})\exp(\beta\hat{a}^{+} - \beta^{*}\hat{a}) =$$
$$= \hat{D}(\alpha + \beta)\exp\tfrac{1}{2}(\alpha\beta^{*} - \alpha^{*}\beta)\,, \qquad (13.34)$$

where (13.3) has been used.

Taking the trace of $\hat{D}$ using (13.18) we arrive at

$$\operatorname{Tr}\hat{D}(\alpha) = \frac{1}{\pi}\int\langle\beta|\,\hat{D}(\alpha)\,|\beta\rangle\,d^{2}\beta =$$
$$= \frac{1}{\pi}\int\exp(-\tfrac{1}{2}|\alpha|^{2})\,\langle\beta|\exp(\alpha\hat{a}^{+})\exp(-\alpha^{*}\hat{a})\,|\beta\rangle\,d^{2}\beta =$$
$$= \frac{1}{\pi}\int\exp(-\tfrac{1}{2}|\alpha|^{2} + \alpha\beta^{*} - \alpha^{*}\beta)\,d^{2}\beta =$$
$$= \exp(-\tfrac{1}{2}|\alpha|^{2})\,\frac{1}{\pi}\int\exp[2i\,\mathrm{Im}\,(\alpha\beta^{*})]\,d^{2}\beta =$$
$$= \exp(-\tfrac{1}{2}|\alpha|^{2})\,\frac{1}{\pi}\iint_{-\infty}^{+\infty}\exp[2i(\mathrm{Im}\,\alpha\,\mathrm{Re}\,\beta - \mathrm{Re}\,\alpha\,\mathrm{Im}\,\beta)]\,.$$
$$.\,d(\mathrm{Re}\,\beta)\,d(\mathrm{Im}\,\beta) = \exp(-\tfrac{1}{2}|\alpha|^{2})\,\pi\,\delta(\mathrm{Re}\,\alpha)\,\delta(\mathrm{Im}\,\alpha) = \pi\,\delta(\alpha)\,, \qquad (13.35)$$

where (13.3), (13.5) and the substitutions $2\,\mathrm{Re}\,\beta = x$ and $2\,\mathrm{Im}\,\beta = y$ have been used; $\delta(\alpha)$ is the two-dimensional Dirac function.

Equations (13.34) and (13.35) give

$$\operatorname{Tr}\{\hat{D}^{+}(\alpha)\,\hat{D}(\beta)\} = \pi\,\delta(\alpha - \beta)\,. \qquad (13.36)$$

From (13.35) it follows that the complex $\delta(\alpha)$-function can be represented by a Fourier integral of the form

$$\delta(\alpha) = \frac{1}{\pi^{2}}\int\exp(\alpha\beta^{*} - \alpha^{*}\beta)\,d^{2}\beta\,. \qquad (13.37)$$

Expectation values of operators in coherent states

Assume an operator $\hat{F}^{(\mathcal{N})}(\hat{a}^+, \hat{a})$ to be in normal form. Then it holds that

$$\langle\alpha|\, \hat{F}^{(\mathcal{N})}(\hat{a}^+, \hat{a})\, |\alpha'\rangle = F^{(\mathcal{N})}(\alpha^*, \alpha')\, \langle\alpha \mid \alpha'\rangle \tag{13.38}$$

and in particular

$$\langle\alpha|\, \hat{F}^{(\mathcal{N})}(\hat{a}^+, \hat{a})\, |\alpha\rangle = F^{(\mathcal{N})}(\alpha^*, \alpha)\,, \tag{13.39}$$

i.e. the expectation value of a normally ordered operator $\hat{F}^{(\mathcal{N})}$ is obtained by the substitutions $\hat{a} \to \alpha$, $\hat{a}^+ \to \alpha^*$ in the operator function. If $\hat{F}^{(\mathcal{N})}(\hat{a}^+, \hat{a}) = \hat{a}^{+k}\hat{a}^l$ (k, l are integers) then $\langle\alpha|\, \hat{a}^{+k}\, \hat{a}^l|\alpha\rangle = \alpha^{*k}\alpha^l$.

Further we can prove using (13.4) that

$$\hat{a}^+|\alpha\rangle = \left(\frac{\alpha^*}{2} + \frac{\partial}{\partial\alpha}\right)|\alpha\rangle\,,$$

$$\langle\alpha|\, \hat{a} = \left(\frac{\alpha}{2} + \frac{\partial}{\partial\alpha^*}\right)\langle\alpha|\,, \tag{13.40}$$

and

$$\hat{a}^+|\alpha\rangle\,\langle\alpha| = \left(\alpha^* + \frac{\partial}{\partial\alpha}\right)|\alpha\rangle\,\langle\alpha|\,,$$

$$|\alpha\rangle\,\langle\alpha|\, \hat{a} = \left(\alpha + \frac{\partial}{\partial\alpha^*}\right)|\alpha\rangle\,\langle\alpha|\,. \tag{13.41}$$

For an operator $\hat{F}(\hat{a}^+, \hat{a})$ we obtain

$$\hat{F}(\hat{a}^+, \hat{a})\,|\alpha\rangle\,\langle\alpha| = F\left(\alpha^* + \frac{\partial}{\partial\alpha,}\,, \alpha\right)|\alpha\rangle\,\langle\alpha|\,,$$

$$|\alpha\rangle\,\langle\alpha|\, \hat{F}(\hat{a}^+, \hat{a}) = F\left(\alpha^*, \alpha + \frac{\partial}{\partial\alpha^*}\right)|\alpha\rangle\,\langle\alpha| \tag{13.42}$$

and multiplying this by $|\alpha\rangle$ from the right and $\langle\alpha|$ from the left we arrive at

$$\langle\alpha|\, \hat{F}(\hat{a}^+, \hat{a})\,|\alpha\rangle = F\left(\alpha^* + \frac{\partial}{\partial\alpha}\,, \alpha\right) = F\left(\alpha^*, \alpha + \frac{\partial}{\partial\alpha^*}\right). \tag{13.43}$$

These rules are useful for solving the equations of motion. As an example we consider the Schrödinger equation for an operator $\hat{S}$

$$i\hbar\, \frac{\partial\hat{S}(\hat{a}^+, \hat{a}, t)}{\partial t} = \hat{H}(\hat{a}^+, \hat{a})\, \hat{S}(\hat{a}^+, \hat{a}, t)\,. \tag{13.44}$$

If we multiply (13.44) by $|\alpha\rangle$ from the right and by $\langle\alpha|$ from the left and use (13.43) we arrive at

$$i\hbar \frac{\partial S\left(\alpha^* + \dfrac{\partial}{\partial\alpha}, \alpha, t\right)}{\partial t} = H\left(\alpha^* + \frac{\partial}{\partial\alpha}, \alpha\right) S\left(\alpha^* + \frac{\partial}{\partial\alpha}, \alpha, t\right), \quad (13.45)$$

which is a classical equation. Solving this equation for S we can obtain

$$S\left(\alpha^* + \frac{\partial}{\partial\alpha}, \alpha, t\right) = S^{(\mathcal{N})}(\alpha^*, \alpha, t) = \langle\alpha|\, \hat{S}^{(\mathcal{N})}(\hat{a}^+, \hat{a}, t)\, |\alpha\rangle\, ; \quad (13.46)$$

his procedure leads directly to the normal form of $\hat{S}$.

Generalized coherent states

Generalized coherent states have been introduced by Titulaer and Glauber [442] and may be written as

$$|\alpha, \{\vartheta_n\}\rangle = \exp\left(-\tfrac{1}{2}|\alpha|^2\right) \sum_{n=0}^{\infty} \frac{\alpha^n \exp(i\vartheta_n)}{\sqrt{n!}} |n\rangle\, , \quad (13.47)$$

where $\{\vartheta_n\}$ is a sequence of real numbers. These states may be regarded as the most general pure fully coherent states (in the sense $m = n$). Some of their properties were studied by Titulaer and Glauber [442], Crosignani, Di Porto and Solimeno [120] and Bialynicka-Birula [82]. Some properties of a subset of the coherent states were investigated by Campagnoli and Zambotti [107].

Putting $\vartheta_n = n\vartheta$ we see that $\hat{a}|\alpha, \{\vartheta_n\}\rangle = \alpha \exp(i\vartheta)\, |\alpha, \{\vartheta_n\}\rangle$ and so $\langle\alpha, \{\vartheta_n\}|\, \hat{a}^{+k}\, \hat{a}^k|\alpha, \{\vartheta_n\}\rangle = \alpha^{*k}\alpha^k$ independently of ϑ. This shows that the field in the coherent state fulfils the full coherence factorization conditions and also that mixtures of the coherent states which differ from one another by phase shifts also describe coherent fields ($m = n$). The difference between the coherent states $|\alpha\rangle$ and generalized coherent states $|\alpha, \{\vartheta_n\}\rangle$ does not manifest (itself in experiments described by even-order correlation functions and consequently we may use only the coherent states $|\alpha\rangle$ in the following (only $|\alpha\rangle$ fulfil (12.34) for all m, n).

Multimode description

The coherent-state formulation will be useful for applications to the electromagnetic field in a finite volume having a finite or at most a countably infinite number of degrees of freedom. (A generalization to fields in an infinite volume having a innumerable number of degrees of freedom is also possible if functional (continuous) integrals are used.) For this purpose we introduce the coherent state

$$|\{\alpha_\lambda\}\rangle \equiv \prod_\lambda |\alpha_\lambda\rangle \tag{13.48}$$

for which

$$|\{\alpha_\lambda\}\rangle = \sum_{\{n_\lambda\}=0}^{\infty} \left\{\prod_{\lambda'} \frac{\alpha_{\lambda'}^{n_{\lambda'}}}{\sqrt{(n_{\lambda'}!)}} \exp\left(-\frac{|\alpha_{\lambda'}|^2}{2}\right) |\{n_\lambda\}\rangle\right\} \tag{13.49}$$

and the completeness condition is

$$\int |\{\alpha_\lambda\}\rangle\langle\{\alpha_\lambda\}|\, d\mu(\{\alpha_\lambda\}) = \hat{1}\,, \tag{13.50}$$

where

$$d\mu(\{\alpha_\lambda\}) \equiv \prod_\lambda \frac{d^2\alpha_\lambda}{\pi}\,, \tag{13.51}$$

and $|\{n_\lambda\}\rangle$ is given by (11.36).

If (11.22) and (13.5a) are used we obtain the following eigen-property of the coherent state

$$\hat{A}^{(+)}(x)\,|\{\alpha_\lambda\}\rangle = V(x)\,|\{\alpha_\lambda\}\rangle \tag{13.52a}$$

and its Hermitian adjoint

$$\langle\{\alpha_\lambda\}|\,\hat{A}^{(-)}(x) = \langle\{\alpha_\lambda\}|\,V^*(x)\,, \tag{13.52b}$$

where

$$V(x) = \frac{\sqrt{(2\pi\hbar c)}}{L^{3/2}} \sum_{k,s} \frac{1}{\sqrt{k}}\, e^{(s)}(k)\,\alpha_{ks} \exp\left[i(k\,.\,x - ckt)\right]. \tag{13.53}$$

Hence, choosing the density matrix in the form $\hat{\varrho} = |\{\alpha_\lambda\}\rangle\langle\{\alpha_\lambda\}|$ the correlation functions (12.5) are factorized in the form (12.34), and the field in the coherent state fulfils the full coherence conditions. One can also see that all frequencies are present in (13.53) so that coherence does not restrict the spectral composition of the field but rather it restricts the statistics of the field.

Time development of the coherent states

Considering a one-mode field for simplicity with the Hamiltonian (11.13) (but with the vacuum fluctuation energy $(1/2)\,\hbar\omega$ subtracted) the time dependence of the coherent state $|\alpha(t)\rangle$ is given by the Schrödinger equation

$$i\hbar \frac{\partial|\alpha(t)\rangle}{\partial t} = \hat{H}|\alpha(t)\rangle . \tag{13.54}$$

If the initial condition is $|\alpha(0)\rangle = |\alpha\rangle$, the solution can be written as

$$|\alpha(t)\rangle = \exp\left(-i\frac{\hat{H}t}{\hbar}\right)|\alpha\rangle = \exp\left(-i\omega t\hat{a}^+\hat{a} - \tfrac{1}{2}|\alpha|^2\right)\sum_{n=0}^{\infty}\frac{\alpha^n}{\sqrt{n!}}|n\rangle =$$
$$= |\alpha \exp(-i\omega t)\rangle . \tag{13.55}$$

Therefore the coherent state remains coherent at all times for the free field Hamiltonian. The complex amplitude $\alpha \exp(-i\omega t)$ describes circles in the complex plane.

It can be shown more generally that a coherent state remains coherent at all times when the Hamiltonian takes the form

$$\hat{H} = \sum_{j,k} f_{jk}(t)\, \hat{a}_j^+ \hat{a}_k + \sum_k (g_k(t)\, \hat{a}_k^+ + g_k^*(t)\, \hat{a}_k) + h(t) , \tag{13.56}$$

where $f_{jk} = f_{kj}^*$, $h^*(t) = h(t)$ and $g_k(t)$ are arbitrary functions of time [178, 305, 302].

It was also shown [305, 302] that the eigenvalues of the annihilation operator for interacting systems are not, in general, analytic signals (they contain both positive and negative frequency components). Further it has been shown by Glauber [174, 176] that the field in a coherent state may be generated by non-random prescribed currents. A Hamiltonian including a third-order non-linear term yet preserving special coherent states has been discussed in [317]. A quantum harmonic oscillator with time-dependent frequency has been considered for the case where the initial state is a coherent state [121, 122].

13.2 Glauber-Sudarshan representation of the density matrix

As states of the field are mixed states rather than pure states we must use the density matrix to describe the fields and it will be useful to express the density matrix in terms of the coherent states. Using the operator expansion (13.21) for the density matrix we obtain [174]

$$\hat{\varrho} = \frac{1}{\pi^2}\iint |\alpha\rangle \langle\alpha|\, \hat{\varrho}|\beta\rangle \langle\beta|\, d^2\alpha\, d^2\beta , \tag{13.57}$$

which is an expansion in terms of dyadic products $|\alpha\rangle\langle\beta|$. By using the Fock form (11.25) of $\hat{\varrho}$ and (13.12a) we obtain for the weight function $\langle\alpha|\hat{\varrho}|\beta\rangle$ in (13.57)

$$\langle\alpha|\hat{\varrho}|\beta\rangle = \sum_{n,m} \varrho(n, m) \frac{\alpha^{*n}\beta^m}{\sqrt{(n!\,m!)}} \exp\left(-\tfrac{1}{2}|\alpha|^2 - \tfrac{1}{2}|\beta|^2\right). \qquad (13.58)$$

Since $\operatorname{Tr}\{\hat{\varrho}^2\} = \sum_{n,m} |\varrho_{nm}|^2 \leqq 1$ $(|\varrho_{nm}|^2 \leqq \varrho_{nn}\varrho_{mm}$, i.e. $\operatorname{Tr}\{\hat{\varrho}^2\} \leqq (\operatorname{Tr}\hat{\varrho})^2 = 1)$, the series in (13.58) converges for all finite values of α^* and β and so $\langle\alpha|\hat{\varrho}|\beta\rangle$ is an entire function in α and β, i.e. it is a well-behaved function.

The advantage of the expansion (13.57) is that the quantum expectation values of normally ordered operators (q-numbers) reduce to expectation values of eigenvalues in the coherent states (c-numbers) with weight function $\langle\alpha|\hat{\varrho}|\beta\rangle$. These expectation values may be calculated simply by integration over the complex planes of α and β. The expansion (13.57) corresponds to the condition of "resolution of unity" in the form

$$\frac{1}{\pi^2}\iint |\alpha\rangle\langle\alpha\,|\,\beta\rangle\langle\beta|\, d^2\alpha\, d^2\beta =$$

$$= \frac{1}{\pi^2}\iint |\alpha\rangle \exp\left(\alpha^*\beta - \tfrac{1}{2}|\alpha|^2 - \tfrac{1}{2}|\beta|^2\right)\langle\beta|\, d^2\alpha\, d^2\beta = \hat{1}, \qquad (13.59)$$

where (13.13a) has been used. This is just equation (13.15) which leads to the simple completeness relation (13.18) using (13.16). We can now ask whether the density matrix $\hat{\varrho}$ can be expanded in the simpler form [172, 436, 174]

$$\hat{\varrho} = \int \Phi_{\mathcal{N}}(\alpha)\, |\alpha\rangle\langle\alpha|\, d^2\alpha \qquad (13.60)$$

as a "mixture" of the projection operators $|\alpha\rangle\langle\alpha|$ onto the coherent states $((|\alpha\rangle\langle\alpha|)^2 = |\alpha\rangle\langle\alpha\,|\,\alpha\rangle\langle\alpha| = |\alpha\rangle\langle\alpha|)$, where $\Phi_{\mathcal{N}}(\alpha)$ is the weight function. This diagonal representation of the density matrix is called the *Glauber-Sudarshan representation*. Since $\operatorname{Tr}\hat{\varrho} = 1$

$$\int \Phi_{\mathcal{N}}(\alpha)\, d^2\alpha = 1 \qquad (13.61)$$

and from the Hermiticity $\hat{\varrho}^+ = \hat{\varrho}$ it follows that $[\Phi_{\mathcal{N}}(\alpha)]^* = \Phi_{\mathcal{N}}(\alpha)$, i.e. $\Phi_{\mathcal{N}}(\alpha)$ is a real function of the complex variable α.

The function $\langle\alpha|\,\hat{\varrho}\,|\beta\rangle$ can be expressed in terms of $\Phi_{\mathcal{N}}(\alpha)$, using (13.60) and (13.13a), in the form

$$\langle\alpha|\,\hat{\varrho}\,|\beta\rangle = \int \langle\alpha\mid\gamma\rangle\,\langle\gamma\mid\beta\rangle\,\Phi_{\mathcal{N}}(\gamma)\,d^2\gamma =$$

$$= \exp\left(-\tfrac{1}{2}|\alpha|^2 - \tfrac{1}{2}|\beta|^2\right)\int \Phi_{\mathcal{N}}(\gamma)\exp\left(-|\gamma|^2 + \alpha^*\gamma + \beta\gamma^*\right)d^2\gamma \quad (13.62a)$$

and for $\alpha = \beta$ we arrive at

$$\langle\alpha|\,\hat{\varrho}\,|\alpha\rangle = \int \Phi_{\mathcal{N}}(\beta)\exp\left(-|\alpha-\beta|^2\right)d^2\beta\,. \quad (13.62b)$$

Another relation between the weight function $\Phi_{\mathcal{N}}$ and the Fock matrix elements $\varrho(n, m)$ of the density matrix can be obtained, using (13.60), [436, 174]

$$\langle n|\,\hat{\varrho}\,|m\rangle = \varrho(n, m) = \int \Phi_{\mathcal{N}}(\alpha)\,\langle n\mid\alpha\rangle\,\langle\alpha|\,m\rangle\,d^2\alpha =$$

$$= \int \Phi_{\mathcal{N}}(\alpha)\,\frac{\alpha^n\alpha^{*m}}{\sqrt{(n!\,m!)}}\exp\left(-|\alpha|^2\right)d^2\alpha\,. \quad (13.63)$$

Equation (13.60) can formally be derived from the Fock form (11.25) of the density matrix as follows: substituting (13.14) into (11.25) and making use of (13.63) we find

$$\hat{\varrho} = \frac{1}{\pi^2}\iiint \exp\left(-\tfrac{1}{2}|\alpha|^2 - \tfrac{1}{2}|\beta|^2 - |\gamma|^2 + \alpha^*\gamma + \beta\gamma^*\right)\Phi_{\mathcal{N}}(\gamma)\,|\alpha\rangle\,\langle\beta|\,d^2\alpha\,d^2\beta\,d^2\gamma \quad (13.64)$$

and using (13.16) we obtain (13.60).

The diagonal form (13.60) of the density matrix is very convenient for calculations of expectation values of normally ordered operators. For example, for the expectation value of the moment operator $\hat{a}^{+k}\hat{a}^l$ (k, l are integers) we obtain

$$\mathrm{Tr}\,\{\hat{\varrho}\hat{a}^{+k}\hat{a}^l\} = \mathrm{Tr}\left\{\int \Phi_{\mathcal{N}}(\alpha)\,|\alpha\rangle\,\langle\alpha|\,d^2\alpha\;\hat{a}^{+k}\hat{a}^l\right\} =$$

$$= \int \Phi_{\mathcal{N}}(\alpha)\,\langle\alpha|\,\hat{a}^{+k}\,\hat{a}^l|\alpha\rangle\,d^2\alpha = \int \Phi_{\mathcal{N}}(\alpha)\,\alpha^{*k}\alpha^l d^2\alpha = \langle\alpha^{*k}\alpha^l\rangle_{\mathcal{N}}\,, \quad (13.65)$$

where we have used the eigenvalue properties (13.5a, b) of the coherent states. The suffix $\mathcal{N}$ on Φ expresses the fact that this function is related to the *normally ordered operator*. Consequently, the quantum expectation value of a normally ordered operator can be expressed as a "classical" expectation value in a generalized phase space with the "probability distribution" $\Phi_{\mathcal{N}}(\alpha)$ if the integration is carried out over the whole complex α-plane. As we have seen, the correlation functions suitable for the description of experiments with photodetectors, yielding information about the coherence properties of light, represent just the expectation values of normally ordered products of the field operators and so they may be expressed (using the Glauber-Sudarshan representation of the density matrix) in a form very close to the classical one studied earlier. This representation provides a basis for a general *formal* equivalence between the classical and quantum descriptions of optical coherence.

The field in the coherent state $|\beta\rangle$ possesses the weight function $\Phi_{\mathcal{N}}(\alpha)$

$$\Phi_{\mathcal{N}}(\alpha) = \delta(\alpha - \beta) \tag{13.66}$$

and the density matrix $\hat{\varrho}$ is just the projection operator $|\beta\rangle\langle\beta|$ onto the coherent state. A general density matrix describing a mixed state of the field can be interpreted, according to (13.60), as a superposition of the projection operators $|\alpha\rangle\langle\alpha|$ with the weight function $\Phi_{\mathcal{N}}(\alpha)$.

Although $\Phi_{\mathcal{N}}(\alpha)$ has some of the properties of a probability distribution (it is a real-valued function fulfilling the normalization (13.61)) it cannot generally be interpreted as a probability distribution since it is not non-negative and it may have singularities stronger than the δ-function. This property is a consequence of the fact that the quantities $\operatorname{Re}\alpha$ and $\operatorname{Im}\alpha$, which are eigenvalues of the non-commuting operators $(\hat{a} + \hat{a}^+)/2$ and $(\hat{a} - \hat{a}^+)/2i$ in the coherent state, are not simultaneously measurable and so $\Phi_{\mathcal{N}}(\alpha)$ cannot be measured directly. We will call the $\Phi_{\mathcal{N}}$-function the *quasi-probability function*; it represents one of a whole family of quasi-probability functions. Another quasi-probability function was introduced by Wigner [468] and $\langle\alpha|\,\hat{\varrho}\,|\alpha\rangle$ is in fact a quasi-probability function. This latter function has many properties in common with a probability density, including positive-definiteness and regularity, but it cannot be measured directly.

Some of the properties of $\langle\alpha|\,\hat{\varrho}\,|\alpha\rangle$ can be derived as follows. First we see that the function $\pi^{-1}\langle\alpha|\,\hat{\varrho}\,|\alpha\rangle$ plays the same role in the averaging of anti-normally ordered products of field operators as $\Phi_{\mathcal{N}}(\alpha)$ plays for normally ordered products, since

$$\operatorname{Tr}\{\hat{\varrho}\hat{a}^l\hat{a}^{+k}\} = \operatorname{Tr}\{\hat{a}^{+k}\hat{\varrho}\hat{a}^l\} = \operatorname{Tr}\left\{\frac{1}{\pi}\int|\alpha\rangle\langle\alpha|\,d^2\alpha\,\hat{a}^{+k}\hat{\varrho}\hat{a}^l\right\} =$$

$$= \frac{1}{\pi}\int\alpha^{*k}\,\alpha^l\langle\alpha|\,\hat{\varrho}\,|\alpha\rangle\,d^2\alpha\,. \tag{13.67}$$

We have carried out a cyclic permutation of the operators, which does not change the trace, and introduced the unity operator (13.18). We may denote $\pi^{-1}\langle\alpha|\,\hat{\varrho}|\alpha\rangle$ as $\Phi_{\mathscr{A}}(\alpha)$, which emphasizes that this function is related to *antinormally ordered operators.* From (13.62b) we have

$$\Phi_{\mathscr{A}}(\alpha) = \frac{1}{\pi}\int \Phi_{\mathscr{N}}(\beta) \exp\left(-|\alpha - \beta|^2\right) d^2\beta\,. \tag{13.68}$$

Putting $k = l = 0$ in (13.67) we arrive at the normalization condition

$$\frac{1}{\pi}\int \langle\alpha|\,\hat{\varrho}|\alpha\rangle\, d^2\alpha = 1\,. \tag{13.69}$$

Introducing the characteristic function [177]

$$C(\beta) = \mathrm{Tr}\,\{\hat{\varrho}\,\hat{D}(\beta)\} = \mathrm{Tr}\,\{\hat{\varrho}\exp(\beta\hat{a}^+ - \beta^*\hat{a})\} \tag{13.70}$$

and making use of (13.3) we obtain

$$C^{(\mathscr{N})}(\beta) = \exp\left(\tfrac{1}{2}|\beta|^2\right) C(\beta) = \exp\left(|\beta|^2\right) C^{(\mathscr{A})}(\beta)\,, \tag{13.71}$$

where

$$C^{(\mathscr{N})}(\beta) = \mathrm{Tr}\,\{\hat{\varrho}\exp(\beta\hat{a}^+)\exp(-\beta^*\hat{a})\} = \int \Phi_{\mathscr{N}}(\alpha)\exp(\beta\alpha^* - \beta^*\alpha)\, d^2\alpha \tag{13.72}$$

and

$$C^{(\mathscr{A})}(\beta) = \mathrm{Tr}\,\{\hat{\varrho}\exp(-\beta^*\hat{a})\exp(\beta\hat{a}^+)\} = \int \Phi_{\mathscr{A}}(\alpha)\exp(\beta\alpha^* - \beta^*\alpha)\, d^2\alpha\,. \tag{13.73}$$

If the complex δ-function (13.37) is employed,

$$\Phi_{\mathscr{N}}(\alpha) = \frac{1}{\pi^2}\int C^{(\mathscr{N})}(\beta)\exp(-\beta\alpha^* + \beta^*\alpha)\, d^2\beta \tag{13.74}$$

and

$$\Phi_{\mathscr{A}}(\alpha) = \frac{1}{\pi^2}\int C^{(\mathscr{A})}(\beta)\exp(-\beta\alpha^* + \beta^*\alpha)\, d^2\beta\,. \tag{13.75}$$

The Wigner function mentioned above can be introduced as

$$\Phi_{\mathrm{Weyl}}(\alpha) = \frac{1}{\pi^2}\int C(\beta)\exp(-\beta\alpha^* + \beta^*\alpha)\, d^2\beta\,, \tag{13.76}$$

where the characteristic function $C(\beta)$ is given by (13.70). The moments of the Wigner function $\Phi_{\text{Weyl}}(\alpha)$ can be calculated as

$$\int \alpha^{*k}\alpha^{l}\Phi_{\text{Weyl}}(\alpha)\, d^2\alpha = \frac{\partial^k}{\partial\beta^k}\frac{\partial^l}{\partial(-\beta^*)^l} C(\beta)\big|_{\beta=0} = \langle\alpha^{*k}\,\alpha^{l}\rangle_{\text{Weyl}}\,, \quad (13.77)$$

where

$$C(\beta) = \int \Phi_{\text{Weyl}}(\alpha) \exp(\beta\alpha^* - \beta^*\alpha)\, d^2\alpha\,. \quad (13.78)$$

Calculating derivatives in (13.77) with the help of (13.70) we can see they are equal to averages of the *symmetrically* (Weyl) ordered products of k factors of $\hat{a}^+$ and l factors of $\hat{a}$. Thus the Wigner function is the quasi-probability function appropriate to symmetric (Weyl) ordering. If $k = l = k$ we have the normalization condition

$$\int \Phi_{\text{Weyl}}(\alpha)\, d^2\alpha = C(0) = 1\,. \quad (13.79)$$

In the same way the moments of $\Phi_{\mathcal{N}}(\alpha)$ and $\Phi_{\mathcal{A}}(\alpha)$ can be calculated from the characteristic functions $C^{(\mathcal{N})}(\beta)$ and $C^{(\mathcal{A})}(\beta)$ respectively. From (13.72) and (13.73) we have

$$\frac{\partial^k}{\partial\beta^k}\frac{\partial^l}{\partial(-\beta^*)^l} C^{(\mathcal{N})}(\beta)\big|_{\beta=0} = \text{Tr}\,\{\hat{\varrho}\hat{a}^{+k}\hat{a}^l\} = \int \Phi_{\mathcal{N}}(\alpha)\, \alpha^{*k}\alpha^l\, d^2\alpha = \langle\alpha^{*k}\alpha^l\rangle_{\mathcal{N}}\,, \quad (13.80)$$

$$\frac{\partial^k}{\partial\beta^k}\frac{\partial^l}{\partial(-\beta^*)^l} C^{(\mathcal{A})}(\beta)\big|_{\beta=0} = \text{Tr}\,\{\hat{\varrho}\hat{a}^l\hat{a}^{+k}\} = \int \Phi_{\mathcal{A}}(\alpha)\, \alpha^{*k}\alpha^l\, d^2\alpha = \langle\alpha^{*k}\alpha^l\rangle_{\mathcal{A}}\,. \quad (13.81)$$

If $|\psi_n\rangle$ are normalized eigenstates of $\hat{\varrho}$ with eigenvalues λ_n then

$$\hat{\varrho} = \sum_n \lambda_n |\psi_n\rangle\,\langle\psi_n|$$

and since $0 \leqq \lambda_n \leqq 1$ ($\lambda_n \geqq 0$, $\text{Tr}\,\hat{\varrho} = \sum_n \lambda_n = 1$) we obtain

$$\Phi_{\mathcal{A}}(\alpha) = \frac{1}{\pi}\sum_n \lambda_n |\langle\alpha\,|\,\psi_n\rangle|^2 \geqq 0\,. \quad (13.82a)$$

Also

$$\Phi_{\mathcal{A}}(\alpha) \leqq \frac{1}{\pi}\sum_n |\langle\alpha\,|\,\psi_n\rangle|^2 = \frac{1}{\pi}\,, \quad (13.82b)$$

i.e. $0 \leqq \Phi_{\mathscr{A}}(\alpha) \leqq 1/\pi$ since $|\alpha\rangle = \sum_n |\psi_n\rangle \langle\psi_n|\alpha\rangle$ and $\langle\alpha|\alpha\rangle = \sum_n |\langle\alpha|\psi_n\rangle|^2 = 1$. Further, since $\hat{D}(\alpha)$ is a unitary operator ($\hat{D}^+\hat{D} = 1$), $|C(\beta)| \leqq 1$ and from (13.71)

$$|C^{(\mathscr{A})}(\beta)| \leqq \exp(-\tfrac{1}{2}|\beta|^2) \tag{13.83a}$$

and

$$|C^{(\mathscr{N})}(\beta)| \leqq \exp(\tfrac{1}{2}|\beta|^2)\,. \tag{13.83b}$$

Consequently $|C^{(\mathscr{A})}(\beta)|$ always decreases at least as fast as $\exp(-|\beta|^2/2)$ for $|\beta| \to \infty$, while $|C^{(\mathscr{N})}(\beta)|$ may diverge as rapidly as $\exp(|\beta|^2/2)$ in that limit.

One can see from (13.58) that $\langle\alpha|\,\hat{\varrho}\,|\alpha\rangle = \pi\,\Phi_{\mathscr{A}}(\alpha)$ represents an entire analytic function

$$\Phi_{\mathscr{A}}(\alpha) = \frac{1}{\pi}\sum_{n,m}\varrho(n, m)\,\frac{\alpha^{*n}\alpha^m}{\sqrt{(n!\,m!)}}\exp(-|\alpha|^2)\,. \tag{13.84}$$

The function $\Phi_{\mathscr{A}}(\alpha)$ was first introduced and its properties were studied by Glauber [176], Kano [223, 224] and Mehta and Sudarshan [304].

Mutual relations between functions $\Phi_{\mathscr{N}}(\alpha)$, $\Phi_{\mathscr{A}}(\alpha)$ and $\Phi_{\mathrm{Weyl}}(\alpha)$ can be derived from the relation (13.71) between the characteristic functions using the convolution theorem (cf. also Sec. 16.2)

$$\Phi_{\mathscr{A}}(\alpha) = \frac{2}{\pi}\int \exp(-2|\alpha - \beta|^2)\,\Phi_{\mathrm{Weyl}}(\beta)\,d^2\beta\,, \tag{13.85a}$$

$$\Phi_{\mathrm{Weyl}}(\alpha) = \frac{2}{\pi}\int \exp(-2|\alpha - \beta|^2)\,\Phi_{\mathscr{N}}(\beta)\,d^2\beta\,; \tag{13.85b}$$

equation (13.68) is the corresponding relation between $\Phi_{\mathscr{A}}(\alpha)$ and $\Phi_{\mathscr{N}}(\alpha)$. In general, the functions $\Phi_{\mathscr{A}}$ and Φ_{Weyl} on the left-hand side represent averages of the functions Φ_{Weyl} and $\Phi_{\mathscr{N}}$ on the right-hand side with exponential weight factors. This averaging process which takes us from $\Phi_{\mathscr{N}}$ to Φ_{Weyl} and $\Phi_{\mathscr{A}}$ tends to smear out any unruly behaviour of $\Phi_{\mathscr{N}}$ and to transform it into a smooth function; this smoothing process ensures that $\Phi_{\mathscr{A}}$ is non-negative and regular. While the functions $\Phi_{\mathscr{A}}$ and Φ_{Weyl} exist for all quantum states, the cases in which $\Phi_{\mathscr{N}}$ cannot be defined in the usual sense are precisely those in which the convolution integrals (13.68) and (13.85b) cannot be inverted.

Ordering of the density matrix and quasi-probabilities

An important rule for obtaining $\Phi_{\mathscr{N}}$ or $\Phi_{\mathscr{A}}$ from the density matrix, ordered in a certain way, can be derived as follows, [251, 247, 193]. Suppose that an operator $\hat{F}$ can be written in the normal and antinormal forms

$$\hat{F}(\hat{a}^+, \hat{a}) = \hat{F}^{(\mathscr{N})}(\hat{a}^+, \hat{a}) = \hat{F}^{(\mathscr{A})}(\hat{a}^+, \hat{a}) = \sum_{r,s} F_{rs}^{(\mathscr{N})} \hat{a}^{+r} \hat{a}^s = \\ = \sum_{r,s} F_{rs}^{(\mathscr{A})} \hat{a}^s \hat{a}^{+r} . \tag{13.86}$$

Assuming similar expansions to be valid for the density matrix

$$\hat{\varrho}(\hat{a}^+, a) = \hat{\varrho}^{(\mathscr{N})}(\hat{a}^+, \hat{a}) = \hat{\varrho}^{(\mathscr{A})}(\hat{a}^+, \hat{a}) = \sum_{r,s} G_{rs}^{(\mathscr{N})} \hat{a}^{+r} \hat{a}^s = \\ = \sum_{r,s} G_{rs}^{(\mathscr{A})} \hat{a}^s \hat{a}^{+r} \tag{13.87}$$

we obtain

$$\mathrm{Tr}\,\{\hat{\varrho}\hat{F}\} = \mathrm{Tr}\,\{\hat{\varrho}^{(\mathscr{A})}\hat{F}^{(\mathscr{N})}\} = \\ = \mathrm{Tr}\left\{\frac{1}{\pi}\int |\alpha\rangle\,\langle\alpha|\; d^2\alpha \sum_{r,s,u,v} G_{rs}^{(\mathscr{A})} \hat{a}^s \hat{a}^{+r} F_{uv}^{(\mathscr{N})} \hat{a}^{+u} \hat{a}^v\right\} = \\ = \frac{1}{\pi}\int \sum_{r,s,u,v} G_{rs}^{(\mathscr{A})} F_{uv}^{(\mathscr{N})} \langle\alpha|\; \hat{a}^{+r} \hat{a}^{+u} \hat{a}^v \hat{a}^s \;|\alpha\rangle\; d^2\alpha = \\ = \frac{1}{\pi}\int \varrho^{(\mathscr{A})}(\alpha^*, \alpha)\; F^{(\mathscr{N})}(\alpha^*, \alpha)\; d^2\alpha , \tag{13.88}$$

where

$$\frac{1}{\pi}\,\varrho^{(\mathscr{A})}(\alpha^*, \alpha) \equiv \Phi_{\mathscr{N}}(\alpha) = \sum_{r,s} G_{rs}^{(\mathscr{A})} \alpha^{*r} \alpha^s \tag{13.89}$$

and

$$F^{(\mathscr{N})}(\alpha^*, \alpha) = \sum_{r,s} F_{rs}^{(\mathscr{N})} \alpha^{*r} \alpha^s . \tag{13.90}$$

Similarly

$$\mathrm{Tr}\,\{\hat{\varrho}\hat{F}\} = \mathrm{Tr}\,\{\hat{\varrho}^{(\mathscr{N})}\hat{F}^{(\mathscr{A})}\} = \frac{1}{\pi}\int \varrho^{(\mathscr{N})}(\alpha^*, \alpha)\; F^{(\mathscr{A})}(\alpha^*, \alpha)\; d^2\alpha , \tag{13.91}$$

i.e.

$$\Phi_{\mathscr{A}}(\alpha) = \varrho^{(\mathscr{N})}(\alpha^*, \alpha)/\pi$$

and we also have

$$\Phi_{\mathscr{A}}(\alpha) = \langle\alpha|\; \hat{\varrho}|\alpha\rangle/\pi = \langle\alpha|\; \hat{\varrho}^{(\mathscr{N})}|\alpha\rangle/\pi = \varrho^{(\mathscr{N})}(\alpha^*, \alpha)/\pi .$$

Further

$$\hat{\varrho} = \int \Phi_{\mathcal{N}}(\alpha) \, |\alpha\rangle \langle\alpha| \, d^2\alpha = \sum_{r,s} G_{rs}^{(\mathcal{A})} \hat{a}^s \hat{a}^{+r} = \sum_{r,s} G_{rs}^{(\mathcal{A})} \hat{a}^s \frac{1}{\pi} \int |\alpha\rangle \langle\alpha| \, d^2\alpha \, \hat{a}^{+r} =$$

$$= \frac{1}{\pi} \int \Big(\sum_{r,s} G_{rs}^{(\mathcal{A})} \alpha^s \alpha^{*r}\Big) |\alpha\rangle \langle\alpha| \, d^2\alpha \, ,$$

i.e. $\Phi_{\mathcal{N}}(\alpha) = \varrho^{(\mathcal{A})}(\alpha^*, \alpha)/\pi$ again. Hence the function $\Phi_{\mathcal{N}}$ ($\Phi_{\mathcal{A}}$) can be calculated from the antinormal (normal) form of the density matrix (obtained using the commutation rules for the operators) with the substitution $\hat{a} \to \alpha$, $\hat{a}^+ \to \alpha^*$; substituting $\alpha \to \hat{a}$, $\alpha^* \to \hat{a}^+$ in $\Phi_{\mathcal{N}}$ ($\Phi_{\mathcal{A}}$) we obtain the antinormal (normal) form of the density matrix.

An example

As an example, consider the case of Gaussian light (the mean number of photons is $\langle n \rangle$) with a coherent component, which is described by the density matrix

$$\hat{\varrho} = \frac{\langle n \rangle^{\hat{b}^+\hat{b}}}{(1 + \langle n \rangle)^{1 + \hat{b}^+\hat{b}}} , \tag{13.92}$$

where $\hat{b} = \hat{a} - \beta$; here β is the complex amplitude of the coherent component. It is obvious that $[\hat{b}^+, \hat{b}] = [\hat{a}^+, \hat{a}] = 1$ and we can write

$$\hat{\varrho} = \frac{1}{\langle n \rangle} \left(1 + \frac{1}{\langle n \rangle}\right)^{-\hat{b}^+\hat{b} - 1} = \frac{1}{\langle n \rangle} \sum_{n=0}^{\infty} \frac{(-1)^n}{n!} \frac{(\hat{b}^+\hat{b} + n)!}{(\hat{b}^+\hat{b})!} \langle n \rangle^{-n} . \tag{13.93}$$

Making use of (11.8b) we obtain

$$\hat{\varrho}^{(\mathcal{A})} = \frac{1}{\langle n \rangle} \sum_{n=0}^{\infty} \frac{(-1)^n}{n!} \frac{\hat{b}^n \hat{b}^{+n}}{\langle n \rangle^n} = \frac{1}{\langle n \rangle} \mathcal{A} \left\{ \exp\left(-\frac{\hat{b}^+\hat{b}}{\langle n \rangle}\right) \right\} , \tag{13.94}$$

which gives by the substitution $\hat{b} \to \alpha - \beta$, $\hat{b}^+ \to \alpha^* - \beta^*$

$$\Phi_{\mathcal{N}}(\alpha) = \frac{1}{\pi \langle n \rangle} \exp\left(-\frac{|\alpha - \beta|^2}{\langle n \rangle}\right) . \tag{13.95}$$

In (13.94) $\mathcal{A}$ denotes the antinormally ordering operator.

Similarly

$$\hat{\varrho} = \frac{1}{1+\langle n\rangle}\left(1 - \frac{1}{1+\langle n\rangle}\right)^{\hat{b}^+\hat{b}} =$$

$$= \frac{1}{1+\langle n\rangle}\sum_{n=0}^{\infty}(-1)^n\binom{\hat{b}^+\hat{b}}{n}(1+\langle n\rangle)^{-n} =$$

$$= \frac{1}{1+\langle n\rangle}\sum_{n=0}^{\infty}\frac{(-1)^n}{n!}\frac{(\hat{b}^+\hat{b})!}{(\hat{b}^+\hat{b}-n)!}(1+\langle n\rangle)^{-n}. \tag{13.96}$$

Making use of (11.8a) we obtain

$$\hat{\varrho}^{(\mathcal{N})} = \frac{1}{1+\langle n\rangle}\sum_{n=0}^{\infty}\frac{(-1)^n}{n!}\frac{\hat{b}^{+n}\hat{b}^n}{(1+\langle n\rangle)^n} = \frac{1}{1+\langle n\rangle}\mathcal{N}\left\{\exp\left(-\frac{\hat{b}^+\hat{b}}{1+\langle n\rangle}\right)\right\} \tag{13.97}$$

and so

$$\Phi_{\mathscr{A}}(\alpha) = \frac{1}{\pi(1+\langle n\rangle)}\exp\left(-\frac{|\alpha-\beta|^2}{1+\langle n\rangle}\right). \tag{13.98}$$

In (13.97) $\mathcal{N}$ denotes the normally ordering operator.

Substituting the expression (11.30) for $\langle n\rangle$ into these equations we can obtain the corresponding results for light with a superimposed coherent component β in thermal equilibrium at the temperature T. Comparing (13.93) with (13.97) and (13.94) we arrive at

$$\hat{\varrho} = (1-e^{-\Theta})\exp\left[-\Theta(\hat{a}^+ - \beta^*)(\hat{a}-\beta)\right] =$$

$$= (1-e^{-\Theta})\mathcal{N}\{\exp\left[-(1-e^{-\Theta})(\hat{a}^+ - \beta^*)(\hat{a}-\beta)\right]\} \equiv \hat{\varrho}^{(\mathcal{N})} \tag{13.99}$$

$$= (e^{\Theta}-1)\mathscr{A}\{\exp\left[-(e^{\Theta}-1)(\hat{a}^+ - \beta^*)(\hat{a}-\beta)\right]\} \equiv \hat{\varrho}^{(\mathscr{A})}. \tag{13.100}$$

An equation of the form (13.99) also follows since ($\beta = 0$)

$$\langle\alpha|\exp(-\Theta\hat{a}^+\hat{a})|\alpha\rangle = \sum_n \exp(-\Theta n)\,|\langle\alpha\mid n\rangle|^2 =$$

$$= \sum_n \frac{(\exp(-\Theta)\,|\alpha|^2)^n}{n!}\exp(-|\alpha|^2) = \exp\left[-|\alpha|^2(1-e^{-\Theta})\right] =$$

$$= \langle\alpha|\,\mathcal{N}\{\exp\left[-\hat{a}^+\hat{a}(1-e^{-\Theta})\right]\}\,|\alpha\rangle, \tag{13.101}$$

where we have used (11.18) and (13.12b) (cf. (10.8a) in the semiclassical treatment).

Superimposition of fields

Sometimes we need to consider superimposed fields. If the complex amplitude α is composed of two components β and γ, i.e. $\alpha = \beta + \gamma$, then the density matrix describing fluctuations of β is

$$\hat{\varrho}_1 = \int \Phi_{\mathcal{N}}^{(1)}(\beta) \, |\beta + \gamma\rangle \langle \beta + \gamma| \, d^2\beta \,, \tag{13.102}$$

while that describing the fluctuations of γ is

$$\hat{\varrho}_2 = \int \Phi_{\mathcal{N}}^{(2)}(\gamma) \, |\beta + \gamma\rangle \langle \beta + \gamma| \, d^2\gamma \,. \tag{13.103}$$

The resulting density matrix will be equal to

$$\hat{\varrho} = \iint \Phi_{\mathcal{N}}^{(1)}(\beta) \, \Phi_{\mathcal{N}}^{(2)}(\gamma) \, |\beta + \gamma\rangle \langle \beta + \gamma| \, d^2\beta \, d^2\gamma = \int \Phi_{\mathcal{N}}(\alpha) \, |\alpha\rangle \langle \alpha| \, d^2\alpha \,, \tag{13.104}$$

where

$$\Phi_{\mathcal{N}}(\alpha) = \int \Phi_{\mathcal{N}}^{(1)}(\alpha - \gamma) \, \Phi_{\mathcal{N}}^{(2)}(\gamma) \, d^2\gamma = \int \Phi_{\mathcal{N}}^{(1)}(\beta) \, \Phi_{\mathcal{N}}^{(2)}(\alpha - \beta) \, d^2\beta =$$
$$= \iint \Phi_{\mathcal{N}}^{(1)}(\beta) \, \Phi_{\mathcal{N}}^{(2)}(\gamma) \, \delta(\alpha - \beta - \gamma) \, d^2\beta \, d^2\gamma \,. \tag{13.105}$$

More generally, if $\alpha = \sum_j \alpha_j$,

$$\Phi_{\mathcal{N}}(\alpha) = \int \delta\Big(\alpha - \sum_j \alpha_j\Big) \prod_j \Phi_{\mathcal{N}}^{(j)}(\alpha_j) \, d^2\alpha_j \,, \tag{13.106}$$

which is in close analogy to the classical convolution law (10.46) for the distributions of statistically independent quantities. This convolution law for the quasi-distributions $\Phi_{\mathcal{N}}^{(j)}$ is a characteristic property of the description of the field by terms of $\Phi_{\mathcal{N}}(\alpha)$ and it is an expression of the general quantum superposition principle

$$\langle\alpha\rangle = \int \Phi_{\mathcal{N}}(\alpha) \, \alpha \, d^2\alpha = \iint \Phi_{\mathcal{N}}^{(1)}(\beta) \, \Phi_{\mathcal{N}}^{(2)}(\gamma) \, (\beta + \gamma) \, d^2\beta \, d^2\gamma = \langle\beta\rangle + \langle\gamma\rangle \,, \tag{13.107a}$$

$$\langle|\alpha|^2\rangle = \langle|\beta|^2\rangle + \langle|\gamma|^2\rangle + \langle\beta\gamma^*\rangle + \langle\beta^*\gamma\rangle \,. \tag{13.107b}$$

13.3 The existence of the Glauber-Sudarshan quasi-probability $\Phi_{\mathcal{N}}$

A criterion for $\Phi_{\mathcal{N}}$ to be an L_2-function (square integrable function) can be derived from the characteristic function $C^{(\mathcal{N})}$ [177, 180]. It can be seen from (13.74) that if $C^{(\mathcal{N})}(\beta) \in L_2$, then $\Phi_{\mathcal{N}}(\alpha) \in L_2$. Since (13.83b) holds, $C^{(\mathcal{N})}(\beta)$ is not in general an L_2-function and consequently $\Phi_{\mathcal{N}}(\alpha)$ need not exist as an ordinary function; nevertheless it may exist as a *generalized function (distribution)*.

Another L_2-criterion was proposed by Mehta [299]. Using (13.60) and (13.13a) we obtain

$$\langle -\alpha|\,\hat{\varrho}|\alpha\rangle = \int \Phi_{\mathcal{N}}(\beta)\langle -\alpha \mid \beta\rangle \langle \beta \mid \alpha\rangle \, d^2\beta =$$

$$= \int \Phi_{\mathcal{N}}(\beta) \exp\left(-|\alpha|^2 - |\beta|^2 - \alpha^*\beta + \alpha\beta^*\right) d^2\beta \quad (13.108)$$

and so

$$\Phi_{\mathcal{N}}(\beta) = \frac{\exp\left(|\beta|^2\right)}{\pi^2} \int \langle -\alpha|\,\hat{\varrho}|\alpha\rangle \exp\left(|\alpha|^2\right) \exp\left(\alpha^*\beta - \alpha\beta^*\right) d^2\alpha\,. \quad (13.109)$$

Hence, if $\langle -\alpha|\,\hat{\varrho}|\alpha\rangle \exp\left(|\alpha|^2\right)$ is an L_2-function, then $\Phi_{\mathcal{N}}(\beta)\exp\left(-|\beta|^2\right)$ is also an L_2-function.

Before going into further details of the existence problems of the Glauber-Sudarshan representation we give some definitions of classes of generalized functions.

Definition and classes of generalized functions

A generalized function f is a continuous linear functional which maps each *test function* φ of a linear space onto a complex number $(f, \varphi) = \int f(x)\,\varphi(x)\,dx$ (see e.g. [12]). Linearity implies

$$(f, u\varphi_1 + v\varphi_2) = u(f, \varphi_1) + v(f, \varphi_2) \quad (13.110)$$

for any two functions φ_1, φ_2 of the considered space of test functions and any complex numbers u, v; continuity implies

$$\lim_{n\to\infty} (f, \varphi_n) = (f, \lim_{n\to\infty} \varphi_n) = (f, \varphi)\,. \quad (13.111)$$

The set $\{\varphi_n\}$ is a sequence of elements of the space which converges to φ in the space.

The class of infinitely differentiable test functions of bounded support (they are zero outside a finite interval) defines the linear space D of test functions. All the continuous linear functionals defined on D form the space of generalized functions (distributions) D'.

The space S of the test functions is composed of infinitely differentiable test functions which together with their derivatives vanish at infinity more rapidly than any negative power of $|x|$. Then S' is the linear space of continuous linear functionals defined on S; these functionals are called *tempered distributions*. If $f(x)$ is a locally integrable function of polynomial growth,

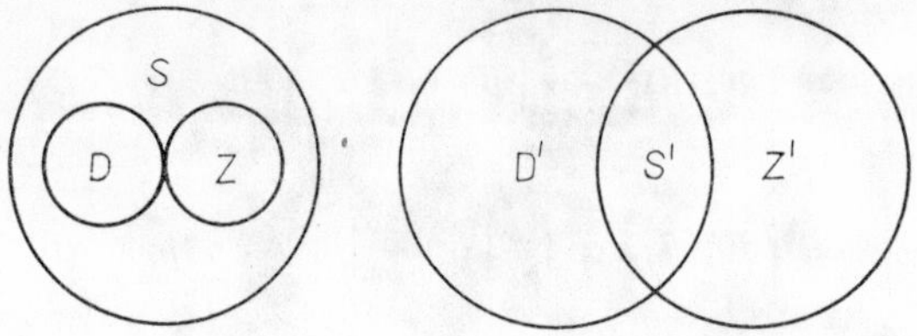

Fig. 13.1. Relationships of the test function spaces and the generalized function spaces.

then the linear functional $\int f(x)\, \varphi(x)\, dx$ converges for $\varphi \in S$ and hence it defines a tempered distribution. It is obvious that $D \subset S$, i.e. every continuous linear functional defined on S is also a continuous functional on D, i.e. $S' \subset D'$. We see that every tempered distribution is a distribution D'. The space of tempered distributions includes the Dirac function, its finite derivatives and all their finite combinations.

The entire functions $\psi(z)$ of the complex variable $z = u + iv$, which satisfy the inequalities

$$|z^n \psi^{(m)}(z)| \leqq c_{n,m} \exp(a|v|), \quad n, m = 0, 1, \ldots, \tag{13.112}$$

where a and $c_{n,m}$ are constants (which may depend on ψ, n and m), form the linear test function space Z. Then Z' is the space of generalized functions defined on Z; they are sometimes called *ultradistributions*. Since $\psi(z)$ is an entire function, it cannot vanish along any finite interval of the real axis (or it would have to be identically zero over the whole complex plane). This means that the spaces D and Z have only the zero element in common. But if $\psi(z) \in Z$, then $\psi(u) \in S$ as follows from (13.112), i.e. the space Z consists of all entire analytic functions of the exponential type which decrease rapidly for real values of their argument and so $Z \subset S$. Therefore $Z' \supset S'$. Relationships between the spaces of test functions and of the corresponding generalized functions are schematically given in Fig. 13.1.

The spaces D and Z as well as D' and Z' are related by means of a Fourier transform. Let $\varphi(x)$ be an arbitrary test function in D, $\varphi(x) = 0$, $|x| > a$. Its Fourier transform $\psi(u)$ can be considered as an entire function of $z = u + iv$ (by using the Paley-Wiener theorem on entire functions [25, 16]), i.e.

$$\psi(z) = \int_{-a}^{+a} \varphi(x) \exp ixz \, dx \, . \tag{13.113}$$

Differentiating (13.113) m times with respect to z and integrating n times by parts with respect to x we obtain, taking the absolute value,

$$\left|z^n \psi^{(m)}(z)\right| \leqq \int_{-a}^{+a} \left| \frac{d^n}{dx^n} [x^m \varphi(x)] \exp(-xv) \right| dx \leqq$$

$$\leqq c_{n,m} \exp(a|v|) \, , \quad n, m = 0, 1, \ldots \, , \tag{13.114}$$

where

$$c_{n,m} = \int_{-a}^{+a} \left| \frac{d^n}{dx^n} [x^m \varphi(x)] \right| dx \, . \tag{13.115}$$

The condition (13.114) is just that required in (13.112). Also every entire function $\psi(z)$ which satisfies (13.114) is the Fourier transform of some infinitely differentiable function $\varphi(x)$ of bounded support $|x| \leqq a$. Thus the Fourier transform establishes a one-to-one correspondence between the D- and Z-spaces and also between the D'- and Z'-spaces.

An interesting example of an element of Z' is

$$\delta(z + c) = \sum_{n=0}^{\infty} \frac{c^n}{n!} \delta^{(n)}(z) \, , \tag{13.116}$$

where $z = u + iv$ and c is a complex constant. This is an infinite series of derivatives of the δ-function.

An important example connected with the existence of $\Phi_{\mathcal{N}}(\alpha)$ is the series

$$f(u) = \sum_{n=0}^{\infty} c_n \, \delta^{(n)}(u) \, , \tag{13.117}$$

which is of the form (10.21). Although it has been shown by Cahill [103] that this series cannot be considered as an element of D' unless $c_n = 0$ for $n > N$ (N is finite) it may represent an element of Z' [316]. If $\psi(z) \in Z$

$$\int f(u) \, \psi(u) \, du = \sum_{n=0}^{\infty} (-1)^n \, c_n \, \psi^{(n)}(0) \, . \tag{13.118}$$

However, from (13.114) it follows that $|\psi^{(n)}(0)| \leqq ca^n$ (c, a are constants) so that (13.117) is an element of Z' if $\sum_{n=0}^{\infty} c_n z^n = g(z)$ is an entire analytic function.

Further details about generalized functions can be found in [12] and about their application to physics in [189].

Explicit determination of $\Phi_{\mathcal{N}}$

An insight into the existence of $\Phi_{\mathcal{N}}(\alpha)$ can be given by inverting (13.63). This inversion provides information about the quasi-distribution of the complex amplitude of the field expressed by means of the Fock matrix elements $\varrho(n, m)$. This problem was first solved by Sudarshan [436] in the form

$$\Phi_{\mathcal{N}}(\alpha) = \sum_{n=0}^{\infty} \sum_{m=0}^{\infty} \frac{\sqrt{(n!\, m!)}\, \varrho(n, m)}{(n + m)!\, \pi r} \exp\left[r^2 + i(m - n)\,\varphi\right] \left(-\frac{\partial}{\partial r}\right)^{n+m} \delta(r), \tag{13.119}$$

where $\alpha = r \exp i\varphi$. This series is a two-dimensional modification of (10.21) and the solution is expressed by means of a double series of derivatives of the δ-function. As we have noted this series does not lead to a distribution on the D-space of test functions whenever an infinite number of photons are present in the state of the field [103]. From this point of view the prescription (13.119) for $\Phi_{\mathcal{N}}(\alpha)$ was criticized by a number of authors (cf. e.g. [199]). However we have also seen that the series (13.119) may give rise to a generalized function on the Z-space of test functions which is the Fourier transform of D.

We can propose another prescription for constructing the function $\Phi_{\mathcal{N}}(\alpha)$ which is based on the decomposition of $\Phi_{\mathcal{N}}(\alpha)$ in terms of the Laguerre polynomials [369, 371]. Consider $\Phi_{\mathcal{N}}(\alpha)$ in the form

$$\Phi_{\mathcal{N}}(\alpha) = \exp\left[-(\zeta - 1)\,|\alpha|^2\right] .$$

$$\cdot \sum_{j=0}^{\infty} \Big\{ \sum_{\mu=0}^{\infty} c_{j\mu} (\zeta\alpha)^\mu L_j^\mu(\zeta|\alpha|^2) + \sum_{\mu=1}^{\infty} c_{j\mu}^* (\zeta\alpha^*)^\mu L_j^\mu(\zeta|\alpha|^2) \Big\}, \tag{13.120}$$

where $\zeta \geqq 1$ is a real number and L_j^μ are the Laguerre polynomials defined by (10.27) and fulfilling the orthogonality condition (10.28). Multiplying (13.120) by $\alpha^{*\mu'} L_k^{\mu'}(\zeta|\alpha|^2) \exp(-|\alpha|^2)$, integrating over $\zeta\alpha$ and using (10.28) we obtain for the decomposition coefficients $c_{j\mu}$ in (13.120) the following expression in terms of the Fock matrix elements $\varrho(n, m)$

$$c_{j\mu} = \frac{j!}{\pi[(j+\mu)!]^3} \int \Phi_{\mathcal{N}}(\alpha) \exp(-|\alpha|^2)\, \alpha^{*\mu} L_j^{\mu}(\zeta|\alpha|^2)\, d^2(\zeta\alpha) =$$

$$= \frac{\zeta}{\pi(j+\mu)!} \sum_{s=0}^{j} (-1)^s \binom{j}{s} \left[\frac{s!}{(s+\mu)!}\right]^{1/2} \zeta^s \varrho(s, s+\mu), \quad (13.121)$$

where we have also used (10.27) and (13.63). This relation can be inverted in the form

$$\varrho(s, s+\mu) = \frac{\pi}{\zeta^{s+1}} \left[\frac{(s+\mu)!}{s!}\right]^{1/2} \sum_{j=0}^{s} (-1)^j \binom{s}{j} (j+\mu)!\, c_{j\mu}\,. \quad (13.122)$$

One can verify that the substitution of (13.120) and (13.4) into (13.60) with the use of (13.122) leads to the Fock form (11.25) of the density matrix again [371].

The moments $\langle \alpha^{*k} \alpha^l \rangle_{\mathcal{N}}$ of $\Phi_{\mathcal{N}}(\alpha)$ can be calculated using (13.120) [371] or (13.119) [199] in the form

$$\langle \alpha^{*k} \alpha^l \rangle_{\mathcal{N}} = \sum_{r=0}^{\infty} \varrho(r+l, r+k) \frac{[(r+l)!\,(r+k)!]^{1/2}}{r!}\,. \quad (13.123)$$

Equations (13.120) and (13.121) with $\mu = 0$ correspond to (10.26) and (10.29).

From (13.120) we obtain (putting $\zeta = 1$ for simplicity)

$$\int \Phi_{\mathcal{N}}^2(\alpha)\, d\mu(\alpha) = \pi \sum_{j=0}^{\infty} \sum_{\mu=0}^{\infty} |c_{j\mu}|^2 \frac{[(j+\mu)!]^3}{j!} + \pi \sum_{j=0}^{\infty} \sum_{\mu=1}^{\infty} |c_{j\mu}|^2 \frac{[(j+\mu)!]^3}{j!}, \quad (13.124)$$

where $d\mu(\alpha) = \exp(-|\alpha|^2)\, d^2\alpha$. Hence, if $\int \Phi_{\mathcal{N}}^2(\alpha)\, d\mu(\alpha)$ (i.e. $\sum_{j=0}^{\infty} \sum_{\mu=0}^{\infty} |c_{j\mu}|^2 [(j+\mu)!]^3 (j!)^{-1}$ is finite), then $\Phi_{\mathcal{N}}(\alpha)$ is an L_2-function in the space with the measure $d\mu$ and the series (13.120) converges to $\Phi_{\mathcal{N}}(\alpha)$ in the norm of L_2, i.e.

$$\lim_{N \to \infty} \int [\Phi_{\mathcal{N}}(\alpha) - \Phi_{\mathcal{N}}^{(N)}(\alpha)]^2\, d\mu(\alpha) = 0\,, \quad (13.125)$$

where $\Phi_{\mathcal{N}}^{(N)}(\alpha)$ is the Nth partial sum of (13.120). In this way a class of physical fields, with $\Phi_{\mathcal{N}}(\alpha)$ as a member of the L_2-space, may be specified by $\varrho(n, m)$ and determined through (13.121).

Examples

We consider first Gaussian light and Gaussian light with a coherent component. In the first case

$$\varrho(n, m) = \delta_{nm} \frac{\langle n \rangle^n}{(1 + \langle n \rangle)^{1+n}}, \tag{13.126}$$

and using (13.121) with $\zeta = 1$ we obtain

$$c_{j0} = \frac{(1 + \langle n \rangle)^{-j-1}}{\pi j!} \tag{13.127}$$

so that

$$\pi^2 \sum_{j=0}^{\infty} c_{j0}^2 (j!)^2 = \sum_{j=0}^{\infty} [(1 + \langle n \rangle)^2]^{-j-1} = \frac{1}{\langle n \rangle^2 + 2\langle n \rangle} < \infty . \tag{13.128}$$

Indeed, the function $\Phi_{\mathcal{N}}(\alpha)$ has the form (13.95) with $\beta = 0$, which is an L_2-function.

In the second case $\varrho(n, m)$ is given by [322] (for a derivation see section 17.3)

$$\varrho(n, m) = \left(\frac{n!}{m!}\right)^{1/2} \exp\left(-\frac{|\beta|^2}{1 + \langle n \rangle}\right) \frac{\langle n \rangle^n}{(1 + \langle n \rangle)^{1+m}} \cdot$$

$$\cdot \beta^{*m-n} \frac{1}{m!} L_n^{m-n}\left(-\frac{|\beta|^2}{\langle n \rangle (\langle n \rangle + 1)}\right), \quad m \geqq n . \tag{13.129}$$

(If $n > m$, then $\varrho(n, m) = \varrho^*(m, n)$.) The coefficients $c_{j\mu}$ are equal to

$$c_{j\mu} = \frac{\zeta}{\pi(1 + \langle n \rangle)^{j+\mu+1}} \exp\left(-\frac{|\beta|^2}{1 + \langle n \rangle}\right) \frac{j!}{[(j + \mu)!]^3} \cdot$$

$$\cdot [1 + \langle n \rangle (1 - \zeta)]^j \beta^{*\mu} L_j^{\mu}\left(\frac{\zeta |\beta|^2}{(1 + \langle n \rangle)[1 + \langle n \rangle (1 - \zeta)]}\right). \tag{13.130}$$

Substituting this result into (13.120) and using the identity ([14], p. 1038)

$$\sum_{n=0}^{\infty} \frac{n!}{[(n + \mu)!]^3} L_n^{\mu}(x) L_n^{\mu}(y) z^n =$$

$$= \frac{(xyz)^{-\mu/2}}{1 - z} \exp\left(-z \frac{x + y}{1 - z}\right) I_\mu\left(2 \frac{\sqrt{(xyz)}}{1 - z}\right), \tag{13.131}$$

(where $I_\mu(x)$ is the modified Bessel function) the property $I_{-\mu}(x) = I_\mu(x)$ and the identity

$$\sum_{n=-\infty}^{+\infty} I_n(x) \exp in\varphi = \exp(x \cos \varphi), \tag{13.132}$$

we arrive at the diagonal representation function (13.95) again. We can easily verify using (13.131) and (13.132) that the coefficients (13.130) are such that

$$\sum_{j,\mu} |c_{j\mu}|^2 \frac{[(j+\mu)!]^3}{j!} < \infty. \tag{13.133}$$

Note that special cases of completely coherent and chaotic fields can be treated in the limits $\langle n \rangle \to 0$ and $\beta \to 0$, respectively.

One can see that while the support of every term in (13.119) is one point, the support of every term in (13.120) is the whole complex plane. This means that if it is ensured *a priori* that the function $\Phi_{\mathcal{N}}(\alpha)$ exists as an ordinary function, then all $\varrho(n, m)$ must be known for the construction of this function from $\varrho(n, m)$ using the prescription (13.119). If only a finite number of $\varrho(n, m)$ are different from zero, then $\Phi_{\mathcal{N}}(\alpha)$ has a one-point support. On the other hand the prescription (13.120) can be used (terminating the series at a finite term) as an approximation to $\Phi_{\mathcal{N}}(\alpha)$ as an ordinary function if only a finite set of $\varrho(n, m)$ is given. If all $\varrho(n, m)$ are given, then (13.120) represents an ordinary function when this series is uniformly convergent (for all α). If this series is divergent, the sequence of partial sums may converge to a generalized function. Some physical examples of the field leading to tempered distributions can be given [371] (putting $\zeta = 1$ again).

a) *The coherent state* $|\beta\rangle$

$$\Phi_{\mathcal{N}}(\alpha) = \delta(\alpha - \beta) =$$
$$= \lim_{M\to\infty} \frac{1}{\pi} \exp(-|\beta|^2) \sum_{j=0}^{M} \left\{ \sum_{\mu=0}^{M} \frac{j!}{[(j+\mu)!]^3} (\beta\alpha^*)^\mu L_j^\mu(|\alpha|^2) L_j^\mu(|\beta|^2) + \right.$$
$$\left. + \sum_{\mu=1}^{M} (\text{c. c.}) \right\} \tag{13.134}$$

(c.c. denotes the complex conjugate to the first expression). This also expresses the completeness condition of the function system $\{\alpha^\mu L_j^\mu\}$ in the L_2-space with the measure $d\mu(\alpha) = \exp(-|\alpha|^2)\, d^2\alpha$. The decomposition (13.134) also follows from (13.120) when (13.121) is substituted with $\varrho(n, m) = \beta^n \beta^{*m} \exp(-|\beta|^2)(n!\, m!)^{-1/2}$ (the last expression follows from (13.63) for $\Phi_{\mathcal{N}}(\alpha) = \delta(\alpha - \beta)$).

In experiments the phase information is very often lost so that averaging (13.134) over the phase we obtain

$$\Phi_{\mathcal{N}}(\alpha) = \frac{1}{2\pi}\int_0^{2\pi} \delta(\alpha - \beta)\, d\varphi = \frac{1}{2\pi}\int_0^{2\pi} \frac{\delta(|\alpha| - |\beta|)}{|\alpha|}\, \delta(\varphi - \psi)\, d\varphi =$$

$$= \frac{\delta(|\alpha| - |\beta|)}{2\pi|\alpha|} = \frac{\delta(|\alpha| - |\beta|)}{\pi\,.\,2|\alpha|} + \frac{\delta(|\alpha| + |\beta|)}{\pi\,.\,2|\alpha|} = \frac{1}{\pi}\,\delta(|\alpha|^2 - |\beta|^2) =$$

$$= \lim_{M\to\infty} \frac{1}{\pi} \exp(-|\beta|^2) \sum_{j=0}^{M} \frac{1}{(j!)^2}\, L_j^0(|\alpha|^2)\, L_j^0(|\beta|^2)\,, \tag{13.135}$$

since $\delta(|\alpha| + |\beta|)/2|\alpha| = 0$ and $\alpha = |\alpha| \exp i\varphi$, $\beta = |\beta| \exp i\psi$.

b) *The Fock state* $|n\rangle$

From (13.119) or from (13.120)

$$\Phi_{\mathcal{N}}(\alpha) = \frac{1}{\pi}\frac{n!}{(2n)!\,|\alpha|} \exp(|\alpha|^2) \left(\frac{\partial}{\partial|\alpha|}\right)^{2n} \delta(|\alpha|) =$$

$$= \frac{1}{\pi}(-1)^n \exp(|\alpha|^2)\, \delta^{(n)}(|\alpha|^2) = \lim_{M\to\infty} \frac{1}{\pi} \sum_{j=n}^{M} \frac{(-1)^n}{n!\,(j-n)!}\, L_j^0(|\alpha|^2)\,, \tag{13.136}$$

where we have used the identity

$$\frac{\delta^{(2n)}(|\alpha|)}{|\alpha|} = \frac{\delta^{(2n)}(|\alpha|) + \delta^{(2n)}(-|\alpha|)}{2|\alpha|} = \left(\frac{\partial}{\partial c}\right)^{2n} \frac{\delta(c + |\alpha|) + \delta(c - |\alpha|)}{2|\alpha|}\bigg|_{c=0} =$$

$$= \left(\frac{\partial}{\partial c}\right)^{2n} \delta(c^2 - |\alpha|^2)\big|_{c=0} = \left(\frac{\partial}{\partial c}\right)^{2n} \sum_{j=0}^{\infty} \frac{(-1)^j}{j!}\, c^{2j}\delta^{(j)}(|\alpha|^2)\big|_{c=0} =$$

$$= (-1)^n \frac{(2n)!}{n!}\, \delta^{(n)}(|\alpha|^2)\,. \tag{13.137}$$

One can conclude that the state $|n\rangle$ has no classical analogue, since for $n > 0$ the function $\Phi_{\mathcal{N}}(\alpha)$ is more singular that the δ-function (it is proportional to the nth derivative of the δ-function).

General solution of the existence problem

Equation (13.68) can yield a complete solution of the existence problem for $\Phi_{\mathcal{N}}(\alpha)$. This *convolution* equation has been solved, making a Fourier transform, by Bonifacio, Narducci and Montaldi [85, 86] who have shown that if a solution of (13.68) exists, it is unique. Taking the Fourier transform of (13.68) we arrive at

$$C^{(\mathscr{A})}(\beta) = C^{(\mathscr{N})}(\beta) \exp(-|\beta|^2), \tag{13.138}$$

where the characteristic functions $C^{(\mathscr{A})}(\beta)$ and $C^{(\mathscr{N})}(\beta)$ are defined by (13.72) and (13.73). Hence, if

$$C^{(\mathscr{N})}(\beta) = C^{(\mathscr{A})}(\beta) \exp |\beta|^2 \tag{13.139}$$

is a tempered distribution, then the Fourier transform, i.e. $\Phi_{\mathscr{N}}(\alpha)$, is also a tempered distribution [86]. Further, Miller and Mishkin [316] have shown that $\Phi_{\mathscr{A}}(\alpha)$ (a function of very regular behaviour as we have seen) is a member of S' and consequently of Z'. Then $C^{(\mathscr{A})}(\beta) \in D'$ and also $C^{(\mathscr{N})}(\beta)$, given by (13.139), lies in D'; therefore the Fourier transform $\Phi_{\mathscr{N}}(\alpha)$ is a member of Z'. It can easily be shown that this solution is unique [316]. Hence, for all physical fields the function $\Phi_{\mathscr{N}}(\alpha)$, the weight function of the Glauber-Sudarshan diagonal representation of the density matrix exists at least as a distribution Z' (ultradistribution). In fact $\Phi_{\mathscr{N}}(\alpha)$ is an element of a subspace of Z' [316]. In this sense it can be said that the function $\Phi_{\mathscr{N}}(\alpha)$ always exists.

Other approaches to the problem of the existence of the diagonal representation of the density matrix have been proposed. In these approaches, developed by Mehta and Sudarshan [304], Klauder, McKenna and Currie [236], Klauder [235] (see also [17], section 8.4) and Rocca [402], the weight function $\Phi_{\mathscr{N}}(\alpha)$ is defined as the limit of an infinite sequence of well-behaved weight functions $\{\Phi_{\mathscr{N}}^{(N)}(\alpha)\}$. Thus all density matrices can be expressed as limits of sequences each member of which has the Glauber-Sudarshan diagonal representation with the weight functions $\Phi_{\mathscr{N}}^{(N)}(\alpha)$ as square integrable functions or tempered generalized functions. For the construction of $\Phi_{\mathscr{N}}^{(N)}$ in the first case, the prescription (13.120) terminated after the Nth term in every series can be used, while in the second case the prescription (13.119) terminated after the Nth term in both the series is suitable. However, since in the process of measuring the density matrices fulfilling the condition

$$\operatorname{Tr}\{\hat{\varrho}\hat{A} - \varrho_N \hat{A}\} \leqq \varepsilon \|\hat{A}\| \tag{13.140}$$

(where $\hat{A}$ is a bounded operator with the norm $\|\hat{A}\|$ and ε is an arbitrarily small number) are not distinguished, we can conclude that the density matrix can be expressed in the diagonal form with arbitrary accuracy. This result applies to the physically important operators $\hat{A} = \mathscr{N}\{\exp(ix\hat{a}^+\hat{a})\}$ and $(1/n!)\,(d^n/d(ix)^n)\,\hat{A}|_{ix=-1} = (1/n!)\,\mathscr{N}\{(\hat{a}^+\hat{a})^n \,.$ $.\exp(-\hat{a}^+\hat{a})\}$ which correspond to the normal characteristic function and the photon-counting distribution respectively. The photon-counting distribution is, for example,

$$p(n) \approx \int \Phi_{\mathcal{N}}^{(N)}(\alpha) \frac{|\alpha|^{2n}}{n!} \exp(-|\alpha|^2)\, d^2\alpha\,, \tag{13.141}$$

where $p(n) \equiv \varrho(n, n)$ and $\varrho(n, n)$ is given by (13.63).

A regularization procedure for the function $\Phi_{\mathcal{N}}(\alpha)$ was proposed by Cahill [104].

13.4 Multimode description

All earlier results and relations can be generalized to fields and systems, having a finite or countably infinite number of degrees of freedom, if the states $|\{\alpha_\lambda\}\rangle$ defined by (13.48) are used. The Glauber-Sudarshan representation has the form

$$\hat{\varrho} = \int \Phi_{\mathcal{N}}(\{\alpha_\lambda\})\, |\{\alpha_\lambda\}\rangle \langle\{\alpha_\lambda\}|\, d^2\{\alpha_\lambda\}\,, \tag{13.142}$$

where $\Phi_{\mathcal{N}}(\{\alpha_\lambda\}) \equiv \Phi_{\mathcal{N}}(\alpha_1, \alpha_2, \ldots)$ is the weight function (or functional) and $d^2\{\alpha_\lambda\} \equiv \prod_\lambda d^2\alpha_\lambda$. For example, we obtain instead of (13.63)

$$\langle\{n_\lambda\}|\hat{\varrho}|\{m_\lambda\}\rangle = \varrho(\{n_\lambda\}, \{m_\lambda\}) = \\ = \int \Phi_{\mathcal{N}}(\{\alpha_\lambda\}) \prod_\lambda \frac{\alpha_\lambda^{n_\lambda} \alpha_\lambda^{*m_\lambda}}{\sqrt{(n_\lambda!\, m_\lambda!)}} \exp(-|\alpha_\lambda|^2)\, d^2\alpha_\lambda\,. \tag{13.143}$$

Further,

$$\Phi_{\mathscr{A}}(\{\alpha_\lambda\}) = \int \Phi_{\mathcal{N}}(\{\beta_\lambda\}) \prod_\lambda \exp(-|\alpha_\lambda - \beta_\lambda|^2) \frac{d^2\beta_\lambda}{\pi}\,, \tag{13.144}$$

which corresponds to (13.68) and

$$\Phi_{\mathscr{A}}(\{\alpha_\lambda\}) = \sum_{\{n_\lambda\}} \sum_{\{m_\lambda\}} \varrho(\{n_\lambda\}, \{m_\lambda\}) \prod_\lambda \frac{\alpha_\lambda^{*n_\lambda} \alpha_\lambda^{m_\lambda}}{\pi \sqrt{(n_\lambda!\, m_\lambda!)}} \exp(-|\alpha_\lambda|^2)\,, \tag{13.145}$$

which corresponds to (13.84). All other relations can be generalized in an identical way.

Using (13.142) and (13.52) we can write the correlation function (12.5) in the form

$$\Gamma_{\mathcal{N}}^{(m,n)}(x_1, \ldots, x_{m+n}) = \int \Phi_{\mathcal{N}}(\{\alpha_\lambda\}) \prod_{j=1}^{m} V^*(x_j) \prod_{k=m+1}^{m+n} V(x_k)\, d^2\{\alpha_\lambda\}\,, \tag{13.146}$$

where $V(x)$ is given by (13.53).

The superposition of two fields described by $\Phi_{\mathcal{N}}^{(1)}(\{\alpha_\lambda\})$ and $\Phi_{\mathcal{N}}^{(2)}(\{\alpha_\lambda\})$ will be described, in analogy to (13.105), by

$$\Phi_{\mathcal{N}}(\{\alpha_\lambda\}) = \iint \Phi_{\mathcal{N}}^{(1)}(\{\beta_\lambda\})\, \Phi_{\mathcal{N}}^{(2)}(\{\gamma_\lambda\}) \prod_\lambda \delta(\alpha_\lambda - \beta_\lambda - \gamma_\lambda)\, d^2\beta_\lambda\, d^2\gamma_\lambda\,. \tag{13.147}$$

This convolution law expresses the quantum superposition principle which leads to the multimode interference law identical with the classical interference law

$$\begin{aligned}
&\langle \hat{A}^{(-)}(x)\, \hat{A}^{(+)}(x)\rangle = \\
&= \int \Phi_{\mathcal{N}}(\{\alpha_\lambda\}) \langle\{\alpha_\lambda\}|\, \hat{A}^{(-)}(x, \{\hat{a}_\lambda^+\})\, \hat{A}^{(+)}(x, \{\hat{a}_\lambda\})|\,\{\alpha_\lambda\}\rangle\, d^2\{\alpha_\lambda\} = \\
&= \iint \Phi_{\mathcal{N}}^{(1)}(\{\beta_\lambda\})\, \Phi_{\mathcal{N}}^{(2)}(\{\gamma_\lambda\}) \left[V^*(x, \{\beta_\lambda\}) + V^*(x, \{\gamma_\lambda\})\right] \left[V(x, \{\beta_\lambda\}) + \right. \\
&\quad \left. + V(x, \{\gamma_\lambda\})\right] d^2\{\beta_\lambda\}\, d^2\{\gamma_\lambda\} = \langle |V(x, \{\beta_\lambda\})|^2\rangle + \langle |V(x, \{\gamma_\lambda\})|^2\rangle + \\
&\quad + \langle V^*(x, \{\beta_\lambda\})\, V(x, \{\gamma_\lambda\})\rangle + \langle V(x, \{\beta_\lambda\})\, V^*(x, \{\gamma_\lambda\})\rangle\,,
\end{aligned} \tag{13.148}$$

since $V(x, \{\alpha_\lambda\})$ is linear in $\{\alpha_\lambda\}$ so that $V(x, \{\beta_\lambda + \gamma_\lambda\}) = V(x, \{\beta_\lambda\}) + V(x, \{\gamma_\lambda\})$. This is in agreement with the rather intuitive formula (12.20).

13.5 Time development of the Glauber-Sudarshan quasi-probability $\Phi_{\mathcal{N}}$

As we have noted in section 13.1 the most general form of the Hamiltonian for which the *coherent state remains coherent* is given by (13.56). This means, writing the density matrix in the form

$$\hat{\varrho}(t) = \int \Phi_{\mathcal{N}}(\alpha, t)\, |\alpha\rangle \langle\alpha|\, d^2\alpha\,, \tag{13.149}$$

which is the form of $\hat{\varrho}$ in the Schrödinger picture, that in this case the weight function $\Phi_{\mathcal{N}}(\alpha) = \delta(\alpha - \beta)$ describing the initial coherent state $|\beta\rangle$ remains a δ-function for all times. More generally it is true [462, 463] that

$$\Phi_{\mathcal{N}}(\{\alpha_\lambda(0)\}, t) = \Phi_{\mathcal{N}}(\{\sum_\mu S^+_{\lambda\mu}(t)\, \alpha_\mu(0) + Q_\lambda(t)\}, 0)\,, \qquad (13.150)$$

where $S_{\lambda\mu}$ are matrix elements of the $\hat{S}$-matrix (11.55) calculated with $\hat{H} \equiv (f_{jk}) = \hat{f}$, that is

$$\hat{S}(t) = 1 + \sum_{n=1}^{\infty} \left(\frac{-i}{\hbar}\right)^n \int_0^t dt'_1 \ldots \int_0^{t'_{n-1}} dt'_n\, \hat{f}(t'_1) \ldots \hat{f}(t'_n) \qquad (13.151)$$

and

$$Q_\lambda(t) = \frac{i}{\hbar} \sum_\mu \int_0^t S^+_{\lambda\mu}(t')\, g_\mu(t')\, dt'\,. \qquad (13.152)$$

Here f_{jk} and g_k are coefficients from (13.56). As a consequence of (13.150) a Gaussian distribution for $t = 0$ remains Gaussian for all times [463, 400, 401, 508].

Some other Hamiltonians have been also investigated in connection with various physical problems. Thus in papers by Mollow and Glauber [322, 323] the time dependence of $\Phi_{\mathcal{N}}(t)$ has been studied for the *parametric amplification process* which is described by the Hamiltonian

$$\hat{H}(t) = \hbar\omega_a\, \hat{a}^+(t)\, \hat{a}(t) + \hbar\omega_b\, \hat{b}^+(t)\, \hat{b}(t) - $$
$$- \hbar\varkappa[\hat{a}^+(t)\, \hat{b}^+(t) \exp(-i\omega t) + \hat{a}(t)\, \hat{b}(t) \exp i\omega t]\,. \qquad (13.153)$$

Here two modes a, b with annihilation and creation operators $\hat{a}$, $\hat{a}^+$ and $\hat{b}$, $\hat{b}^+$ are coupled by a harmonic field $\exp i\omega t$ which is assumed to be strong enough so that it may be described by a classical function with a coupling constant $\varkappa$. The frequencies ω_a, ω_b and the pumping frequency ω satisfy $\omega = \omega_a + \omega_b$ (cf. stationary condition (2.21)). This Hamiltonian also describes the frequency splitting of light beams and coherent Raman and Brillouin scattering. The time dependence of $\Phi_{\mathcal{N}}(\alpha, \beta, t)$ for the light of the modes a and b has been investigated. In the Heisenberg picture [322, 323] characteristic functions were employed and in the Schrödinger picture in the Fourier frequency domain [318] the characteristics of the corresponding partial differential equation for the characteristic function were used. Other approaches have been proposed in [188, 187, 453] and the quantum statistics of coupled oscillator systems has been investigated in [320].

Considering for simplicity the statistics of light of the mode a we can obtain for the time development of $\Phi_{\mathcal{N}}(\alpha, t)$ the equation [322]

$$\Phi_{\mathcal{N}}(\alpha, t) = \frac{1}{\pi(\sinh \varkappa t)^2} \iint \Phi_{\mathcal{N}}(\alpha_0', \beta_0') \exp\left[- \frac{|\alpha - \bar{\alpha}(\alpha_0', \beta_0', t)|^2}{(\sinh \varkappa t)^2} \right] d^2\alpha_0' \, d^2\beta_0' , \tag{13.154}$$

where $\Phi_{\mathcal{N}}(\alpha_0, \beta_0)$ describes the initial state for $t = 0$ and $\bar{\alpha}(\alpha_0, \beta_0, t) = \alpha_0 \cosh(\varkappa t) \exp(-i\omega_a t) + i\beta_0^* \sinh(\varkappa t) \exp(-i\omega_a t)$. If the initial state is coherent, i.e. $\Phi_{\mathcal{N}}(\alpha_0', \beta_0') = \delta(\alpha_0' - \alpha_0)\, \delta(\beta_0' - \beta_0)$ we obtain

$$\Phi_{\mathcal{N}}(\alpha, t) = \frac{1}{\pi(\sinh \varkappa t)^2} \exp\left[- \frac{|\alpha - \bar{\alpha}(\alpha_0, \beta_0, t)|^2}{(\sinh \varkappa t)^2} \right]. \tag{13.155}$$

The time evolution of this function is shown in Fig. 13.2 for $\beta_0 = 0$.

Some further examples of the time development of $\Phi_{\mathcal{N}}$ can also be solved in a closed form – in particular the case of the *randomly modulated harmon-*

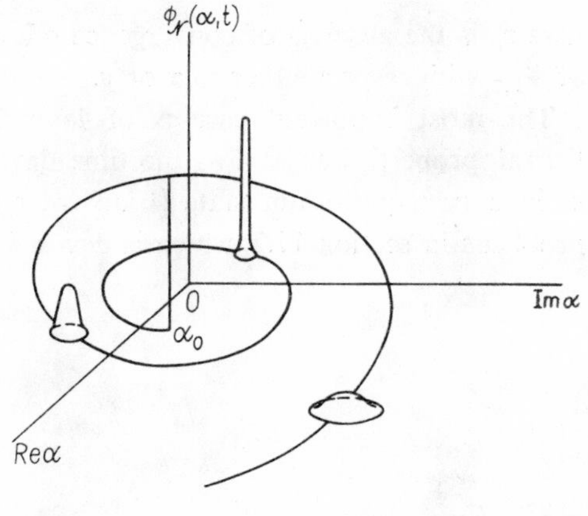

Fig. 13.2. Illustration of the way in which the function $\Phi_{\mathcal{N}}(\alpha, t)$ varies with time if the system is initially in a coherent state.

ic oscillator [123]. Assuming short-range frequency correlations, strong perturbation and that the process is Markoffian, the Heisenberg equation (11.48a) for the density matrix leads (after some complicated mathematics) to the following differential equation

$$\frac{\partial}{\partial t} \Phi_{\mathcal{N}}(\alpha, t) = \tfrac{1}{4} D \left(\frac{\partial}{\partial x} - 4x \right)^2 \left(\frac{\partial}{\partial y} \right)^2 \Phi_{\mathcal{N}}(\alpha, t) , \tag{13.156}$$

where D is a constant. This can be solved in the form

$$\Phi_{\mathcal{N}}(\alpha, t) =$$

$$= \exp(2x^2) \int_{-\infty}^{+\infty} dx' \int_{\tau_0 - i\infty}^{\tau_0 + i\infty} dy' \exp(-2x'^2)\, \Phi_{\mathcal{N}}(\alpha', 0)\, \mathcal{G}_1(x, y; t \mid x', y'; 0) ,$$

$$y > \operatorname{Re} y' \tag{13.157a}$$

or

$$\Phi_{\mathcal{N}}(\alpha, t) =$$

$$= \exp(2x^2) \int_{\tau_0 - i\infty}^{\tau_0 + i\infty} dx' \int_{-\infty}^{+\infty} dy' \exp(-2x'^2)\, \Phi_{\mathcal{N}}(\alpha', 0)\, \mathcal{G}_2(x, y; t \mid x', y'; 0)\,,$$

$$x > \operatorname{Re} x'\,, \tag{13.157b}$$

where $\alpha = x + iy$ and the Green's functions are given by

$$\mathcal{G}_{1,2}(x, y; t \mid x', y'; 0) =$$

$$= [2i\pi^{3/2} D^{1/2} t^{1/2}]^{-1} \int_0^\infty \exp\left[-\frac{(x - x')^2}{\nu^2 Dt} - \nu(y - y')\right] \frac{d\nu}{\nu} = \tag{13.158a}$$

$$= [2i\pi^{3/2} D^{1/2} t^{1/2}]^{-1} \int_0^\infty \exp\left[-\frac{(y - y')^2}{\mu^2 Dt} - \mu(x - x')\right] \frac{d\mu}{\mu}\,; \tag{13.158b}$$

here τ_0 is the abscissa of convergence relevant to the analytic continuation of $\mathcal{G}_{1,2}$ with respect either to x or y.

The most important case is of laser light and in particular its statistical properties as well as the time development of the beam when it is in a nonequilibrium state (transient effects). We shall touch on these problems in section 17.2 which is devoted to laser light.

Chapter 14

RELATION BETWEEN THE QUANTUM AND CLASSICAL THEORIES OF COHERENCE

14.1 Quantum and classical correlation functions

Introducing the probability distribution $P_n(V_1, \ldots, V_n)$, $V_j \equiv V(x_j)$, by the relation [271]

$$P_n(V_1, \ldots, V_n) = \int \Phi_{\mathcal{N}}(\{\alpha'_\lambda\}) \prod_{j=1}^{n} \delta(V_j - V'_j)\, d^2\{\alpha'_\lambda\} = \langle \prod_{j=1}^{n} \delta(V_j - V'_j) \rangle , \tag{14.1}$$

where the functions V'_j are given by (13.53) with $\alpha \to \alpha'$, we can rewrite (13.146) in the form

$$\Gamma_{\mathcal{N}}^{(m,n)}(x_1, \ldots, x_{m+n}) = \int \ldots \int P_{m+n}(V_1, \ldots, V_{m+n}) \prod_{j=1}^{m} V_j^* \prod_{k=m+1}^{m+n} V_k \prod_{l=1}^{m+n} d^2 V_l =$$
$$= \langle \prod_{j=1}^{m} V^*(x_j) \prod_{k=m+1}^{m+n} V(x_k) \rangle , \tag{14.2}$$

which is just the form (8.4) of the classical correlation function. Thus there is a very close correspondence between the classical and the normally ordered quantum correlation functions. In this correspondence the Glauber-Sudarshan diagonal representation of the density matrix serves as a bridge between the classical and quantum descriptions. More generally, if $\hat{F}\{\hat{A}^{(-)}(x), \hat{A}^{(+)}(x)\}$ is a functional of the field operators $\hat{A}^{(+)}$ and $\hat{A}^{(-)}$ and if we define the displacement operator $\hat{D}(\{\alpha_\lambda\})$ as the product of one-mode displacement operators (13.2), we obtain using the property (13.32)

$$\hat{D}^{-1}(\{\alpha_\lambda\})\, \hat{A}^{(+)}(x)\, \hat{D}(\{\alpha_\lambda\}) = \hat{A}^{(+)}(x) + V(x) ,$$
$$\hat{D}^{-1}(\{\alpha_\lambda\})\, \hat{A}^{(-)}(x)\, \hat{D}(\{\alpha_\lambda\}) = \hat{A}^{(-)}(x) + V^*(x) , \tag{14.3}$$

where $V(x)$ is given by (13.53). Hence, one obtains

$$\mathrm{Tr}\,\{\hat{\varrho}\,\hat{F}\{\hat{A}^{(-)}(x), \hat{A}^{(+)}(x)\}\} =$$

$$= \int \Phi_{\mathcal{N}}(\{\alpha_\lambda\})\,\langle 0|\hat{D}^{-1}(\{\alpha_\lambda\})\,\hat{F}\{\hat{A}^{(-)}(x), \hat{A}^{(+)}(x)\}\,\hat{D}(\{\alpha_\lambda\})\,|0\rangle\, d^2\{\alpha_\lambda\} =$$

$$= \int \Phi_{\mathcal{N}}(\{\alpha_\lambda\})\,\langle 0|\hat{F}\{\hat{A}^{(-)}(x) + V^*(x), \hat{A}^{(+)}(x) + V(x)\}\,|0\rangle\, d^2\{\alpha_\lambda\}\,, \tag{14.4}$$

since $|\{\alpha_\lambda\}\rangle = \hat{D}(\{\alpha_\lambda\})\,|0\rangle$. The operators $\hat{A}^{(+)}$ and $\hat{A}^{(-)}$ stand on the right-hand side to give the contribution of the physical vacuum fluctuations. However, if $\hat{F}$ is a normally ordered functional, then the vacuum expectation value of a normally ordered operator is zero since $\hat{A}^{(+)}|0\rangle = \langle 0|\,\hat{A}^{(-)} = 0$ (i.e. the vacuum fluctuations give no contribution) and we have

$$\mathrm{Tr}\,\{\hat{\varrho}\,\hat{F}\{\hat{A}^{(-)}(x), \hat{A}^{(+)}(x)\}\} = \int \Phi_{\mathcal{N}}(\{\alpha_\lambda\})\,F\{V^*(x), V(x)\}\, d^2\{\alpha_\lambda\}\,. \tag{14.5}$$

This result holds no matter how weak the field, i.e. it holds for arbitrarily low intensities. This result also gives a justification for the procedure used in the calculations of classical optics [286]. We can see that there is a formal equivalence between the quantum and classical descriptions in this sense; of course, this formal equivalence does not imply a physical equivalence between the classical and quantum theories and we have seen, for example, that the Fock state $|n\rangle$ has no classical analogue. However equation (14.2) shows that all results given in earlier chapters devoted to the classical theory of the space-time-polarization behaviour of the correlation functions are also correct for the quantum correlation functions.

14.2 Photon-number and photon-counting distributions

The success of the classical (semiclassical) theory is a consequence of the fact that the photodetection equation (10.3a), derived by the semiclassical method, follows simply from the properties of the quantized electromagnetic field [169, 353, 357, 176, 231, 520, 523].

First consider the probability distribution $p(n)$ of the number of photons n within the normalization volume L^3 at time t. It is clear that

$$p(m) = \sum_{\{n_\lambda\}} \varrho(\{n_\lambda\}, \{n_\lambda\})\,\delta_{nm}\,, \tag{14.6}$$

where $\sum_\lambda n_\lambda = n$. The moment $\langle \hat{n}^k \rangle$ (k is integer), where $\hat{n}$ is defined in (11.24b), is, with the help of the Fock and Glauber-Sudarshan representations of the density matrix,

$$\langle \hat{n}^k \rangle = \sum_{\{n_\lambda\}} \varrho(\{n_\lambda\}) \langle \{n_\lambda\} | \hat{n}^k | \{n_\lambda\} \rangle = \int \Phi_{\mathcal{N}}(\{\alpha_\lambda\}) \langle \{\alpha_\lambda\} | \hat{n}_k | \{\alpha_\lambda\} \rangle \, d^2\{\alpha_\lambda\} , \tag{14.7}$$

where $\varrho(\{n_\lambda\}) \equiv \varrho(\{n_\lambda\}, \{n_\lambda\})$. Substituting the unit operator (11.37) into the right-hand side of (14.7) and using the multimode analogue of (13.12b),

$$|\langle \{n_\lambda\} | \{\alpha_\lambda\} \rangle|^2 = \prod_\lambda \frac{|\alpha_\lambda|^{2n_\lambda}}{n_\lambda!} \exp(-|\alpha_\lambda|^2) , \tag{14.8}$$

we obtain

$$\sum_{n=0}^{\infty} p(n) \, n^k = \sum_{\{n_\lambda\}} \left\{ \int \Phi_{\mathcal{N}}(\{\alpha_\lambda\}) \prod_\lambda \frac{|\alpha_\lambda|^{2n_\lambda}}{n_\lambda!} \exp(-|\alpha_\lambda|^2) \, d^2\{\alpha_\lambda\} \, n^k \right\} , \tag{14.9}$$

where $n = \sum_\lambda n_\lambda$ is the eigenvalue of the operator $\hat{n}$ in the Fock state $|\{n_\lambda\}\rangle$ and $p(n)$ is given by (14.6). As (14.9) must hold for every k we arrive at [169, 357]

$$p(n) = \int \Phi_{\mathcal{N}}(\{\alpha_\lambda\}) \frac{W^n}{n!} \exp(-W) \, d^2\{\alpha_\lambda\} , \tag{14.10}$$

where we have used the multinomial theorem

$$\sum_{\{n_\lambda\}}' \prod_\lambda \frac{x_\lambda^{n_\lambda}}{n_\lambda!} = \frac{\left(\sum_\lambda x_\lambda\right)^n}{n!} , \tag{14.11}$$

where Σ' is taken under the condition $\sum_\lambda n_\lambda = n$; the quantity W is equal to

$$W = \langle \{\alpha_\lambda\} | \hat{n} | \{\alpha_\lambda\} \rangle = \int_{L^3} \mathscr{A}^*(x) \, . \, \mathscr{A}(x) \, d^3x = \sum_\lambda |\alpha_\lambda|^2 , \tag{14.12}$$

where $\mathscr{A}(x)$ is the eigenvalue of $\hat{A}(x)$ (given by (11.23)) in the coherent state. If we consider a detector with a plane cathode of surface S on which plane waves are normally incident, then $L^3 = ScT$ and (14.12) is proportional to (10.3b). The substitution

$$P_{\mathcal{N}}(W) = \int \Phi_{\mathcal{N}}(\{\alpha'_\lambda\}) \, \delta(W - \sum_\lambda |\alpha'_\lambda|^2) \, d^2\{\alpha'_\lambda\} \tag{14.13}$$

finally gives the photodetection equation with the photoefficiency $\eta = 1$

$$p(n) = \int_0^\infty P_{\mathcal{N}}(W) \frac{W^n}{n!} \exp(-W)\, dW. \tag{14.14}$$

Thus we have obtained the same relation for the distribution of the number of photons in the volume L^3 at time t as for the distribution of emitted photoelectrons within the time interval $(t, t + T)$ and it can be said that in photoelectric detection measurements the statistical properties of photons are directly reflected in the statistical properties of photoelectrons. The use of normally ordered operators, ensuring that the contribution of the vacuum is zero, is responsible for this result, as we have pointed out.

We can also show that equation (14.14) holds for an arbitrary volume V (the linear dimensions of which are large compared with the wavelength) which need not coincide with the normalization volume L^3. Considering (V, t') conjoint with (x, t) and using the commutation rule (11.44a) we obtain [277]

$$\langle \mathcal{N}\{\hat{n}_{Vt}^k\}\rangle \equiv \langle \hat{n}_{Vt}^k \rangle_{\mathcal{N}} = \langle \hat{n}_{Vt}(\hat{n}_{Vt} - 1) \dots (\hat{n}_{Vt} - k + 1)\rangle, \tag{14.15}$$

which is a multimode analogue of (11.8a). Thus the normal characteristic function is equal to

$$\langle \exp ix\hat{n}_{Vt}\rangle_{\mathcal{N}} = \sum_{r=0}^{\infty} \frac{(ix)^r}{r!} \langle \hat{n}_{Vt}^r \rangle_{\mathcal{N}} = \langle (1 + ix)^{\hat{n}_{Vt}} \rangle, \tag{14.16}$$

which corresponds to (10.11) and substituting $1 + ix = e^{iy}$ we have

$$\langle \exp iy\hat{n}_{Vt}\rangle = \langle \exp [\hat{n}_{Vt}(e^{iy} - 1)]\rangle_{\mathcal{N}}, \tag{14.17}$$

which corresponds to (10.8) and (13.101); the inverse Fourier transform leads to (14.14) at once. As a consequence the coefficients in the expansion of moments $\langle \hat{n}_{Vt}^k \rangle$ in terms of the normal moments $\langle \hat{n}_{Vt}^j \rangle_{\mathcal{N}}$ $(j = 1, \dots, k)$ are the same as in (10.4), (10.17) or (10.19); this can also be verified directly using the commutation rules (11.42) [275]. As an example, considering one mode for simplicity and using the commutation rule (11.3), $\langle \hat{n}^2 \rangle = \langle \hat{a}^+ \hat{a} \hat{a}^+ \hat{a}\rangle = \langle \hat{a}^+ \hat{a}^+ \hat{a} \hat{a}\rangle + \langle \hat{a}^+ \hat{a}\rangle = \langle \hat{n}^2 \rangle_{\mathcal{N}} + \langle \hat{n} \rangle_{\mathcal{N}}$, in agreement with (10.4b) or the second equation in (10.17). All these results again confirm the close formal analogy between the quantum and classical descriptions.

Note that the normalized factorial moments of the photon-counting distribution

$$\frac{\langle W^k \rangle_{\mathcal{N}}}{\langle W \rangle_{\mathcal{N}}^k} = \frac{\left\langle \dfrac{n!}{(n - k)!} \right\rangle}{\langle n \rangle^k}, \tag{14.18}$$

which follows from (10.13), are independent of the photoefficiency η so that they equal the factorial moments of the photon-number distribution. Thus the photon-counting distribution can be considered as the photon-number distribution independently of the detector efficiency if the factorial moments are used for the interpretation of experiments. Denoting the photon-number distribution by $p(n)$ and the photon-counting distribution for this moment by $p_\eta(n)$ their connection is given by Bernoulli's distribution

$$p_\eta(n) = \sum_{m=n}^{\infty} \binom{m}{n} \eta^n (1 - \eta)^{m-n} \, p(m) . \tag{14.19}$$

We can also derive the photon-counting distribution in the form (10.3a) with the help of quantum methods [176, 177, 180, 231, 520, 523, 17]. In order to find the photon-counting distribution we assume that the sensitive element of a photon counter consists of many atoms, any of which may undergo a photoabsorption process and give rise to a detected photon count.† Let $\hat{n}(T) = \sum_j^N \hat{n}_j(T)$ be interpreted as the number of counts in $(0, T)$, where $\hat{n}_j(T)$ represent the contributions of the individual atoms. The eigenvalues of $\hat{n}_j$ are one or zero; one for final states to which the jth atom has contributed a photon count and zero for states to which it has not contributed. Therefore we can write

$$(1 + ix)^{\hat{n}(T)} = (1 + ix)^{\sum_j^N \hat{n}_j(T)} = \prod_j^N (1 + ix)^{\hat{n}_j(T)} = \prod_j^N [1 + ix \, \hat{n}_j(T)] , \tag{14.20}$$

since the eigenvalues of $\hat{n}_j$ are 1 or 0. Expanding (14.20), expressing $\langle \hat{n}_1(T) \ldots \hat{n}_m(T) \rangle = p^{(m)}(T, \ldots, T)$ from (12.13) and approximating the combination number $N!/(N - m)!\, m!$ appearing in the expansion of (14.20)

† Also *two-photon* and *multi-photon* absorptions can be considered. The perturbation theory of quantum mechanics leads to the result [243, 111, 112] that the probability of transition for the s-quantum absorption is proportional to $I^s(t)$, i.e. all results concerning the detection of the field given here are correct if $I(t)$ is replaced by $I^s(t)$. The relation between the two-quantum photon-counting distribution and the intensity fluctuations of the radiation incident on the detector has been studied and compared with the one-quantum results for laser and chaotic radiation by Teich and Diament [438]. Some further investigations of the two- and multi-photon processes were performed in [439, 102, 242, 130, 321, 491–493, 511, 526–528, 531, 532, 536]. Some experimental results have been referred to in [422, 502, 521, 535].

by $1/m!$ (since N is large and $p^{(m)}$ are non zero only for $m \ll N$) we obtain

$$\langle (1 + ix)^{\hat{n}(T)} \rangle = 1 + \sum_{m=1}^{\infty} \frac{(ix)^m}{m!} F_m , \tag{14.21}$$

where

$$F_m = \int_0^T \cdots \int_0^T \underbrace{\int \cdots \int}_{\substack{\text{Volume} \\ \text{of detector}}} \Gamma_{\mathcal{N}}^{(m,m)}(x_1', \ldots, x_m', x_m'', \ldots, x_1'') \, .$$

$$\cdot \prod_{j=1}^{m} \mathscr{S}(x_j'' - x_j', t_j'' - t_j') \, d^4x_j' \, d^4x_j'' \, .$$

If we now use the Glauber-Sudarshan representation, we obtain for the characteristic function

$$\langle (1 + ix)^{\hat{n}(T)} \rangle = \int \Phi_{\mathcal{N}}(\{\alpha_\lambda\}) \exp\left[ix \, \overline{W}(\{\alpha_\lambda\})\right] d^2\{\alpha_\lambda\} \, , \tag{14.22}$$

where

$$\overline{W}(\{\alpha_\lambda\}) = \int_0^T \int_0^T \underbrace{\iint}_{\substack{\text{Volume} \\ \text{of detector}}} \mathscr{S}(x'' - x', t'' - t') \, V^*(x'') \, V(x') \, d^4x' \, d^4x'' \, . \tag{14.23}$$

Remember that in (14.23) as well as in (14.21) one must understand the integrals over x as also implying a sum over the polarization indices. If the analogous substitution to (14.13) is used,

$$P_{\mathcal{N}}(\overline{W}) = \int \Phi_{\mathcal{N}}(\{\alpha_\lambda'\}) \, \delta(\overline{W} - \overline{W}(\{\alpha_\lambda'\})) \, d^2\{\alpha_\lambda'\} \, , \tag{14.24}$$

we have the final form of the characteristic function

$$\langle (1 + ix)^{\hat{n}(T)} \rangle = \int_0^\infty P_{\mathcal{N}}(\overline{W}) \exp ix\overline{W} \, d\overline{W} \, . \tag{14.25}$$

The required photon-counting distribution is equal to

$$p(n, T) = \frac{1}{n!} \frac{d^n}{d(ix)^n} \langle (1 + ix)^{\hat{n}(T)} \rangle \big|_{ix=-1} = \int_0^\infty P_{\mathcal{N}}(\overline{W}) \frac{\overline{W}^n}{n!} \exp(-\overline{W}) \, d\overline{W} \tag{14.26}$$

and the factorial moments are from (14.21)

$$\left\langle \frac{\hat{n}!}{(\hat{n}-k)!} \right\rangle = \frac{d^k}{d(ix)^k} \langle (1+ix)^{\hat{n}(T)} \rangle \Big|_{ix=0} =$$

$$= \int_0^T \cdots \int_0^T \underbrace{\int \cdots \int}_{\substack{\text{Volume} \\ \text{of detector}}} \prod_{j=1}^{k} \mathcal{S}(x_j'' - x_j', t_j'' - t_j')\, \Gamma_{\mathcal{N}}^{(k,k)}(x_1', \ldots, x_k', x_k'', \ldots, x_1'') \,.$$

$$\cdot \prod_{j=1}^{k} d^4x_j'\, d^4x_j'' \,. \tag{14.27}$$

For a broad-band detector and plane quasi-monochromatic waves normally incident in the z-direction on a plane photocathode of surface S we can write $\mathcal{S}(\mathbf{x}'' - \mathbf{x}', t'' - t') = (\eta/ScT)\, \delta(x'' - x') = (\eta/ScT) \,.$ $\delta(\mathbf{x}'' - \mathbf{x}')\, \delta(t'' - t')$ and obtain

$$\overline{W} = \frac{\eta}{cST} \int_0^T dt' \int_0^{cT} dz' \int_S d^2\mathbf{x}' I(x') = \eta \int_0^T I(t')\, dt' = \eta W \,, \tag{14.28}$$

where $I(x) = \sum_{\mu} V_\mu^*(x)\, V_\mu(x) = \boldsymbol{V}^*(x) \,.\, \boldsymbol{V}(x)$ is explicitly independent of x, y and z. This leads, together with (14.26), to the photodetection equation (10.3a) derived by the semiclassical method. Under these simplifications the factorial moments (14.27) are equal to

$$\left\langle \frac{\hat{n}!}{(\hat{n}-k)!} \right\rangle = \eta^k \int_0^T \cdots \int_0^T \Gamma_{\mathcal{N}}^{(k,k)}(t_1', \ldots, t_k', t_k', \ldots, t_1') \prod_{j=1}^{k} dt_j' \tag{14.29}$$

(cf. (12.11) and (10.92)). It is a formal matter to generalize these results to multiple detectors in analogy with (10.87)–(10.95).

Finally we can conclude that there is indeed a very close similarity between the quantum and classical descriptions of the coherence properties of the electromagnetic field. If we interpret the Glauber-Sudarshan representation of the density matrix as implying the latter's existence as a Z'-generalized function (ultradistribution), there is a complete formal equivalence between the two descriptions. This may be regarded as a consequence of the fact that vacuum expectation values of normally ordered operators are zero and thus the physical vacuum gives no contribution. This says nothing about the physical equivalence, since the classical description is applicable to strong fields where energies comparable with the field quantum $\hbar\omega$ are neglected whereas the quantum description is valid in all cases having typical quantum-mechanical properties (including arbitrarily weak fields). The physical inequivalence of the quantum and the classical descriptions reflects itself in the fact that classical distributions are well-behaved non-negative functions while the function $\Phi_{\mathscr{N}}(\{\alpha_\lambda\})$ can take on negative values and can be very singular (e.g. for the Fock state $|n\rangle$ it is proportional to the nth derivative of the δ-function, which has no classical analogue; it may be much more singular since it is in general a Z'-distribution). Only in the limit of strong fields where the ordering of the field operators plays no role and $\Phi_{\mathscr{N}} \to \Phi_{\mathscr{A}}$, which is a well-behaved function, may the complete physical equivalence of the quantum and the classical descriptions be reached.

Chapter 15

STATIONARY CONDITIONS OF THE FIELD

15.1 Time invariance properties of the correlation functions

It is well known that the quantum stationary condition is expressed by

$$[\hat{H}, \hat{\varrho}] = 0 \tag{15.1}$$

which follows from (11.48a) since $\partial\hat{\varrho}/\partial t = 0$ in this case; here $\hat{H}$ is Hamiltonian of the system. We show that the earlier definition of a stationary field (Chapter 2) (the $(m + n)$th-order correlation function are invariant with respect to translations of the time origin or that they depend on $(m + n - 1)$ time differences $(t_j - t_1)$, $j = 2, \ldots, m + n$ only) is a simple consequence of (15.1) [222, 176, 181].

We can write, in the Heisenberg picture,

$$\Gamma_{\mathcal{N}}^{(m,n)}(x_1, \ldots, x_{m+n}) = \mathrm{Tr}\,\{\hat{\varrho} \prod_{j=1}^{m} \hat{A}^{(-)}(\boldsymbol{x}_j, t_j) \prod_{k=m+1}^{m+n} \hat{A}^{(+)}(\boldsymbol{x}_k, t_k)\} =$$

$$= \mathrm{Tr}\left\{\exp\left(i\frac{\hat{H}\tau}{\hbar}\right)\hat{\varrho}\exp\left(-i\frac{\hat{H}\tau}{\hbar}\right).\right.$$

$$.\exp\left(i\frac{\hat{H}\tau}{\hbar}\right)\hat{A}^{(-)}(\boldsymbol{x}_1, t_1)\exp\left(-i\frac{\hat{H}\tau}{\hbar}\right)\ldots$$

$$\left.\ldots\exp\left(i\frac{\hat{H}\tau}{\hbar}\right)\hat{A}^{(+)}(\boldsymbol{x}_{m+n}, t_{m+n})\exp\left(-i\frac{\hat{H}\tau}{\hbar}\right)\right\}, \tag{15.2}$$

where we have used the cyclic property of the trace and substituted the unit operator $\exp[-i(\hat{H}\tau/\hbar)]\exp[i(\hat{H}\tau/\hbar)]$. Using (11.46a) and assuming

(15.1) we arrive at

$$\Gamma_{\mathcal{N}}^{(m,n)}(x_1, \ldots, x_{m+n}) =$$
$$= \mathrm{Tr}\,\{\hat{\varrho} \prod_{j=1}^{m} \hat{A}^{(-)}(\boldsymbol{x}_j, t_j + \tau) \prod_{k=m+1}^{m+n} \hat{A}^{(+)}(\boldsymbol{x}_k, t_k + \tau)\} =$$
$$= \Gamma_{\mathcal{N}}^{(m,n)}(\boldsymbol{x}_1, \ldots, \boldsymbol{x}_{m+n}; t_1 + \tau, \ldots, t_{m+n} + \tau)\,. \tag{15.3}$$

Putting $\tau = -t_1$ we have

$$\Gamma_{\mathcal{N}}^{(m,n)}(x_1, \ldots, x_{m+n}) = \Gamma_{\mathcal{N}}^{(m,n)}(\boldsymbol{x}_1, \ldots, \boldsymbol{x}_{m+n}; 0, t_2 - t_1, \ldots, t_{m+n} - t_1)\,, \tag{15.4}$$

which is just the above-mentioned stationary condition. From (15.3) or (15.4) the condition (15.1) can also be obtained.

Another stationary condition which makes use of the time derivatives of $\Gamma_{\mathcal{N}}^{(m,n)}$ can be obtained as follows [206]. In the Heisenberg picture

$$i\hbar \frac{\partial \hat{A}^{(+)}(\boldsymbol{x}_l, t_l)}{\partial t_l} = [\hat{A}^{(+)}(\boldsymbol{x}_l, t_l), \hat{H}]\,. \tag{15.5}$$

Multiplying this equation by $\prod_{j=1}^{m} \hat{A}^{(-)}(\boldsymbol{x}_j, t_j) \prod_{k=m+1}^{l-1} \hat{A}^{(+)}(\boldsymbol{x}_k, t_k)$ from the left and by $\prod_{k=l+1}^{m+n} \hat{A}^{(+)}(\boldsymbol{x}_k, t_k)$ from the right and taking the average we arrive at

$$i\hbar \langle \prod_{j=1}^{m} \hat{A}^{(-)}(\boldsymbol{x}_j, t_j) \prod_{k=m+1}^{l-1} \hat{A}^{(+)}(\boldsymbol{x}_k, t_k) \frac{\partial \hat{A}^{(+)}(\boldsymbol{x}_l, t_l)}{\partial t_l} \prod_{k=l+1}^{m+n} \hat{A}^{(+)}(\boldsymbol{x}_k, t_k)\rangle =$$
$$= \langle \prod_{j=1}^{m} \hat{A}^{(-)}(\boldsymbol{x}_j, t_j) \prod_{k=m+1}^{l-1} \hat{A}^{(+)}(\boldsymbol{x}_k, t_k)\, \hat{A}^{(+)}(\boldsymbol{x}_l, t_l)\, \hat{H} \prod_{k=l+1}^{m+n} \hat{A}^{(+)}(\boldsymbol{x}_k, t_k)\rangle -$$
$$- \langle \prod_{j=1}^{m} \hat{A}^{(-)}(\boldsymbol{x}_j, t_j) \prod_{k=m+1}^{l-1} \hat{A}^{(+)}(\boldsymbol{x}_k, t_k)\, \hat{H}\, \hat{A}^{(+)}(\boldsymbol{x}_l, t_l) \prod_{k=l+1}^{m+n} \hat{A}^{(+)}(\boldsymbol{x}_k, t_k)\rangle\,. \tag{15.6}$$

Summing all these equations for $l = m + 1, \ldots, m + n$ together with the other set of the equations obtained with the help of the Hermitian conjugate to (15.5) we obtain the stationary condition (cf. (2.21) in the frequency region)

$$i\hbar \sum_{l=1}^{m+n} \frac{\partial}{\partial t_l} \Gamma_{\mathcal{N}}^{(m,n)}(x_1, \ldots, x_{m+n}) =$$
$$= \mathrm{Tr}\left\{\hat{\varrho} \prod_{j=1}^{m} \hat{A}^{(-)}(x_j) \prod_{k=m+1}^{m+n} \hat{A}^{(+)}(x_k)\, \hat{H}\right\} - \mathrm{Tr}\left\{\hat{\varrho}\hat{H} \prod_{j=1}^{m} \hat{A}^{(-)}(x_j) \prod_{k=m+1}^{m+n} \hat{A}^{(+)}(x_k)\right\} =$$
$$= \mathrm{Tr}\left\{[\hat{H}, \hat{\varrho}] \prod_{j=1}^{m} \hat{A}^{(-)}(x_j) \prod_{k=m+1}^{m+n} \hat{A}^{(+)}(x_k)\right\} = 0\,, \tag{15.7}$$

provided (15.1) holds. For $m = n = 1$ we have $\partial \Gamma_{\mathcal{N}}^{(1,1)}/\partial t_1 = -\partial \Gamma_{\mathcal{N}}^{(1,1)}/\partial t_2$ as must be since $\Gamma_{\mathcal{N}}^{(1,1)}(t_1, t_2) = \Gamma_{\mathcal{N}}^{(1,1)}(t_2 - t_1)$.

15.2 Stationary conditions in phase space

First let us consider a free one-mode field. Using the renormalized Hamiltonian (11.13), $\hat{H} = \hbar\omega \hat{a}^+ \hat{a}$, and the form (11.25) of the density matrix we have for (15.1)

$$[\hat{H}, \hat{\varrho}] = \hbar\omega \sum_{n,m} \varrho(n, m)(n - m) |n\rangle \langle m| = 0\,, \tag{15.8}$$

i.e.

$$\langle k| [\hat{H}, \hat{\varrho}] |l\rangle = \hbar\omega\, \varrho(k, l)(k - l) = 0 \tag{15.9}$$

for all k, l. Consequently $\varrho(k, l)$ must have the form

$$\varrho(k, l) = \varrho(k, k)\, \delta_{kl}\,. \tag{15.10}$$

Thus we can see from (13.119) or (13.121) and (13.120) that $\Phi_{\mathcal{N}}(\alpha)$ is independent of the phase φ of α. On the other hand, if $\Phi_{\mathcal{N}}(\alpha)$ is independent of the phase φ, then $\varrho(k, l)$ has the form (15.10) and (15.1) holds. Therefore for a one-mode free field the stationary condition (15.1) is necessary and sufficient for $\Phi_{\mathcal{N}}(\alpha)$ to be independent of the phase φ of the complex amplitude α.

For a multimode free field we again consider the renormalized Hamiltonian (11.13) and, using the multimode form (11.38) of the density matrix, obtain

$$[\hat{H}, \hat{\varrho}] = \sum_{\{n_\lambda\}} \sum_{\{m_\lambda\}} \varrho(\{n_\lambda\}, \{m_\lambda\}) [\sum_\lambda \hbar\omega_\lambda (n_\lambda - m_\lambda)] |\{n_\lambda\}\rangle \langle\{m_\lambda\}| = 0\,, \tag{15.11}$$

which leads to

$$\varrho(\{n_\lambda\}, \{m_\lambda\}) = \varrho_0(\{n_\lambda\}, \{m_\lambda\})\, \delta_{NM}\,, \tag{15.12}$$

where

$$N = \sum_\lambda \hbar\omega_\lambda n_\lambda\,, \quad M = \sum_\lambda \hbar\omega_\lambda m_\lambda \tag{15.13}$$

and ϱ_0 is equal to ϱ when $N = M$. Substituting (15.12) into the multimode form of (13.119),

$$\Phi_{\mathcal{N}}(\{\alpha_\lambda\}) = \sum_{\{n_\lambda\}} \sum_{\{m_\lambda\}} \varrho(\{n_\lambda\}, \{m_\lambda\}) \prod_\lambda \frac{\sqrt{(n_\lambda!\, m_\lambda!)}}{(n_\lambda + m_\lambda)!\, \pi r_\lambda}\,.$$
$$.\exp[r_\lambda^2 + i(m_\lambda - n_\lambda)\varphi_\lambda] \left(-\frac{\partial}{\partial r_\lambda}\right)^{n_\lambda + m_\lambda} \delta(r_\lambda)\,, \tag{15.14}$$

or (13.120),

$$\Phi_{\mathcal{N}}(\{\alpha_\lambda\}) = \exp\left[-(\zeta - 1)\sum_\lambda |\alpha_\lambda|^2\right].$$

$$\sum_{\{j_\lambda\}=0}^{\infty} \left\{ \sum_{\{\mu_\lambda\}=0}^{\infty} c_{\{j_\lambda\},\{\mu_\lambda\}} \prod_\lambda (\zeta\alpha_\lambda)^{\mu_\lambda} L_{j_\lambda}^{\mu_\lambda}(\zeta|\alpha_\lambda|^2) + \sum_{\{\mu_\lambda\}=1}^{\infty} (c.\,c.)\right\}, \quad (15.15)$$

where

$$c_{\{j_\lambda\},\{\mu_\lambda\}} = \sum_{\{s_\lambda\}=0}^{\{j_\lambda\}} \varrho(\{s_\lambda\}, \{s_\lambda + \mu_\lambda\}) \prod_\lambda \frac{\zeta^{s_\lambda+1}(-1)^{s_\lambda}}{\pi(j_\lambda + \mu_\lambda)!} \binom{j_\lambda}{s_\lambda} \left[\frac{s_\lambda!}{(s_\lambda + \mu_\lambda)!}\right]^{1/2}, \quad (15.16)$$

we can obtain the corresponding function $\Phi_{\mathcal{N}}(\{\alpha_\lambda\})$ for the stationary field. This function will in general depend upon the phases $\{\varphi_\lambda\}$ of $\{\alpha_\lambda\}$.

If the additional condition [222, 225]

$$[\hat{\varrho}, \hat{a}_\lambda^+ \hat{a}_\lambda] = 0\,, \quad (\text{all } \lambda) \quad (15.17)$$

holds, then

$$\varrho(\{n_\lambda\}, \{m_\lambda\}) = \varrho(\{n_\lambda\}, \{n_\lambda\})\, \delta_{\{n_\lambda\},\{m_\lambda\}} \quad (15.18)$$

and consequently $\Phi_{\mathcal{N}}(\{\alpha_\lambda\})$ is independent of the phases $\{\varphi_\lambda\}$. The Fock form of the density matrix (11.38) is diagonal in this case.

Analogous results for the space dependence of the correlation functions can be obtained from the condition $[\hat{\boldsymbol{P}}, \hat{\varrho}] = 0$ [134], where $\hat{\boldsymbol{P}} = \sum_\lambda \hbar \boldsymbol{k}_\lambda \hat{a}_\lambda^+ \hat{a}_\lambda$ is the total momentum of the system. This condition leads to correlation functions which are stationary in space (homogeneity and isotropy). One can see that if (15.17) holds, then $[\hat{H}, \hat{\varrho}] = [\hat{\boldsymbol{P}}, \hat{\varrho}] = 0$ and the field is stationary in time and in space as well. This can also be seen from the form of the correlation functions calculated under the assumption that $\Phi_{\mathcal{N}}(\{\alpha_\lambda\})$ is independent of $\{\varphi_\lambda\}$ (cf. Sec. 17.1, equation (17.26)). In general, if a field is stationary in time, it need not be stationary in space, so that for example the function $\Gamma_{\mathcal{N}}^{(1,1)}(\boldsymbol{x}, \boldsymbol{x}; t_2 - t_1)$ will generally depend on $\boldsymbol{x}$.

Hence, the condition (15.17) is a necessary and sufficient condition for $\Phi_{\mathcal{N}}(\{\alpha_\lambda\})$ to be independent of phases $\{\varphi_\lambda\}$; in this case the phases $\{\varphi_\lambda\}$ are distributed uniformly in the interval $(0, 2\pi)$. The condition (15.1) is a necessary but not a sufficient condition for $\Phi_{\mathcal{N}}(\{\alpha_\lambda\})$ to be independent of the phases $\{\varphi_\lambda\}$.

If $\Phi_{\mathcal{N}}(\{\alpha_\lambda\})$ is independent of $\{\varphi_\lambda\}$ (the field is stationary in time and in space), then $\mathrm{Tr}\,\{\hat{\varrho}\hat{a}_\lambda\} = \mathrm{Tr}\,\{\hat{\varrho}\hat{a}_\lambda^+\} = 0$ and so

$$\mathrm{Tr}\,\{\hat{\varrho}\, \hat{A}(x)\} = 0\,; \quad (15.19)$$

and also

$$\mathrm{Tr}\,\{\hat{\varrho}\,\hat{A}^{(-)}(x_1)\ldots\hat{A}^{(-)}(x_m)\,\hat{A}^{(+)}(x_{m+1})\ldots\hat{A}^{(+)}(x_{m+n})\} = 0\,, \quad (15.20)$$

if $m \neq n$, since the integrals over the phases (the Glauber-Sudarshan representation is used) must equal zero. However, if the field is stationary in time only, then $\Gamma_{\mathcal{N}}^{(m,n)}$ for $m \neq n$ need not be zero, since (15.12) holds in this case and so $\Phi_{\mathcal{N}}(\{\alpha_\lambda\})$ will depend in general on the phases $\{\varphi_\lambda\}$. Also the conditions (2.21) must be fulfilled in the frequency domain for the non-vanishing of $\Gamma_{\mathcal{N}}^{(m,n)}$ for $m \neq n$.

Some possibilities of measuring the correlation functions for $m \neq n$ have been discussed in Sec. 12.4. However, in usual experiments with stationary fields the correlation functions $\Gamma_{\mathcal{N}}^{(m,n)}$ with $m \neq n$ are equal to zero [303, 273].

The idea of a *quasi-stationary* field was recently introduced by Picinbono [384] (see also [533]).

Chapter 16

ORDERING OF FIELD OPERATORS IN QUANTUM OPTICS

In the preceding chapters devoted to the quantum theory of coherence, although antinormal and Weyl orderings were considered, we have dealt mainly with normal orderings and with their correspondence to *c*-number functions. In this chapter we develop a systematic approach to the correspondence between *q*-number and *c*-number functions.

The correspondence between functions of *q*-numbers (operators) and functions of *c*-numbers (classical functions) was first investigated by Wigner [468]. He introduced a generalized phase space quasi-distribution function in connection with quantum-mechanical corrections to thermodynamic equilibrium. In this formulation the problem of expressing functions of *q*-numbers according to a prescribed rule of ordering (and the closely related problem of the calculation of quantum expectation values as averages with respect to a quasi-distribution function in some generalized phase space) arose. Later Moyal [327] used the Wigner formulation as a statistical interpretation of quantum mechanics and showed that the Wigner quasi-distribution function is related to a symmetrized (Weyl) ordering of boson field operators. Among other papers devoted to this problem we mention one by Mehta [296].

The problem of the ordering of field operators in quantum optics plays an important role in connection with generalized phase space quasi-distributions of optical fields [435, 436, 173 – 176, 285, 17, 362, 301, 304, 223, 224], laser theory [251, 247, 248, 19, 193, 250, 252] and some investigations of non-linear processes such as parametric amplification [322, 323, 188, 187, 318] and non-linear optics [421]. A systematic way of treating problems of the ordering of field operators and the correspondence between functions of *q*-numbers and functions of *c*-numbers has recently been found by Agarwal and Wolf [37, 484, 40]. Another general approach to this problem was proposed by Lax [250] (see also [259]); this theory is also valid for Fermi particles. These methods were used in [38, 36] to handle quantum dynamics problems in phase space and to obtain an expression for the time-ordered product of Heisenberg operators in terms of products arranged according to a prescribed rule of ordering (Wick's theorem can be deduced as a special case) [39]. Cahill and Glauber [105, 106] have given a general treatment of the ordering of operators showing the inter-relationships between the normal, symmetric (Weyl) and antinormal orderings; they also show the correspondence between q-number and c-number functions and how quantum-mechanical expectation values may be calculated using a parametrization making it possible to interpolate between normal and antinormal orderings (a special case of the theory of Agarwal and Wolf). These methods also provide a way of seeing when and how singularities appear in quasi-probability distributions including the weight function of the Glauber-Sudarshan representation. Such a theory for multimode fields has been developed in [365, 367] and has been demonstrated for the example of a superimposition of coherent and chaotic fields in [366].

Such results have particular significance for correlation and photon-counting experiments; photoelectric detectors measure observables connected with the normally ordered products of field operators and the quantum counters operating by stimulated emission rather than by absorption, measure observables connected with the antinormal ordering of field operators [278, 359, 360]. Another relevant field is the scattering of light from fluctuations in liquids and turbulent fluids; this is described by the correlation functions of the field in a symmetrical form and these are closely connected to correlation functions of the fluctuating medium (e.g. the fluctuating density of particles [77, 79, 80, 118, 119, 133]).

16.1 Definition of Ω-ordering and the general decompositions of operators

The property (13.36) of the displacement operator $\hat{D}(\alpha)$ enables us to decompose any operator $\hat{A}$ in terms of the displacement operator $\hat{D}(\alpha)$ in the form

$$\hat{A} = \frac{1}{\pi} \int \tilde{a}(\beta)\, \hat{D}^{+}(\beta)\, d^2\beta\,, \tag{16.1}$$

where

$$\tilde{a}(\beta) = \mathrm{Tr}\,\{\hat{D}(\beta)\,\hat{A}\}\,. \tag{16.2}$$

Considering another operator $\hat{B}$ we can write

$$\mathrm{Tr}\,\{\hat{A}\hat{B}\} = \frac{1}{\pi} \int \tilde{a}(-\beta)\, \tilde{b}(\beta)\, d^2\beta \tag{16.3}$$

and for $\hat{B} \equiv \hat{A}^+$

$$\|\hat{A}\|^2 = \mathrm{Tr}\,\{\hat{A}^+\hat{A}\} = \frac{1}{\pi}\int |\tilde{a}(\beta)|^2\, d^2\beta\,. \tag{16.4}$$

Thus, if $\tilde{a}(\beta)$ is a square integrable function, then the Hilbert-Schmidt norm $\|\hat{A}\| = [\mathrm{Tr}\,\{\hat{A}^+\hat{A}\}]^{1/2}$ is finite and vice versa. In particular for the density matrix $\hat{\varrho} = \hat{\varrho}^+ \equiv \hat{A}$, $\mathrm{Tr}\,\{\hat{\varrho}^2\} = \int |C(\beta)|^2\, d^2\beta/\pi \leqq 1$, where $C(\beta)$ is the characteristic function (13.70); thus we conclude that $0 \leqq |C(\beta)| \leqq 1$ again.

If $\hat{A}(\hat{a}^+, \hat{a}) \equiv \hat{\varrho}(\hat{a}^+, \hat{a})$, then we have obtained a representation of the density matrix in terms of $\hat{D}(\beta)$. It is obvious that $\tilde{a}(\beta)$ is identical to the Wigner characteristic function $C(\beta)$ given by (13.70) and that $\varrho(\alpha^*, \alpha)/\pi$ obtained from (16.1) is just the Wigner function (13.76). Some further details about the completeness properties of the displacement operators $\hat{D}(\alpha)$ can be found in [105, 40].

We introduce the operator $\Delta(\alpha, \Omega)$ as

$$\Delta(\alpha, \Omega) = \frac{1}{\pi}\int \Omega(\beta^*, \beta)\, \hat{D}(\beta) \exp(-\beta\alpha^* + \beta^*\alpha)\, d^2\beta =$$

$$= \frac{1}{\pi}\int \Omega(\beta^*,\beta) \exp[\beta(\hat{a}^+ - \alpha^*) - \beta^*(\hat{a} - \alpha)]\, d^2\beta\,, \tag{16.5}$$

i.e. $\Delta(\alpha, \Omega)/\pi$ is equal to the Fourier transform of the operator

$$\hat{D}(\beta, \Omega) = \hat{\Omega}\, \hat{D}(\beta) = \hat{\Omega} \exp(\beta\hat{a}^+ - \beta^*\hat{a}) = \Omega(\beta^*, \beta) \exp(\beta\hat{a}^+ - \beta^*\hat{a}) \equiv$$
$$\equiv \{\hat{D}(\beta)\}_\Omega\,, \tag{16.6}$$

where $\hat{\Omega}$ is an operator arranging the expression in an Ω-ordered form denoted by $\{\,\}_\Omega$. The function $\Omega(\beta^*, \beta)$ may be called the *filter function*. Introducing (16.6) into (16.5) we can write

$$\Delta(\alpha, \Omega) = \frac{1}{\pi}\int \hat{D}(\beta, \Omega) \exp(\beta^*\alpha - \beta\alpha^*)\, d^2\beta\,. \tag{16.7}$$

We assume that the function $\Omega(\alpha^*, \alpha)$ has the following properties: $\Omega(\alpha^*, \alpha)$ is a function obtained for $\beta = \alpha^*$ from a function $\Omega(\beta, \alpha)$, which is an entire analytic function of two complex variables and $\Omega(\beta, \alpha)$ has no zeros (minimal filter), $\Omega(0, 0) = 1$ and $\Omega(-\beta, -\alpha) = \Omega(\beta, \alpha)$ (symmetric filter). By using the Weierstrass theorem for entire functions the function $\Omega(\beta, \alpha)$ has the form $\exp[\omega(\beta, \alpha)]$, where $\omega(\beta, \alpha)$ is an entire analytic function with $\omega(0, 0) = 0$. It can be seen by taking the above properties into account that

$$\omega(\beta, \alpha) = \frac{r}{2}\alpha^2 + \frac{t}{2}\beta^2 + \frac{s}{2}\alpha\beta + \text{higher-order terms in } \alpha \text{ and } \beta\,, \quad (16.8)$$

where r, t and s are parameters.

As an example we give the cases of normal, symmetric (Weyl) and antinormal orderings. Making use of the Baker-Hausdorff identity (13.3) we obtain for (16.6) identifying Ω with symmetric, normal and antinormal orderings

$$\hat{D}(\beta, \Omega) \equiv \hat{D}(\beta, s) = \exp\left(\frac{s}{2}|\beta|^2\right)\hat{D}(\beta) = \exp\left(\frac{s}{2}|\beta|^2\right)\exp(\beta\hat{a}^+ - \beta^*\hat{a}) =$$

$$= \exp\left(\frac{s-1}{2}|\beta|^2\right)\exp(\beta\hat{a}^+)\exp(-\beta^*\hat{a}) =$$

$$= \exp\left(\frac{s+1}{2}|\beta|^2\right)\exp(-\beta^*\hat{a})\exp(\beta\hat{a}^+)\,, \quad (16.9)$$

i.e. $r = t = 0$ and other higher-order terms in (16.8) vanish. Hence, $s = 0$ for symmetric ordering, $s = 1$ for normal ordering and $s = -1$ for antinormal ordering.

Obviously $\hat{D}^+(\beta, \Omega) = \hat{D}(-\beta, \Omega^*) = \hat{D}^{-1}(\beta, \Omega^{*-1}) = \hat{D}^{-1}(\beta, \tilde{\Omega}^*)$, where the $\tilde{\Omega}$-ordering is defined as the Ω-ordering with $\Omega(\beta^*, \beta) \to \Omega^{-1}(\beta^*, \beta)$. Therefore it follows that $\Delta^+(\alpha, \Omega) = \Delta(\alpha, \Omega^*)$. For the s-ordering (as a special case) $\hat{D}^+(\beta, s) = \hat{D}(-\beta, s^*) = \hat{D}^{-1}(\beta, -s^*)$ and $\Delta^+(\alpha, s) = \Delta(\alpha, s^*)$.

If we express the operators $\hat{a}$ and $\hat{a}^+$ by means of the canonical operators $\hat{p}$, $\hat{q}$ given by (11.15) we can arrange $\hat{D}(\alpha)$ is such a way that all $\hat{q}$ are to the left of all $\hat{p}$ (the standard ordering), or all $\hat{q}$ are to the right of all $\hat{p}$ (the anti-standard ordering). For the first case we obtain $r = 1/2$, $t = -1/2$, $s = 0$ and for the second case $r = -1/2$, $t = 1/2$, $s = 0$ in (16.8).

From (16.7) and (16.6) $\Delta(\alpha, \Omega) = \pi\{\delta(\hat{a} - \alpha)\}_\Omega$ (the δ-function is defined by (13.37)), i.e. the operator $\pi^{-1}\,\Delta(\alpha, \Omega)$ represents the Ω-ordered form of the operator $\delta(\hat{a} - \alpha)$.

Further we obtain

$$\mathrm{Tr}\,\{\hat{D}(\beta, \Omega)\,\hat{D}(\gamma, \tilde{\Omega})\} = \mathrm{Tr}\,\{\hat{D}(\beta, s)\,\hat{D}(\gamma, -s)\} = \mathrm{Tr}\,\{\hat{D}(\beta)\,\hat{D}(\gamma)\} =$$

$$= \pi\,\delta(\beta + \gamma)\,. \quad (16.10)$$

Thus an operator $\hat{A}$ can be decomposed in the form

$$\hat{A} = \frac{1}{\pi} \int \tilde{a}(\beta, \Omega)\, \hat{D}^{-1}(\beta, \Omega)\, d^2\beta\,, \tag{16.11}$$

which is a more general decomposition than (16.1); here

$$\tilde{a}(\beta, \Omega) = \mathrm{Tr}\,\{\hat{D}(\beta, \Omega)\, \hat{A}\}\,. \tag{16.12}$$

If $\hat{A} \equiv \hat{\varrho}$ and $\hat{D}(\beta, \Omega) \equiv \hat{D}(\beta, s)$ we regain for $s = 1$ and -1 the characteristic functions (13.72), $C^{(\mathcal{N})}(\beta) \equiv \tilde{a}(\beta, 1)$, and (13.73), $C^{(\mathcal{A})}(\beta) \equiv \equiv \tilde{a}(\beta, -1)$. Thus $\varrho(\alpha^*, \alpha)/\pi$ obtained from (16.11) by the substitutions $\hat{a} \to \alpha$ and $\hat{a}^+ \to \alpha^*$ is equal to $\Phi_{\mathcal{N}}(\alpha)$ [(13.74)] for $s = 1$ and to $\Phi_{\mathcal{A}}(\alpha)$ [(13.75)] for $s = -1$. For $s = 0$ we have (16.1) and (16.2), which corresponds to the symmetric ordering.

By using (16.7) and (16.10) we arrive at the orthogonality relation

$$\mathrm{Tr}\,\{\Delta(\beta, \Omega)\, \Delta(\gamma, \tilde{\Omega})\} = \pi\, \delta(\beta - \gamma)\,. \tag{16.13}$$

Other interesting properties of the operator $\Delta(\beta, \Omega)$ can also be given: thus from (16.7)

$$\int \Delta(\beta, \Omega)\, d^2\beta = \pi\, \hat{D}(0, \Omega) = \pi \tag{16.14}$$

and

$$\mathrm{Tr}\,\{\Delta(\beta, \Omega)\} = 1\,, \tag{16.15}$$

where (16.5) and (13.35) have been used. Making use of (16.7) and (16.9) for $s = -1$ we arrive at

$$\Delta(\beta, \mathcal{A}) \equiv \Delta(\beta, -1) =$$

$$= \frac{1}{\pi} \int \exp(-\gamma^* \hat{a}) \left(\frac{1}{\pi} \int |\alpha\rangle \langle\alpha|\, d^2\alpha \right) \exp(\gamma \hat{a}^+) \exp(\gamma^* \beta - \gamma \beta^*)\, d^2\gamma =$$

$$= \frac{1}{\pi^2} \int |\alpha\rangle \langle\alpha|\, d^2\alpha \int \exp[\gamma(\alpha^* - \beta^*) - \gamma^*(\alpha - \beta)]\, d^2\gamma = |\beta\rangle \langle\beta|\,, \tag{16.16}$$

i.e. $\Delta(\beta, -1)$ is the projection operator onto the coherent state $|\beta\rangle$.

Making use of the completeness property (16.13) of the Δ-operator we can also decompose any operator $\hat{A}$ in terms of Δ

$$\hat{A} = \frac{1}{\pi} \int a(\beta, \tilde{\Omega})\, \Delta(\beta, \Omega)\, d^2\beta\,, \tag{16.17}$$

where

$$a(\beta, \tilde{\Omega}) = \mathrm{Tr}\,\{\Delta(\beta, \tilde{\Omega})\,\hat{A}\}\,. \tag{16.18}$$

As $\Delta(\beta, \Omega) = \pi\{\delta(\hat{a} - \beta)\}_\Omega$, equation (16.17) provides the Ω-ordered form of the operator $\hat{A}$. Using (16.13) again we have for the trace of the product of two operators $\hat{A}$ and $\hat{B}$

$$\mathrm{Tr}\,\{\hat{A}\hat{B}\} = \frac{1}{\pi^2}\iint a(\beta, \Omega)\, b(\gamma, \tilde{\Omega})\, \mathrm{Tr}\,\{\Delta(\beta, \tilde{\Omega})\, \Delta(\gamma, \Omega)\}\, d^2\beta\, d^2\gamma =$$

$$= \frac{1}{\pi}\int a(\beta, \Omega)\, b(\beta, \tilde{\Omega})\, d^2\beta\,. \tag{16.19}$$

If $\hat{B}$ is identified with the density matrix $\hat{\varrho}$ we can write

$$\mathrm{Tr}\,\{\hat{\varrho}\hat{A}\} = \frac{1}{\pi}\int \Phi(\beta, \tilde{\Omega})\, a(\beta, \Omega)\, d^2\beta\,, \tag{16.20a}$$

where

$$\Phi(\beta, \tilde{\Omega}) = \mathrm{Tr}\,\{\hat{\varrho}\, \Delta(\beta, \tilde{\Omega})\} = \pi\langle\{\delta(\hat{a} - \beta)\}_{\tilde{\Omega}}\rangle \tag{16.21a}$$

and from (16.17)

$$\hat{\varrho} = \frac{1}{\pi}\int \Phi(\beta, \tilde{\Omega})\, \Delta(\beta, \Omega)\, d^2\beta\,. \tag{16.22a}$$

Considering the s-ordering as a special case of the Ω-ordering we then have

$$\mathrm{Tr}\,\{\hat{\varrho}\hat{A}\} = \frac{1}{\pi}\int \Phi(\beta, -s)\, a(\beta, s)\, d^2\beta\,, \tag{16.20b}$$

where

$$\Phi(\beta, -s) = \mathrm{Tr}\,\{\hat{\varrho}\, \Delta(\beta, -s)\} \tag{16.21b}$$

and

$$\hat{\varrho} = \frac{1}{\pi}\int \Phi(\beta, -s)\, \Delta(\beta, s)\, d^2\beta\,. \tag{16.22b}$$

Hence, the expectation value of an operator $\hat{A}$ can be calculated as a classical expectation value of the classical function $a(\beta, \Omega)$ corresponding to $\hat{A}$ via the $\tilde{\Omega}$-ordering with the quasi-probability $\Phi(\beta, \tilde{\Omega})$ corresponding to $\hat{\varrho}$ via the Ω-ordering as follows from (16.17) and (16.22a). This agrees with our special results discussed in section 13.2.

Putting $s = -1$ in (16.22b) and making use of (16.16) we obtain the

Glauber-Sudarshan representation of the density matrix

$$\hat{\varrho} = \int \Phi_{\mathcal{N}}(\beta) \, |\beta\rangle \langle\beta| \, d^2\beta \,, \tag{16.23}$$

where $\Phi_{\mathcal{N}}(\beta) \equiv \Phi(\beta, 1)/\pi$.

If $\hat{A} \equiv 1$, then $a(\beta, \Omega) = 1$ from (16.18) with the use of (16.15), and equation (16.20a) gives us the normalization condition

$$\mathrm{Tr}\,\{\hat{\varrho}\} = 1 = \frac{1}{\pi} \int \Phi(\beta, \tilde{\Omega}) \, d^2\beta \tag{16.24}$$

and $\hat{\varrho} = \hat{\varrho}^+$ implies $\Phi^*(\beta, \tilde{\Omega}) = \Phi(\beta, \tilde{\Omega}^*)$ $[\Phi^*(\beta, s) = \Phi(\beta, s^*)]$.

Writing (16.21a) in the form

$$\Phi(\alpha, \tilde{\Omega}) = \frac{1}{\pi} \int \mathrm{Tr}\,\{\hat{\varrho}\, \hat{D}(\beta, \tilde{\Omega})\} \exp\,(\beta^*\alpha - \beta\alpha^*) \, d^2\beta \,, \tag{16.25}$$

where we have used (16.7), it is obvious that $\mathrm{Tr}\,\{\hat{\varrho}\, \hat{D}(\beta, \tilde{\Omega})\}$ is the characteristic function $C(\beta, \tilde{\Omega})$ of the quasi-distribution $\Phi(\alpha, \tilde{\Omega})$ and so

$$\Phi(\alpha, \tilde{\Omega}) = \frac{1}{\pi} \int C(\beta, \tilde{\Omega}) \exp\,(\beta^*\alpha - \beta\alpha^*) \, d^2\beta \,, \tag{16.26a}$$

$$C(\beta, \tilde{\Omega}) = \frac{1}{\pi} \int \Phi(\alpha, \tilde{\Omega}) \exp\,(-\beta^*\alpha + \beta\alpha^*) \, d^2\alpha \,. \tag{16.26b}$$

The characteristic function $C(\beta, \tilde{\Omega})$ is bounded,

$$|C(\beta, \tilde{\Omega})| = |\tilde{\Omega}| \, |\mathrm{Tr}\,\{\hat{\varrho}\, \hat{D}(\beta)\}| \leq |\tilde{\Omega}| \,.$$

The Ω-ordered moments of $\Phi(\alpha, \Omega)$ can be calculated as follows

$$\frac{\partial^{k+l}}{\partial\beta^k \, \partial(-\beta^*)^l} C(\beta, \Omega)\bigg|_{\beta=\beta^*=0} = \mathrm{Tr}\left\{\hat{\varrho} \frac{\partial^{k+l}}{\partial\beta^k \, \partial(-\beta^*)^l} \hat{D}(\beta, \Omega)\right\}\bigg|_{\beta=\beta^*=0} =$$

$$= \mathrm{Tr}\,\{\hat{\varrho}\{\hat{a}^{+k}\hat{a}^l\}_\Omega\} = \frac{1}{\pi} \int \Phi(\alpha, \Omega) \, \alpha^{*k}\alpha^l \, d^2\alpha = \langle\alpha^{*k}\alpha^l\rangle_\Omega \,, \tag{16.27}$$

where we have used (16.6) and (16.26b). If the inverse relation to (16.7) is used we also obtain

$$\{\hat{a}^{+k}\hat{a}^l\}_\Omega = \frac{\partial^{k+l}}{\partial\beta^k \, \partial(-\beta^*)^l} \hat{D}(\beta, \Omega)\bigg|_{\beta=\beta^*=0} =$$

$$= \frac{\partial^{k+l}}{\partial\beta^k\,\partial(-\beta^*)^l}\frac{1}{\pi}\int\Delta(\alpha,\Omega)\exp(-\beta^*\alpha+\beta\alpha^*)\,d^2\alpha\bigg|_{\beta=\beta^*=0} =$$

$$= \frac{1}{\pi}\int\alpha^{*k}\alpha^l\,\Delta(\alpha,\Omega)\,d^2\alpha\,. \tag{16.28}$$

Considering these equations for the s-ordering specified by $\Omega(\alpha^*, \alpha) = \exp(s|\alpha|^2/2)$ we obtain from (16.26b) [(16.26a)] equations (13.72), (13.73) and (13.78) [(13.74), (13.75) and (13.76)] and from (16.27) equations (13.80), (13.81) and (13.77) for $s = 1, -1$ and 0 respectively with the replacements $\Phi(\alpha, 1)/\pi \equiv \Phi_{\mathcal{N}}(\alpha)$, $\Phi(\alpha, -1)/\pi \equiv \Phi_{\mathcal{A}}(\alpha)$, $\Phi(\alpha, 0)/\pi \equiv \Phi_{\mathrm{Weyl}}(\alpha)$, $C(\beta, 1) \equiv C^{(\mathcal{N})}(\beta)$, $C(\beta, -1) \equiv C^{(\mathcal{A})}(\beta)$, and $C(\beta, 0) \equiv C(\beta)$.

The present general scheme of the correspondence between functions of q-numbers and functions of c-numbers can provide an insight into the problem of the singularities appearing in the quasi-distributions. Consider the s-ordering only. We can decompose the operator $\hat{A}$ in (16.11) into an s-ordered series by expanding the operator $\hat{D}^{-1}$

$$\hat{A} = \frac{1}{\pi}\int\mathrm{Tr}\,\{\hat{D}(\beta, -s)\,\hat{A}\}\exp\left(\frac{s}{2}|\beta|^2\right)\exp(-\beta\hat{a}^+ + \beta^*\hat{a})\,d^2\beta =$$

$$= \sum_n\sum_m\{\hat{a}^{+n}\hat{a}^m\}_s\,A_{nm}\,, \tag{16.29}$$

where we have used (16.9), substituted $s \to -s$ and where

$$A_{nm} = \frac{1}{\pi n!\,m!}\int\mathrm{Tr}\,\{\hat{D}(\beta, -s)\,\hat{A}\}\,(-\beta)^n\,(\beta^*)^m\,d^2\beta =$$

$$= \frac{1}{\pi n!\,m!}\int\mathrm{Tr}\,\{\hat{D}(\beta)\,\hat{A}\}\exp\left(-\frac{s}{2}|\beta|^2\right)(-\beta)^n\,(\beta^*)^m\,d^2\beta\,. \tag{16.30}$$

By applying the Schwarz inequality to (16.30),

$$\left|\frac{1}{\pi}\int\tilde{a}(\beta)\,\tilde{b}(\beta)\,d^2\beta\right| \leq \left\{\frac{1}{\pi}\int|\tilde{a}(\beta)|^2 d^2\beta\right\}^{1/2}\left\{\frac{1}{\pi}\int|\tilde{b}(\beta)|^2\,d^2\beta\right\}^{1/2}, \tag{16.31}$$

where (16.2) is used and $\tilde{b}(\beta) = \exp(-s|\beta|^2/2)\,(-\beta)^n\,(\beta^*)^m$, we see that, as a consequence of the presence of the factor $\exp(-s|\beta|^2/2)$, the coefficients (16.30) are finite for all bounded operators $\hat{A}$ ($\|\hat{A}\|^2 = \mathrm{Tr}\,\{\hat{A}^+\hat{A}\} < \infty$, cf. (16.4)) when $\mathrm{Re}\,s > 0$. Therefore it can be said that these coefficients are finite for orderings which are closer to the normal than to the antinormal order. A further investigation of the convergence of the series (16.29) [105] leads to the result that all bounded operators possess convergent s-ordered power series for $\text{Re}\, s > 1/2$, i.e. when the ordering is closer to normal than to symmetric. Note that in a $(\hat{q}, \hat{p})$-representation the exponential factor $\exp(i\hat{q}\hat{p}/2)$ replaces the factor $\exp(-s|\beta|^2/2)$, as a consequence of the fact that the commutator $[\hat{q}, \hat{p}] = i\hbar$ is purely imaginary. Thus the integrals for various orderings differ from one another only by unimodular factors in their integrands and a bounded operator that possesses an expansion in one ordering is likely to possess it in another. Such a class of operators is approximately the same as the class for which the symmetrically ordered expansion (16.29) with $s = 0$ is appropriate. For some further details concerning the $(\hat{q}, \hat{p})$-representation in this connection we refer the reader to the papers [105, 40]. In general it is seen from (16.30) that the broadest class of operators possess normally ordered expansions while a relatively smaller class of operators possess antinormally ordered expansions. According to our earlier results this means that, for the density matrix $\hat{\varrho} \equiv \hat{A}$, the class of physical fields which possess the diagonal representation of the density matrix with the weight function $\Phi_{\mathscr{N}}(\alpha) = \pi^{-1}\,\varrho^{(\mathscr{A})}(\alpha^*, \alpha)$ as an ordinary function is smaller than the class of fields for which the function $\Phi_{\mathscr{A}}(\alpha) = \pi^{-1}\,\varrho^{(\mathscr{N})}(\alpha^*, \alpha)$ is an ordinary function (in fact for all physical fields $\Phi_{\mathscr{A}}(\alpha)$ is an ordinary function).

A characteristic feature of the expansion (16.22b) is that both the numbers s and $-s$ appear in it. We have seen that $\Delta(\alpha, -1) = |\alpha\rangle\langle\alpha|$, while one can easily verify by a similar calculation such as in (16.16) that the operator $\Delta(\alpha, 1)$ (which can also be used for decompositions of the density matrix with the weight function $\Phi_{\mathscr{A}}(\alpha)$) is singular. In the first case, when $\Delta(\alpha, -1)$ is regular, $\Phi_{\mathscr{N}}(\alpha)$ can have singularities, in the second case, when $\Phi_{\mathscr{A}}(\alpha)$ is regular, $\Delta(\alpha, 1)$ has singularities. Thus extreme smoothness of one quantity leads to singular behaviour of the other. Only for the symmetric ordering for which $s = 0$, are both the quantities $\Delta(\alpha, 0)$ and $\Phi_{\text{Weyl}}(\alpha) = \pi^{-1}\,\Phi(\alpha, 0)$ regular (although the operator $\Delta(\alpha, 0)$ is not bounded; it is finite in the sense $\operatorname{Max}_{\beta} \langle\beta|\,\Delta(\alpha, 0)\,|\beta\rangle$ is finite for β finite).

16.2 Connecting relations for different orderings

If we consider two orderings Ω_1 and Ω_2 we can write on the basis of (16.5)

$$\Delta(\alpha, \Omega_1) = \frac{1}{\pi}\int \Omega_1(\beta^*, \beta)\,\hat{D}(\beta)\exp(-\beta\alpha^* + \beta^*\alpha)\,d^2\beta\,, \quad (16.32a)$$

$$\Delta(\alpha, \Omega_2) = \frac{1}{\pi}\int \Omega_2(\beta^*, \beta)\,\hat{D}(\beta)\exp(-\beta\alpha^* + \beta^*\alpha)\,d^2\beta\,. \quad (16.32b)$$

Expressing the operator $\hat{D}(\beta)$ by means of the inverse relation to (16.32a) and substituting it into (16.32b) we arrive at

$$\Delta(\alpha, \Omega_2) = \int \Delta(\beta, \Omega_1) K_{21}(\alpha - \beta) \, d^2\beta \,, \tag{16.33}$$

where

$$K_{21}(\beta) = \frac{1}{\pi^2} \int \tilde{\Omega}_1(\gamma^*, \gamma) \, \Omega_2(\gamma^*, \gamma) \exp(-\gamma\beta^* + \gamma^*\beta) \, d^2\gamma \,. \tag{16.34}$$

Multiplying (16.33) by the density operator $\hat{\varrho}$, taking the trace and making use of (16.21a) with $\tilde{\Omega} \to \Omega$ we obtain

$$\Phi(\alpha, \Omega_2) = \int \Phi(\beta, \Omega_1) K_{21}(\alpha - \beta) \, d^2\beta \,. \tag{16.35}$$

The corresponding relation between the characteristic functions $C(\alpha, \Omega_1)$ and $C(\alpha, \Omega_2)$ is obviously

$$C(\alpha, \Omega_2) = C(\alpha, \Omega_1) \, \tilde{\Omega}_1(\alpha^*, \alpha) \, \Omega_2(\alpha^*, \alpha) \,. \tag{16.36}$$

Some differential relations between different orderings can also be derived. Writing

$$\Delta(\alpha, \Omega_2) = \frac{1}{\pi} \int \tilde{\Omega}_1(\beta^*, \beta) \, \Omega_2(\beta^*, \beta) \, \Omega_1(\beta^*, \beta) \, \hat{D}(\beta) \exp(-\beta\alpha^* + \beta^*\alpha) \, d^2\beta =$$

$$= \mathscr{L}_{21}\left(\frac{\partial}{\partial \alpha}, -\frac{\partial}{\partial \alpha^*}\right) \Delta(\alpha, \Omega_1) \,, \tag{16.37}$$

where

$$\mathscr{L}_{21}(\gamma^*, \gamma) = \tilde{\Omega}_1(\gamma^*, \gamma) \, \Omega_2(\gamma^*, \gamma) \,, \tag{16.38}$$

we have

$$\Phi(\alpha, \Omega_2) = \mathscr{L}_{21}\left(\frac{\partial}{\partial \alpha}, -\frac{\partial}{\partial \alpha^*}\right) \Phi(\alpha, \Omega_1) \,. \tag{16.39}$$

These general relations can be used for the s-ordering and as a special case relations between normal, symmetric and antinormal orderings can be derived. Since $\Omega(\alpha^*, \alpha) = \exp(s|\alpha|^2/2)$ in this case we obtain from (16.36)

$$C(\alpha, s_2) = C(\alpha, s_1) \exp\left[\tfrac{1}{2}(s_2 - s_1) |\alpha|^2\right], \tag{16.40a}$$

$$\hat{D}(\alpha, s_2) = \hat{D}(\alpha, s_1) \exp\left[\tfrac{1}{2}(s_2 - s_1) |\alpha|^2\right] \tag{16.40b}$$

(recalling that $C(\alpha, s) = \text{Tr}\{\hat{\varrho}\, \hat{D}(\alpha, s)\}$). Calculating $K_{21}(\beta)$ from (16.34),

$$K_{21}(\beta) = \frac{1}{\pi^2} \int \exp\left[\tfrac{1}{2}(s_2 - s_1)|\gamma|^2\right] \exp(-\gamma\beta^* + \gamma^*\beta)\, d^2\gamma$$

and using the integral

$$\int \exp(-s|\gamma|^2 + \alpha\gamma^* + \alpha'\gamma)\, d^2\gamma = \sum_n \sum_m \frac{\alpha^n \alpha'^m}{n!\, m!} \int \exp(-s|\gamma|^2)\, \gamma^{*n}\gamma^m\, d^2\gamma =$$

$$= \frac{\pi}{s} \exp\left(\frac{\alpha\alpha'}{s}\right), \quad \text{Re}\, s > 0 \tag{16.41}$$

(where we have used the integral (13.17) with $s = (s_1 - s_2)/2$, $\alpha = \beta$ and $\alpha' = -\beta^*$) we obtain

$$K_{21}(\beta) = \frac{2}{\pi(s_1 - s_2)} \exp\left[-\frac{2|\beta|^2}{s_1 - s_2}\right], \quad \text{Re}\, s_1 > \text{Re}\, s_2\,. \tag{16.42}$$

Thus from (16.35)

$$\Phi(\alpha, s_2) = \frac{2}{\pi(s_1 - s_2)} \int \Phi(\beta, s_1) \exp\left[-\frac{2|\alpha - \beta|^2}{s_1 - s_2}\right] d^2\beta\,, \quad \text{Re}\, s_1 > \text{Re}\, s_2\,. \tag{16.43}$$

As $\Phi(\alpha, s) = \text{Tr}\{\hat{\varrho}\, \Delta(\alpha, s)\}$ the same relation holds between $\Delta(\alpha, s_2)$ and $\Delta(\alpha, s_1)$ (this follows from (16.33)). Relations (13.85a, b) and (13.68) follow from (16.43) with $s_1 = 0$, $s_2 = -1$, $s_1 = 1$, $s_2 = 0$ and $s_1 = 1$, $s_2 = -1$ respectively. It can be seen from the form of (16.43) that this Gaussian convolution tends to smooth out any unruly behaviour of the function $\Phi(\beta, s_1)$. For example for the coherent state and $s_1 = 1$, $\pi^{-1}\Phi(\beta, 1) \equiv \Phi_{\mathcal{N}}(\beta) = \delta(\beta - \gamma)$ and $\Phi(\alpha, s) = [2/(1 - s)] \exp[-2|\alpha - \gamma|^2/(1 - s)]$, which is a regular function for $s \neq 1$. For $s = -1$, $\pi^{-1}\Phi(\alpha, -1) \equiv \Phi_{\mathcal{A}}(\alpha) = \pi^{-1} \exp(-|\alpha - \gamma|^2)$, for $s = 0$, $\pi^{-1}\Phi(\alpha, 0) \equiv \Phi_{\text{Weyl}}(\alpha) = 2\pi^{-1}$. $\exp(-2|\alpha - \gamma|^2)$.

The relation (16.43) can in principle be inverted (generally in a space of generalized functions) as a convolution integral or, in some cases, in terms of a series of Hermite polynomials H_n [367]

$$\Phi(\beta, s_1) = \sum_{n,m} \frac{1}{n!\, m!} \left(\tfrac{1}{2}\right)^{n+m} \left(\frac{s_1 - s_2}{2}\right)^{(n+m)/2} .$$

$$. H_n\left(\left[\frac{2}{s_1 - s_2}\right]^{1/2} \beta_1\right) H_m\left(\left[\frac{2}{s_1 - s_2}\right]^{1/2} \beta_2\right) \frac{\partial^n}{\partial \alpha_1^n} \frac{\partial^m}{\partial \alpha_2^m} \Phi(\alpha, s_2)\Big|_{\alpha=0},$$

$$\text{Re}\, s_1 > \text{Re}\, s_2\,, \tag{16.44}$$

where $\alpha = \alpha_1 + i\alpha_2$, $\beta = \beta_1 + i\beta_2$ ($\alpha_1, \alpha_2, \beta_1, \beta_2$ are real) under the assumption that

$$\int |\Phi(\beta, s_1)|^2 \exp\left(-\frac{2|\beta|^2}{s_1 - s_2}\right) d^2\beta = \pi \frac{s_1 - s_2}{2} \sum_{n,m} 2^{n+m} n!\, m!\, |c_{nm}|^2 < \infty\,; \tag{16.45}$$

here

$$c_{nm} = \frac{1}{n!\, m!} \left(\tfrac{1}{2}\right)^{n+m} \left(\frac{s_1 - s_2}{2}\right)^{(n+m)/2} \frac{\partial^n}{\partial \alpha_1^n} \frac{\partial^m}{\partial \alpha_2^m} \Phi(\alpha, s_2)\bigg|_{\alpha=0}. \tag{16.46}$$

More generally, making use of (16.39) it follows that

$$\Phi(\alpha, s_1) = \exp\left(-\frac{s_1 - s_2}{2} \frac{\partial^2}{\partial \alpha\, \partial \alpha^*}\right) \Phi(\alpha, s_2) \tag{16.47}$$

and, as special cases,

$$\Phi_{\text{Weyl}}(\alpha) = \exp\left(-\frac{1}{2} \frac{\partial^2}{\partial \alpha\, \partial \alpha^*}\right) \Phi_{\mathscr{A}}(\alpha) \tag{16.48a}$$

for $s_1 = 0$, $s_2 = -1$,

$$\Phi_{\mathscr{N}}(\alpha) = \exp\left(-\frac{1}{2} \frac{\partial^2}{\partial \alpha\, \partial \alpha^*}\right) \Phi_{\text{Weyl}}(\alpha) \tag{16.48b}$$

for $s_1 = 1$, $s_2 = 0$, and

$$\Phi_{\mathscr{N}}(\alpha) = \exp\left(-\frac{\partial^2}{\partial \alpha\, \partial \alpha^*}\right) \Phi_{\mathscr{A}}(\alpha) \tag{16.48c}$$

for $s_1 = 1$, $s_2 = -1$; these are just the inverse relations to (13.85a, b) and (13.68).

We can also derive the corresponding relations between different s-ordered moments $\langle\{\hat{a}^{+k}\hat{a}^l\}\rangle_s \equiv \langle\hat{a}^{+k}\hat{a}^l\rangle_s$ and obtain from (16.43) for $\operatorname{Re} s_1 > \operatorname{Re} s_2$

$$\langle\hat{a}^{+k}\hat{a}^l\rangle_{s_2} = \langle\alpha^{*k}\alpha^l\rangle_{s_2} =$$

$$= \frac{1}{\pi} \int \Phi(\beta, s_1)\, d^2\beta\, \frac{2}{\pi(s_1 - s_2)} \int \alpha^{*k}\alpha^l \exp\left(-\frac{2|\alpha - \beta|^2}{s_1 - s_2}\right) d^2\alpha =$$

$$= \frac{1}{\pi} \int \Phi(\beta, s_1)\, d^2\beta\, \frac{2}{\pi(s_1 - s_2)} \int (\gamma^* + \beta^*)^k (\gamma + \beta)^l \exp\left(-\frac{2|\gamma|^2}{s_1 - s_2}\right) d^2\gamma\,, \tag{16.49}$$

where the substitution $\alpha - \beta = \gamma$ has been used. Making use of the binomial theorem we obtain

$$\langle \hat{a}^{+k}\hat{a}^l \rangle_{s_2} = \frac{1}{\pi}\int \Phi(\beta, s_1)\, d^2\beta \sum_{m=0}^{k}\sum_{n=0}^{l} \frac{k!\, l!\, \beta^{*m}\beta^n}{m!\, n!\, (k-m)!\, (l-n)!}\, \cdot$$
$$\cdot \frac{2}{\pi(s_1 - s_2)} \int_0^\infty \exp\left(-\frac{2|\gamma|^2}{s_1 - s_2}\right) |\gamma|^{k+l-m-n+1}\, d|\gamma|\, \cdot$$
$$\cdot \int_0^{2\pi} \exp\left[i\vartheta(l - n - k + m)\right] d\vartheta\,, \qquad (16.50)$$

where $\vartheta = \arg\gamma$. The calculation of the integrals leads at once to the result

$$\langle \hat{a}^{+k}\hat{a}^l \rangle_{s_2} = \frac{k!}{l!}\left(\frac{s_1 - s_2}{2}\right)^k \left\langle \beta^{l-k} L_k^{l-k}\left(\frac{2|\beta|^2}{s_2 - s_1}\right)\right\rangle_{s_1} =$$
$$= \frac{k!}{l!}\left(\frac{s_1 - s_2}{2}\right)^k \left\langle \hat{a}^{l-k} L_k^{l-k}\left(\frac{2\hat{a}^+\hat{a}}{s_2 - s_1}\right)\right\rangle_{s_1}, \quad l \geqq k\,, \qquad (16.51a)$$
$$= \frac{l!}{k!}\left(\frac{s_1 - s_2}{2}\right)^l \left\langle \hat{a}^{+k-l} L_l^{k-l}\left(\frac{2\hat{a}^+\hat{a}}{s_2 - s_1}\right)\right\rangle_{s_1}, \quad l \leqq k\,; \qquad (16.51b)$$

the moments calculated with the formally inverted relation (16.43) leads to (16.51) again with $s_1 \rightleftarrows s_2$ so that (16.51) holds generally. The same relation can be derived with the help of (16.27). First

$$\hat{D}(\beta, s_2) = \hat{D}(\beta, s_1) \exp\left[\tfrac{1}{2}(s_2 - s_1)|\beta|^2\right] =$$
$$= \{\exp(\beta\hat{a}^+ - \beta^*\hat{a} + \tfrac{1}{2}(s_2 - s_1)|\beta|^2\}_{s_1} =$$
$$= \left\{\sum_{m=0}^{\infty} \frac{\beta^{*m}}{m!}\left(-\hat{a} + \tfrac{1}{2}(s_2 - s_1)\beta\right)^m \exp(\beta\hat{a}^+)\right\}_{s_1} = \qquad (16.52a)$$
$$= \left\{\sum_{n=0}^{\infty} \frac{\beta^n}{n!}\left(\hat{a}^+ + \tfrac{1}{2}(s_2 - s_1)\beta^*\right)^n \exp(-\beta^*\hat{a})\right\}_{s_1}. \qquad (16.52b)$$

Making use of the identity ([22], Chapter 6)

$$(1 + t)^\mu \exp(-xt) = \sum_{n=0}^{\infty} \frac{t^n}{\Gamma(\mu + 1)} L_n^{\mu-n}(x)\,, \qquad (16.53)$$

which leads to

$$\hat{D}(\beta, s_2) = \left\{\sum_{m=0}^{\infty}\sum_{n=0}^{\infty} (-1)^m \frac{\beta^{*m}}{m!}\frac{\hat{a}^{m-n}}{m!}\left[\tfrac{1}{2}(s_1 - s_2)\beta\right]^n L_n^{m-n}\left(\frac{2\hat{a}^+\hat{a}}{s_2 - s_1}\right)\right\}_{s_1} \qquad (16.54a)$$

$$= \left\{ \sum_{n=0}^{\infty} \sum_{m=0}^{\infty} (-1)^m \frac{\beta^n}{n!} \frac{\hat{a}^{+n-m}}{n!} \left[\tfrac{1}{2}(s_1 - s_2)\beta^*\right]^m L_m^{n-m}\left(\frac{2\hat{a}^+\hat{a}}{s_2 - s_1}\right) \right\}_{s_1}, \tag{16.54b}$$

and applying (16.27) we arrive at (16.51) again.

The relations between moments corresponding to normal, symmetric and antinormal orderings can be obtained from (16.51) for $s = 1, 0$ and -1. For example when $s_2 = 1$ and $s_1 = -1$ we have

$$\langle \hat{a}^{+k}\hat{a}^l \rangle_{\mathcal{N}} = \frac{k!}{l!} (-1)^k \langle \hat{a}^{l-k} L_k^{l-k}(\hat{a}^+\hat{a}) \rangle_{\mathcal{A}}, \qquad l \geqq k, \tag{16.55}$$

when $s_2 = -1$ and $s_1 = 1$

$$\langle \hat{a}^{+k}\hat{a}^l \rangle_{\mathcal{A}} = \frac{k!}{l!} \langle \hat{a}^{l-k} L_k^{l-k}(-\hat{a}^+\hat{a}) \rangle_{\mathcal{N}}, \qquad l \geqq k, \tag{16.56}$$

and when $s_2 = 0$ and $s_1 = 1$

$$\langle \hat{a}^{+k}\hat{a}^l \rangle_{\mathrm{Weyl}} = \frac{k!}{l!} (\tfrac{1}{2})^k \langle \hat{a}^{l-k} L_k^{l-k}(-2\hat{a}^+\hat{a}) \rangle_{\mathcal{N}}, \quad l \geqq k, \tag{16.57}$$

etc.

Some further results concerning the Ω-ordering and s-ordering can be found in papers (previously quoted) by Agarwal and Wolf [40] and Cahill and Glauber [105, 106].

16.3 Multimode description

The general theory just developed may easily be extended to systems with any number of degrees of freedom (finite or countably infinite) in the same way as in Chapter 13. For example a multimode operator $\hat{A}(\{\hat{a}_\lambda\}) \equiv \hat{A}(\{\hat{a}_\lambda^+\}, \{\hat{a}_\lambda\})$ can be decomposed in analogy with (16.17) and (16.18), as

$$\hat{A}(\{\hat{a}_\lambda\}) = \int a(\{\beta_\lambda\}, \tilde{\Omega})\, \Delta(\{\beta_\lambda\}, \Omega) \prod_\lambda \frac{d^2\beta_\lambda}{\pi}, \tag{16.58}$$

where

$$a(\{\beta_\lambda\}, \tilde{\Omega}) = \mathrm{Tr}\, \{\Delta(\{\beta_\lambda\}, \tilde{\Omega})\, \hat{A}\} \tag{16.59}$$

and

$$\Delta(\{\alpha_\lambda\}, \Omega) = \{\prod_\lambda \pi\delta(\hat{a}_\lambda - \alpha_\lambda)\}_\Omega =$$

$$= \int \Omega(\{\beta_\lambda^*\}, \{\beta_\lambda\}) \prod_\lambda \exp\left[\beta_\lambda(\hat{a}_\lambda^+ - \alpha_\lambda^*) - \beta_\lambda^*(\hat{a}_\lambda - \alpha_\lambda)\right] \frac{d^2\beta_\lambda}{\pi} \tag{16.60}$$

and the counterpart to (16.20a) is

$$\mathrm{Tr}\,\{\hat{\varrho}\hat{A}\} = \int \Phi(\{\beta_\lambda\}, \tilde{\Omega})\, a(\{\beta_\lambda\}, \Omega) \prod_\lambda \frac{d^2\beta_\lambda}{\pi}\,, \tag{16.61}$$

where Φ is given by (16.59) with $\hat{A}$ substituted by $\hat{\varrho}$. The density operator $\hat{\varrho}$ can be decomposed in the form (16.58) with $a(\{\beta_\lambda\}, \tilde{\Omega})$ replaced by $\Phi(\{\beta_\lambda\}, \tilde{\Omega})$, etc.

Next we shall develop another generalization of the one-mode results [365, 367] in a manner consistent with description of the field in terms of the quantities $\hat{\mathbf{A}}(x)$ and $\hat{n}_{Vt}$ defined by (11.23) and (11.24).

First we need to derive the s-ordered form of the operator $\exp(ix\hat{a}^+\hat{a})$ by substituting this operator into (16.30) giving

$$A_{nm} = \frac{1}{\pi n!\, m!} \int \mathrm{Tr}\,\{\exp(\beta\hat{a}^+ - \beta^*\hat{a}) \exp(ix\hat{a}^+\hat{a})\} \exp\left(-\frac{s}{2}|\beta|^2\right).$$

$$.\,(-\beta)^n \beta^{*m}\, d^2\beta = \frac{1}{\pi^2 n!\, m!} \iint \langle\alpha|\exp(\beta\hat{a}^+)\exp(ix\hat{a}^+\hat{a})\exp(-\beta^*\hat{a})|\alpha\rangle\,.$$

$$.\,\exp\left(-\frac{s-1}{2}|\beta|^2\right)(-\beta)^n \beta^{*m}\, d^2\beta\, d^2\alpha\,, \tag{16.62}$$

where we have used (16.9) and the cyclic property of the trace. If we use the normal form of the operator $\exp(ix\hat{a}^+\hat{a})$ which is given by (13.99) or (14.17), we obtain

$$A_{nm} = \frac{1}{\pi^2 n!\, m!} \iint \exp(\beta\alpha^*) \langle\alpha| \mathcal{N}\{\exp[(e^{ix}-1)\,\hat{a}^+\hat{a}]\}|\alpha\rangle\,.$$

$$.\,\exp(-\beta^*\alpha)\exp\left(-\frac{s-1}{2}|\beta|^2\right)(-\beta)^n \beta^{*m}\, d^2\beta\, d^2\alpha =$$

$$= \frac{1}{\pi^2 n!\, m!} \iint \exp\left(\beta\alpha^* - \beta^*\alpha + (e^{ix}-1)|\alpha|^2 - \frac{s-1}{2}|\beta|^2\right)(-\beta)^n \beta^{*m}\, d^2\beta\, d^2\alpha =$$

$$= \frac{(-1)^n}{n!}\,\delta_{nm}\,\frac{1}{1-\exp ix}\,\frac{1}{\left(\dfrac{s-1}{2} + \dfrac{1}{1-\exp ix}\right)^{n+1}}\,, \tag{16.63}$$

where the integrals (16.41) and (13.17) have been used. Substituting this result into (16.29) we arrive at

$$\exp(ix\hat{a}^+\hat{a}) =$$

$$= \sum_n \{\hat{a}^{+n}\hat{a}^n\}_s \frac{(-1)^n}{n!} \frac{1}{\left(\dfrac{s-1}{2} + \dfrac{1}{1-\exp ix}\right)^n} \frac{1}{\dfrac{s-1}{2}(1-\exp ix)+1} =$$

$$= \frac{2}{1+e^{ix}+s-se^{ix}} \left\{\exp\left[\frac{2(1-e^{ix})\,\hat{a}^+\hat{a}}{1+e^{ix}+s-se^{ix}}\right]\right\}_s. \qquad (16.64)$$

Let us note that, because the operator $\hat{b} = \hat{a} - \beta$ (β is a complex amplitude), describing the superposition of chaotic and coherent fields, obeys with $\hat{b}^+$ the same commutation rule as the operators $\hat{a}$ and $\hat{a}^+$, the relation (16.64) also holds for the operator $\hat{b}$.

First consider the normalization volume of the field; we obtain, introducing the number operator $\hat{n}$ given by (11.24b),

$$\langle \exp ix\hat{n}\rangle = \langle\prod_\lambda^M \exp(ix\,\hat{a}_\lambda^+\hat{a}_\lambda)\rangle =$$

$$= \left[\frac{2}{1+e^{ix}+s-se^{ix}}\right]^M \left\langle \exp\left[\frac{2\hat{n}(e^{ix}-1)}{1+e^{ix}+s-se^{ix}}\right]\right\rangle_s =$$

$$= \left[1 - \frac{1-s}{2}(1-e^{ix})\right]^{-M} \left\langle \exp\left[\frac{-\dfrac{2\hat{n}}{1-s}\dfrac{1-s}{2}(1-e^{ix})}{1-\dfrac{1-s}{2}(1-e^{ix})}\right]\right\rangle_s, \qquad (16.65)$$

where M is the number of modes. This result is valid for an arbitrary volume V (whose linear dimensions are much larger than the wavelength) by the same argument which led to (14.15), (14.16) and (14.17).

The substitution

$$\frac{2(e^{ix}-1)}{1+e^{ix}+s-se^{ix}} = iy \qquad (16.66)$$

in (16.65) gives us the s-ordered characteristic function

$$\langle \exp iy\hat{n} \rangle_s \equiv \langle \exp iyW \rangle_s = \left(1 - \frac{1-s}{2} iy\right)^{-M} \left\langle \left[\frac{1 + \frac{1+s}{2} iy}{1 - \frac{1-s}{2} iy} \right]^{\hat{n}} \right\rangle =$$

$$= \left(1 - \frac{1-s}{2} iy\right)^{-M} \left\langle \left[\frac{2}{(1-s)\left(1 - \frac{1-s}{2} iy\right)} - \frac{1+s}{1-s} \right]^{\hat{n}} \right\rangle . \tag{16.67}$$

Considering two orderings with $s = s_1$ and $s = s_2$ we can write

$$\left[\frac{2}{1 + e^{ix} + s_1 - s_1 e^{ix}}\right]^M \left\langle \exp\left[\frac{2\hat{n}(e^{ix} - 1)}{1 + e^{ix} + s_1 - s_1 e^{ix}}\right]\right\rangle_{s_1} =$$

$$= \left[\frac{2}{1 + e^{ix} + s_2 - s_2 e^{ix}}\right]^M \left\langle \exp\left[\frac{2\hat{n}(e^{ix} - 1)}{1 + e^{ix} + s_2 - s_2 e^{ix}}\right]\right\rangle_{s_2} , \tag{16.68}$$

from which, using the substitution (16.66) with $s \to s_2$, it follows that

$$\langle \exp iy\hat{n} \rangle_{s_2} = \left(1 + \frac{s_2 - s_1}{2} iy\right)^{-M} \left\langle \exp\left[\frac{iy\hat{n}}{1 + \frac{s_2 - s_1}{2} iy}\right]\right\rangle_{s_1} , \tag{16.69}$$

which is the generating function for the Laguerre polynomials L_r^{M-1} ([22], Chapter 6) so that

$$\langle \exp iy\hat{n} \rangle_{s_2} = \sum_{r=0}^{\infty} \frac{(iy)^r}{\Gamma(r + M)} \left[\frac{s_1 - s_2}{2}\right]^r \left\langle L_r^{M-1}\left(\frac{2\hat{n}}{s_2 - s_1}\right)\right\rangle_{s_1} . \tag{16.70}$$

From (16.69) we can obtain the relation between the quasi-distributions $P(W, s_2)$ and $P(W, s_1)$ using the Fourier integral and the residuum theorem

$$P(W, s_2) = \frac{1}{2\pi} \int_{-\infty}^{+\infty} \langle \exp iy\hat{n} \rangle_{s_2} \exp(-iyW)\, dy =$$

$$= \frac{1}{2\pi} \int_0^{\infty} P(W', s_1)\, dW' \int_{-\infty}^{+\infty} \left(1 - \frac{s_1 - s_2}{2} iy\right)^{-M} .$$

$$. \exp\left[\frac{iyW'}{1 - \frac{s_1 - s_2}{2} iy} - iyW\right] dy =$$

$$= \frac{2}{s_1 - s_2} \int_0^\infty \left(\frac{W}{W'}\right)^{(M-1)/2} \exp\left[-\frac{2(W+W')}{s_1 - s_2}\right] .$$

$$\cdot I_{M-1}\left(4\frac{\sqrt{WW'}}{s_1 - s_2}\right) P(W', s_1)\, dW', \quad \operatorname{Re} s_1 > \operatorname{Re} s_2, \tag{16.71}$$

where $I_{M-1}(x)$ is the modified Bessel function.

An inverse relation expressing $P(W, s_1)$ in terms of $P(W, s_2)$ ($\operatorname{Re} s_1 > \operatorname{Re} s_2$) has a singular character because in this case the above integral (16.71) is non-zero in the complex y-plane only for $W = 0$, which is a one-point support of a generalized function. Such a relation can be written in the form of a series of the derivatives of the δ-functions or, in some cases, in the form of a series of the Laguerre polynomials [365, 367].

From (16.67) one can obtain in the same way the relation between the quasi-probability $P(W, s)$ and the photon-number distribution $p(n)$†

$$P(W, s) = \frac{\exp\left(-\frac{2W}{1-s}\right)}{W} \left(\frac{2W}{1-s}\right)^M .$$

$$\cdot \sum_{n=0}^{\infty} \frac{n!\, p(n)}{[(n+M-1)!]^2} \left(\frac{s+1}{s-1}\right)^n L_n^{M-1}\left(\frac{4W}{1-s^2}\right), \tag{16.72}$$

which can be inverted, using the orthogonality relation (10.28), in the form

$$p(n) = \frac{1}{(n+M-1)!} \left(\frac{2}{1+s}\right)^M \left(\frac{s-1}{s+1}\right)^n .$$

$$\cdot \int_0^\infty P(W, s)\, L_n^{M-1}\left(\frac{4W}{1-s^2}\right) \exp\left(-\frac{2W}{1+s}\right) dW. \tag{16.73}$$

For $s \to 1$ if the asymptotic formula

$$L_n^{M-1}(x) \underset{x\to\infty}{\simeq} \frac{\Gamma(n+M)}{n!} (-x)^n \tag{16.74}$$

is used, equation (16.73) gives the standard photodetection equation (14.14).

---

† If M is not integer one must replace the factorials containing M in the following equations by the Γ-functions.

The relation between the moments $\langle \hat{n}^k \rangle_{s_2}$ and $\langle \hat{n}^l \rangle_{s_1}$ follows from (16.70)

$$\langle \hat{n}^k \rangle_{s_2} = \frac{d^k}{d(iy)^k} \langle \exp iy\hat{n} \rangle_{s_2} \Big|_{iy=0} =$$

$$= \frac{k!}{\Gamma(k+M)} \left(\frac{s_1 - s_2}{2}\right)^k \left\langle L_k^{M-1}\left(\frac{2\hat{n}}{s_2 - s_1}\right)\right\rangle_{s_1} . \tag{16.75}$$

From (16.65), which is also the generating function for the Laguerre polynomials, one obtains

$$\langle \hat{n}^k \rangle = \frac{d^k}{d(ix)^k} \langle \exp ix\hat{n} \rangle \Big|_{ix=0} =$$

$$= \frac{d^k}{d(ix)^k} \left\langle \sum_{j=0}^{\infty} \frac{1}{\Gamma(j+M)} \left(\frac{1-s}{2}\right)^j (1 - e^{ix})^j L_j^{M-1}\left(\frac{2\hat{n}}{1-s}\right)\right\rangle_s =$$

$$= \sum_{j=0}^{k} \frac{1}{\Gamma(j+M)} \left(\frac{1-s}{2}\right)^j \left[\sum_{r=0}^{j} \binom{j}{r} (-1)^r r^k\right] \left\langle L_j^{M-1}\left(\frac{2\hat{n}}{1-s}\right)\right\rangle_s , \tag{16.76}$$

where the identity used in connection with (10.19) has been used again, and (16.67) gives

$$\langle \hat{n}^k \rangle_s \equiv \langle W^k \rangle_s = \frac{d^k}{d(iy)^k} \langle \exp iy\hat{n} \rangle_s \Big|_{iy=0} =$$

$$= \left\langle \sum_{\hat{\jmath}=\hat{0}}^{\hat{n}} \frac{\hat{n}!\,(\hat{\jmath} + M + k - 1)!}{(\hat{n} - \hat{\jmath})!\,\hat{\jmath}!\,(\hat{\jmath} + M - 1)!} (-2)^{\hat{\jmath}-k} (s+1)^{\hat{n}-\hat{\jmath}} (s-1)^{k-\hat{n}} \right\rangle . \tag{16.77}$$

A number of special cases for normal, symmetric and antinormal orderings can be obtained in the same way as for the one-mode case by putting $s = 1, 0$ and -1 respectively. For example from (16.71) we have for $s_2 = -1$, $s_1 = 0$,

$$P_{\mathscr{A}}(W) = 2\int_0^{\infty} \left(\frac{W}{W'}\right)^{(M-1)/2} \exp[-2(W + W')]\, I_{M-1}(4\sqrt{WW'})\,.$$
$$.\, P_{\mathrm{Weyl}}(W')\, dW' \tag{16.78a}$$

and for $s_2 = -1$, $s_1 = 1$,

$$P_{\mathscr{A}}(W) = \int_0^{\infty} \left(\frac{W}{W'}\right)^{(M-1)/2} \exp\left[-(W + W')\right] I_{M-1}[2\sqrt{WW'}]\, P_{\mathscr{N}}(W')\, dW' . \tag{16.78b}$$

The corresponding relation between the moments is obtained from (16.75)

$$\langle \hat{n}^k \rangle_{\mathscr{A}} = \frac{k!}{2^k \, \Gamma(k+M)} \langle L_k^{M-1}(-2\hat{n}) \rangle_{\text{Weyl}} \tag{16.79a}$$

$$= \frac{k!}{\Gamma(k+M)} \langle L_k^{M-1}(-\hat{n}) \rangle_{\mathscr{N}} \tag{16.79b}$$

and between the characteristic functions from (16.69)

$$\langle \exp iy\hat{n} \rangle_{\mathscr{A}} = \left(1 - \frac{iy}{2}\right)^{-M} \left\langle \exp\left[\frac{iy\hat{n}}{1 - \dfrac{iy}{2}}\right] \right\rangle_{\text{Weyl}} \tag{16.80a}$$

$$= (1 - iy)^{-M} \left\langle \exp\left[\frac{iy\hat{n}}{1 - iy}\right] \right\rangle_{\mathscr{N}} . \tag{16.80b}$$

Similarly we have from (16.76) putting $s = 1$ and using the asymptotic formula (16.74)†

$$\langle \hat{n}^k \rangle = \sum_{j=0}^{k} \langle \hat{n}^j \rangle_{\mathscr{N}} \sum_{r=0}^{j} \frac{(-1)^{r+j} \, r^k}{r! \, (j-r)!} = \sum_{j=0}^{k} \langle \hat{n}^j \rangle_{\mathscr{N}} \sum_{m=0}^{j} \frac{(-1)^m \, (j-m)^k}{m! \, (j-m)!} , \tag{16.81}$$

which corresponds to the semiclassical relations (10.4) and (10.19). For $s = 0$ we can obtain the corresponding relation between the moments $\langle \hat{n}^k \rangle$ and $\langle \hat{n}^j \rangle_{\text{Weyl}}$ and for $s = -1$ we have

$$\langle \hat{n}^k \rangle = \sum_{j=0}^{k} \langle \hat{n}^j \rangle_{\mathscr{A}} \sum_{r=0}^{j} \frac{(-1)^{k+r} \, (M+r)^k}{r! \, (j-r)!} . \tag{16.82}$$

† More generally for a function $\hat{F}(\hat{a}^+ \hat{a})$ of the operator $\hat{a}^+ \hat{a}$,

$$\langle \alpha | \, \hat{F}(\hat{a}^+ \hat{a}) \, | \alpha \rangle = \sum_n |\langle \alpha \,|\, n \rangle|^2 \, F(n) = \sum_n \frac{|\alpha|^{2n}}{n!} \exp(-|\alpha|^2) \, F(n) =$$

$$= \sum_{n=0}^{\infty} \sum_{m=0}^{\infty} \frac{|\alpha|^{2(n+m)} (-1)^m}{n! \, m!} F(n) = \sum_{m=0}^{\infty} \sum_{j=m}^{\infty} \frac{(-1)^m \, |\alpha|^{2j}}{(j-m)! \, m!} F(j-m) =$$

$$= \langle \alpha | \, \mathscr{N} \left\{ \sum_{j=0}^{\infty} (\hat{a}^+ \hat{a})^j \sum_{m=0}^{j} \frac{(-1)^m \, F(j-m)}{(j-m)! \, m!} \right\} | \alpha \rangle ,$$

i.e.

$$\hat{F}(\hat{a}^+ \hat{a}) = \sum_{j=0}^{\infty} \{ (\hat{a}^+ \hat{a})^j \}_{\mathscr{N}} \sum_{m=0}^{j} \frac{(-1)^m \, F(j-m)}{m! \, (j-m)!} .$$

From (16.77) it follows, for $s = -1$, that

$$\langle \hat{n}^k \rangle_{\mathscr{A}} = \langle (\hat{n} + M)(\hat{n} + M + 1) \dots (\hat{n} + M + k - 1) \rangle = \\ = \left\langle \frac{(\hat{n} + M + k - 1)!}{(\hat{n} + M - 1)!} \right\rangle, \tag{16.83}$$

which can also be derived directly with the help of the commutation rules (11.44) using the relation $\int_V \hat{\mathbf{A}}(\boldsymbol{x}, t) \, . \, \hat{\mathbf{A}}^+(\boldsymbol{x}, t) \, d^3\boldsymbol{x} = \hat{n}_{Vt} + VL^{-3} \sum_{\boldsymbol{k},s} \boldsymbol{e}^{(s)}(\boldsymbol{k}) \, . \, \boldsymbol{e}^{(s)*}(\boldsymbol{k}) = \hat{n}_{Vt} + VL^{-3}\mu = \hat{n}_{Vt} + M$, where (11.4a) has been used; μ is the number of modes in the volume L^3 and so M is the number of modes in the volume V. Note that successive factors in (14.15) are decreased by unity, whereas those in (16.83) are increased by unity. The difference may be regarded as a reflection of the fact that normally ordered correlations correspond to photon *absorption*, whereas antinormally ordered correlations correspond to photon *emission* (this point will be discussed in greater detail in the next section).

Equation (16.67) gives us the following characteristic functions

$$\langle \exp iy\hat{n} \rangle_{\mathscr{N}} = \langle (1 + iy)^{\hat{n}} \rangle \,, \tag{16.84}$$

$$\langle \exp iy\hat{n} \rangle_{\mathrm{Weyl}} = (1 - \tfrac{1}{2}iy)^{-M} \left\langle \left[\frac{1 + \frac{1}{2}iy}{1 - \frac{1}{2}iy} \right]^{\hat{n}} \right\rangle \tag{16.85}$$

and

$$\langle \exp iy\hat{n} \rangle_{\mathscr{A}} = \langle (1 - iy)^{-\hat{n} - M} \rangle \tag{16.86}$$

for $s = 1, 0$ and -1 respectively and from (16.65)

$$\langle \exp ix\hat{n} \rangle = \langle \exp [\hat{n}(e^{ix} - 1)] \rangle_{\mathscr{N}} \tag{16.87a}$$

$$= \left[\frac{2}{1 + e^{ix}} \right]^M \left\langle \exp \left[\frac{2\hat{n}(e^{ix} - 1)}{1 + e^{ix}} \right] \right\rangle_{\mathrm{Weyl}} \tag{16.87b}$$

$$= \langle \exp [-ixM + \hat{n}(1 - e^{-ix})] \rangle_{\mathscr{A}} \, . \tag{16.87c}$$

Equation (16.87a) is in agreement with (14.17) and (16.87c) for $M = 1$ with (13.100).

Finally from (16.72), for $s = -1$,

$$P_{\mathscr{A}}(W) = \exp(-W) \sum_{n=0}^{\infty} \frac{W^{n+M-1}}{(n + M - 1)!} \, p(n) \,, \tag{16.88}$$

so that

$$p(n) = \frac{d^{n+M-1}}{dW^{n+M-1}} \left[P_{\mathscr{A}}(W) \exp W \right] \Big|_{W=0} \, . \tag{16.89}$$

The corresponding formulae for the description of the one-mode stationary fields can be gained by putting $M = 1$ in the above equations.

Some further details concerning the present formulation can be found in papers [365, 367].

In the classical limit, when the average photon number per mode becomes large, the commutator of the field operators $[\hat{a}, \hat{a}^+] = \hat{a}\hat{a}^+ - \hat{a}^+\hat{a}$ is practically zero since $\hat{a}\hat{a}^+ = \hat{a}^+\hat{a} + 1 \approx \hat{a}^+\hat{a}$ (this is also true for the operators $\hat{\mathbf{A}}(x)$ and $\hat{A}(x)$) and the distinction between different orderings of the field operators vanishes as a consequence of the correspondence principle. Thus all distributions $P(W, s)$ for various s and $p(n)$ must be equal and consequently all their moments are also equal. This can be seen as follows: for example for the photodetection equation (14.14) (or alternatively for (16.88)) we can conclude that the function under the integral (summation) sign gives the main contribution to the integral (sum) in the neighbourhood of the point $W = n$ where the function $W^n \exp(-W)/n!$ has its maximum; however this function tends to the function $\delta(W - n)$ for large W. Thus $p(n) \approx P_{\mathcal{N}}(n)$ for n large ($n \approx W$) and $\langle \hat{n}^k \rangle \approx \langle \hat{n}^k \rangle_{\mathcal{N}}$, etc.

Finally we mention a general formulation (suitable for boson as well as fermion fields) for the ordering of field operators; this was given by Lax [250] (see also [259]).

Consider a complete set of operators $\{\hat{a}_\lambda\}$ in the Schrödinger picture obeying some set of commutation or anti-commutation relations (the anti-commutator is defined as $[\hat{A}, \hat{B}]_+ = \hat{A}\hat{B} + \hat{B}\hat{A}$) with an associated set of c-numbers $\{\alpha_\lambda\}$. An operator $\hat{A}^{(Q)}(\{\hat{a}_\lambda\})$ in a Q-ordered form can be expressed as

$$\hat{A}^{(Q)}(\{\hat{a}_\lambda\}) = \int A^{(Q)}(\{\alpha_\lambda\}) \prod_\lambda \delta(\hat{a}_\lambda - \alpha_\lambda)\, d^2\alpha_\lambda\,, \tag{16.90}$$

where the δ-functions are in the chosen order. If the $\{\hat{a}_\lambda\}$ are not Hermitian and $\hat{a}_\lambda^+$ are also present, the δ-functions are defined, for example, by

$$\delta(\hat{a} - \alpha) = \frac{1}{\pi^2} \int \exp[-\beta^*(\hat{a} - \alpha)] \exp[\beta(\hat{a}^+ - \alpha^*)]\, d^2\beta \tag{16.91}$$

if the chosen order is $\hat{a}$, $\hat{a}^+$. The expectation value of (16.90) then becomes

$$\operatorname{Tr}\{\hat{\varrho}\hat{A}^{(Q)}\} = \int A^{(Q)}(\{\alpha_\lambda\}) \langle \prod_\lambda \delta(\hat{a}_\lambda - \alpha_\lambda) \rangle\, d^2\{\alpha_\lambda\} =$$

$$= \int A^{(Q)}(\{\alpha_\lambda\})\, \Phi(\{\alpha_\lambda\})\, d^2\{\alpha_\lambda\}\,, \tag{16.92}$$

where

$$\Phi(\{\alpha_\lambda\}) = \langle\prod_\lambda \delta(\hat{a}_\lambda - \alpha_\lambda)\rangle = \int C(\{\beta_\lambda\}) \prod_\lambda \exp(-\beta_\lambda\alpha_\lambda^* + \beta_\lambda^*\alpha_\lambda)\,\frac{d^2\beta_\lambda}{\pi^2} \tag{16.93}$$

and the characteristic function is equal to

$$C(\{\beta_\lambda\}) = \mathrm{Tr}\,\{\hat{\varrho}\prod_\lambda \exp(-\beta_\lambda^*\hat{a}_\lambda)\exp(\beta_\lambda\hat{a}_\lambda^+)\}\,. \tag{16.94}$$

The advantage of this formulation is that the Q-order of operators explicitly occurs in equation (16.90) while in the Agarwal-Wolf formulation the various orders of operators are specified by the function $\Omega(\alpha^*, \alpha)$ and it is not always clear which order of operators corresponds to a given $\Omega(\alpha^*, \alpha)$.

16.4 Measurements corresponding to antinormally ordered products of field operators – quantum counters

In this section we show that there exists in principle an interesting method of measuring the statistics of optical fields in which the photoelectric detectors normally used are replaced by atomic counting devices, called *quantum counters* [278]. Such detectors operate by *stimulated emission* rather than by the absorption of photons in an external field and consequently correlation measurements by means of quantum counters correspond to antinormally ordered products of field operators. The quantum counter is in principle a device useful for measurements in high degeneracy fields only (it is insensitive to fields with low degeneracy, i.e. with a low mean number of photons per mode) and a practical realization is not a simple matter. However, it is interesting to compare this device with the photoelectric detector and to compare the way the two devices measure fields.

A schematic outline of a quantum counter is given in Fig. 16.1. In this figure a represents a terminal energy level and c is a metastable level which is radiatively coupled to a broad energy band d, corresponding to a very short-lived state. The system will make spontaneous radiative transitions from d to a. We shall assume that this system is prepared in the state c by optical pumping from a to the broad energy level b from which it will make non-radiative transitions to c. Further let the interval $(E_c - E_d)/\hbar c$, defined by the energy levels E_c and E_d, be of the same order as the wave number of a typical mode of the external field, and let $(E_d - E_a) > (E_c - E_d)$.

Under the interaction of the system placed at a space point $\boldsymbol{x}$ at time $\boldsymbol{t}$ with the external field the system can make a stimulated transition from c to d with the emission of a photon. As level d is very short lived, it will decay spontaneously from d to a with the further emission of a photon. The latter photon is distinguishable from the former since $(E_d - E_a) > (E_c - E_d)$ and a photodetector placed in a neighbourhood of $\boldsymbol{x}$ with sufficiently high photoelectric threshold will register the second photon

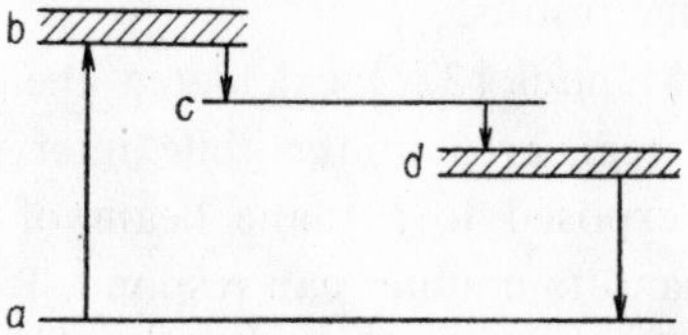

Fig. 16.1.
Energy levels of the quantum counter.

alone. The combination of the photodetector with a large number of such atomic systems acts as a quantum counter of the external field functioning by stimulated emission of radiation. It should be noted that two photons are emitted into the field by the system but only one is absorbed by the photodetector; the photodetector plays an auxiliary role only, the field is actually measured by means of the first induced transition.

Making use of arguments similar to those used in connection with the measurement of the field by means of photoelectric detectors (Sec. 12.1) we find that the probability that one count is registered at $(\boldsymbol{x}, t)$ is proportional to

$$\{\sum_f |\langle f| \hat{A}^{(-)}(x) |i\rangle|^2\}_{\text{average}} =$$
$$= \{\sum_f \langle i| \hat{A}^{(+)}(x) |f\rangle \langle f| \hat{A}^{(-)}(x) |i\rangle\}_{\text{average}} =$$
$$= \text{Tr}\,\{\hat{\varrho}\, \hat{A}^{(+)}(x)\, \hat{A}^{(-)}(x)\} = \Gamma_{\mathscr{A}}^{(1,1)}(x, x)\,, \tag{16.95}$$

i.e. this probability is proportional to the antinormally ordered correlation function $\Gamma_{\mathscr{A}}^{(1,1)}(x, x)$. Similarly the joint probability that counts will be registered by N quantum counters at space-time points $x_1, \ldots, x_N$ is proportional to

$$\Gamma_{\mathscr{A}}^{(N,N)}(x_1, \ldots, x_N, x_N, \ldots, x_1) =$$
$$= \text{Tr}\,\{\hat{\varrho}\, \hat{A}^{(+)}(x_1) \ldots \hat{A}^{(+)}(x_N)\, \hat{A}^{(-)}(x_N) \ldots \hat{A}^{(-)}(x_1)\}\,, \tag{16.96}$$

which is the antinormal correlation function of the $2N$th order.

From (16.83) it is clear that measurements by quantum counters will be sensitive to the field only if the mean number of photons $\langle n\rangle$ is much larger than the number of modes, i.e. if the degeneracy parameter $\langle n \rangle / M = \langle n \rangle L^3 / \mu V = \delta$ is much greater than one. This is the case for laser fields but for non-degenerate fields, such as thermal fields, $\delta \ll 1$ and the quantum counter will not be a useful measuring device. However for sufficiently strong fields (large δ) the difference between normally and antinormally ordered correlations vanishes as we have seen and $\langle \hat{n}^k \rangle_{\mathscr{A}} \approx \langle \hat{n}^k \rangle_{\mathscr{N}} \approx \langle \hat{n}^k \rangle$, because the field approaches the classical limit and consequently both photoelectric detectors and quantum counters will give practically the same results.

Mandel [278] calculated the probability $p'(n, T, t)$ that n counts will be registered in the time interval $(t, t + T)$ when the quantum counter is exposed to a plane beam of quasi-monochromatic light to which the quantum counter can respond. By an argument identical to that used in the derivation of (10.3a) he obtained

$$p'(n, T, t) = \int_0^\infty P_{\mathscr{A}}(\overline{W}) \frac{\overline{W}^n}{n!} \exp(-\overline{W})\, d\overline{W}, \tag{16.97}$$

which corresponds to (10.3a) or to (14.26) with $\overline{W}$ given by (14.28). Consequently all the properties of this relation are the same as those of (10.3a) and they are summarized in Chapter 10. However in this case $\langle (\Delta \hat{n})^2 \rangle_{\mathscr{A}} \geqq 0$ always (we have seen that this is not the case for the photoelectric detector, e.g. for the $|n\rangle$-state $\langle (\Delta \hat{n})^2 \rangle_{\mathscr{N}} = -n$ and $\langle (\Delta \hat{n})^2 \rangle = 0$); thus the variance of n will always be greater than or equal to the variance corresponding to a Poisson distribution. This fact is a consequence of the non-negativeness of the quasi-distribution $\Phi_{\mathscr{A}}(\{\alpha_\lambda\})$ (cf. equation (13.82a)). Hence the antinormal correlations are always positive semi-definite. From (13.68) it can be seen that for strong fields, for which the quasi-probability $\Phi_{\mathscr{N}}(\beta)$ is non-zero for large $|\beta|$ only and the function $\exp[-|\alpha - \beta|^2]$ is sharply peaked, the principal contribution to $\Phi_{\mathscr{A}}(\alpha)$ will come from values of β in a neighbourhood of α and if $\Phi_{\mathscr{N}}(\alpha)$ is sufficiently smooth then $\Phi_{\mathscr{A}}(\alpha) \approx \Phi_{\mathscr{N}}(\alpha)$, and for multimode field $\Phi_{\mathscr{A}}(\{\alpha_\lambda\}) \approx \Phi_{\mathscr{N}}(\{\alpha_\lambda\})$. Thus we conclude again that the normal and antinormal correlations are equal in this classical limit.

We have seen in Chapter 14 that the forms of the photon-number distribution and of the photon-counting distribution are identical (they differ from one another by a scale change given by the photoefficiency η of the detector). This can be regarded as a consequence of the fact that photoelectric detection measurements are related to normally ordered products of field operators for which the vacuum expectation values are zero, so that the vacuum fluctuations give no contribution to the normal correlation functions. We have pointed out that another situation occurs for other orderings. For example for antinormally ordered products of field operators connected with quantum counter measurements the photon-counting distribution is given by (16.97) while the photon-number distribution is given by (16.89) and so they substantially differ from one another. This is caused by the contribution of the physical vacuum through the commutator $\hat{a}\hat{a}^+ - \hat{a}^+\hat{a} = 1$, i.e. $\hat{a}\hat{a}^+ = \hat{a}^+\hat{a} + 1$. This generates the convolution law (13.68) by the correspondence $\hat{a}\hat{a}^+ \rightleftarrows \Phi_{\mathscr{A}}(\alpha)$, $\hat{a}^+\hat{a} \rightleftarrows \Phi_{\mathscr{N}}(\alpha)$ and $1 \rightleftarrows \exp(-|\alpha|^2) = \langle\alpha \mid 0\rangle \langle 0 \mid \alpha\rangle = \langle\alpha| \mathscr{N}\{\exp(-\hat{a}^+\hat{a})\} |\alpha\rangle$ (i.e. $|0\rangle\langle 0| = \mathscr{N}\{\exp(-\hat{a}^+\hat{a})\}$). For M-mode fields the vacuum fluctuations contribute through the number of modes M since $\int_V \hat{\mathbf{A}}(x) \cdot \hat{\mathbf{A}}^+(x)\, d^3\boldsymbol{x} = \int_V \hat{\mathbf{A}}^+(x) \cdot \hat{\mathbf{A}}(x)\, d^3\boldsymbol{x} + M$. The connection between $p'(n, T, t)$ and $p(n)$ is determined by substitution of (16.88) into (16.97), which gives [360]

$$p'(n, T, t) = \sum_{m=0}^{\infty} p(m)\, 2^{-m-n-M} \binom{m + n + M - 1}{n}. \tag{16.98}$$

Finally it can be said that different orderings of field operators correspond to different types of experiments and each type is characteristically different.

Chapter 17

SPECIAL STATES OF THE ELECTROMAGNETIC FIELD

The general methods developed in preceding chapters will be demonstrated now by applying them to the most typical state in nature – the chaotic (Gaussian) field (and as a special case to the thermal field). In addition we apply them to laser light and to the superposition of chaotic and coherent fields.

17.1 Chaotic (Gaussian) radiation

Distributions and characteristic functions

It is well known that the state for which the entropy

$$H = -\mathrm{Tr}\,\{\hat{\varrho} \log \hat{\varrho}\} \tag{17.1}$$

has a maximum under the assumptions that $\mathrm{Tr}\,\hat{\varrho} = 1$ and $\mathrm{Tr}\,\{\hat{\varrho}\hat{a}^+\hat{a}\} = \langle n \rangle$ is the chaotic state for which the density matrix possesses the diagonal Fock form (11.27) with $\varrho(n)$ as members of the Bose-Einstein distribution (11.32), i.e.

$$\hat{\varrho} = \frac{\langle n \rangle^{\hat{a}^+\hat{a}}}{(1 + \langle n \rangle)^{\hat{a}^+\hat{a}+1}}\,. \tag{17.2}$$

For thermal (blackbody) radiation (11.30) holds for $\langle n \rangle$ and we obtain (11.29) for $\hat{\varrho}$.

The density matrix for a multimode field will be, in analogy to (11.29),

$$\hat{\varrho} = \frac{\prod\limits_{\lambda} \exp(-\Theta_\lambda \hat{a}_\lambda^+ \hat{a}_\lambda)}{\mathrm{Tr}\,\{\prod\limits_{\lambda} \exp(-\Theta_\lambda \hat{a}_\lambda^+ \hat{a}_\lambda)\}} = \frac{\exp\left(-\dfrac{1}{KT}\hat{H}\right)}{\mathrm{Tr}\left\{\exp\left(-\dfrac{1}{KT}\hat{H}\right)\right\}}, \tag{17.3}$$

where the Hamiltonian $\hat{H}$ is given by (11.13) without the vacuum energy term $\sum_\lambda \hbar\omega_\lambda/2$.

We must still determine what kind of function $\Phi_{\mathcal{N}}(\{\alpha_\lambda\})$ corresponds to (17.2) and (17.3). Because of the statistical independence of different modes the multimode form of (17.2) will be

$$\hat{\varrho} = \prod_\lambda \frac{\langle n_\lambda\rangle^{\hat{a}^+_\lambda \hat{a}_\lambda}}{(1 + \langle n_\lambda\rangle)^{\hat{a}^+_\lambda \hat{a}_\lambda + 1}} = \sum_{\{n_\lambda\}} \prod_\lambda \frac{\langle n_\lambda\rangle^{n_\lambda}}{(1 + \langle n_\lambda\rangle)^{n_\lambda + 1}} |\{n_\lambda\}\rangle \langle\{n_\lambda\}| \,. \tag{17.4}$$

Making use of (13.50) and (13.49) we can write

$$\hat{\varrho} = \int |\{\alpha_\lambda\}\rangle \langle\{\alpha_\lambda\}| \prod_\lambda \frac{d^2\alpha_\lambda}{\pi} \prod_{\lambda'} \frac{\langle n_{\lambda'}\rangle^{\hat{a}^+_{\lambda'}\hat{a}_{\lambda'}}}{(1 + \langle n_{\lambda'}\rangle)^{\hat{a}^+_{\lambda'}\hat{a}_{\lambda'} + 1}} =$$

$$= \int \sum_{\{n_\lambda\}} \sum_{\{m_\lambda\}} \prod_\lambda \exp(-|\alpha_\lambda|^2) \frac{\alpha_\lambda^{n_\lambda}}{\sqrt{n_\lambda!}} \frac{\alpha_\lambda^{*m_\lambda}}{\sqrt{m_\lambda!}} |\{n_\lambda\}\rangle \langle\{m_\lambda\}| \,.$$

$$\cdot \prod_{\lambda'} \frac{\langle n_{\lambda'}\rangle^{\hat{a}^+_{\lambda'}\hat{a}_{\lambda'}}}{(1 + \langle n_{\lambda'}\rangle)^{\hat{a}^+_{\lambda'}\hat{a}_{\lambda'} + 1}} \frac{d^2\alpha_{\lambda'}}{\pi} = \int \sum_{\{n_\lambda\}} \sum_{\{m_\lambda\}} |\{n_\lambda\}\rangle \langle\{m_\lambda\}| \,.$$

$$\cdot \prod_\lambda \frac{\exp(-|\alpha_\lambda|^2)}{1 + \langle n_\lambda\rangle} \frac{\alpha_\lambda^{n_\lambda}}{\sqrt{n_\lambda!}} \frac{\alpha_\lambda^{*m_\lambda}}{\sqrt{m_\lambda!}} \left(\frac{\langle n_\lambda\rangle}{1 + \langle n_\lambda\rangle}\right)^{m_\lambda} \frac{d^2\alpha_\lambda}{\pi} \,. \tag{17.5}$$

As only the terms of the series with $n_\lambda = m_\lambda$ do not vanish (by integration over the phases), we arrive at

$$\hat{\varrho} = \int \prod_\lambda \frac{\exp\left[-|\alpha_\lambda|^2 + \dfrac{\langle n_\lambda\rangle}{1 + \langle n_\lambda\rangle}|\alpha_\lambda|^2\right]}{\pi(1 + \langle n_\lambda\rangle)} |\beta_\lambda\rangle \langle\beta_\lambda| \, d^2\{\alpha_\lambda\} =$$

$$= \int \prod_\lambda \frac{\exp\left[-\dfrac{|\beta_\lambda|^2}{\langle n_\lambda\rangle}\right]}{\pi\langle n_\lambda\rangle} |\{\beta_\lambda\}\rangle \langle\{\beta_\lambda\}| \, d^2\{\beta_\lambda\} \,, \tag{17.6}$$

where the substitution $\alpha_\lambda[\langle n_\lambda\rangle/(1 + \langle n_\lambda\rangle)]^{1/2} = \beta_\lambda$ has been used. If we write α instead of β again, we obtain the result

$$\Phi_{\mathcal{N}}(\{\alpha_\lambda\}) = \prod_\lambda \frac{\exp\left(-\dfrac{|\alpha_\lambda|^2}{\langle n_\lambda\rangle}\right)}{\pi\langle n_\lambda\rangle} \,. \tag{17.7}$$

This is a Gaussian distribution in the complex amplitude α_λ and this distribution is independent of the phases of α_λ; thus the chaotic field is stationary in time and in space. It is easy to verify that the substitution of (17.7) and (13.49) into (13.142) leads to (17.4) again.

This result can also be derived using the quantum analogue of the central limit theorem [174, 17]. The distribution (17.7) for chaotic field is consistent with the classical distribution (10.39). This follows taking into account the fact that the complex amplitude of the field is a linear superposition of the type (13.53) and that the convolution of two Gaussian distributions is again a Gaussian distribution. Thus we obtain (10.39) with $\langle I\rangle = \langle V^*(x) . V(x)\rangle = 2\pi\hbar c L^{-3} \sum_{k,s} k^{-1}\langle n_{ks}\rangle$, since $\langle \alpha_{ks}^* \alpha_{k's'}\rangle = \delta_{kk'}\delta_{ss'} . \langle |\alpha_{ks}|^2\rangle = \delta_{kk'}\delta_{ss'}\langle n_{ks}\rangle$. If $\langle n_{ks}\rangle$ is independent of the polarization s, which is true for blackbody radiation, then $\langle I\rangle = 2\pi\hbar c L^{-3} \sum_{k,s} k^{-1}\langle n_k\rangle$; for blackbody radiation $\langle n_k\rangle \equiv \langle n_{k}\rangle$ is given by (11.30).

A calculation of the characteristic functional (12.48) for chaotic radiation will enable us to determine the correlation functions according to (12.49) and also the distribution (14.1). Writing

$$\varepsilon_\mu(\lambda, x) = \frac{(2\pi\hbar c)^{1/2}}{L^{3/2}} \frac{1}{k^{1/2}} e_\mu^{(s)}(\mathbf{k}) \exp\left[i(\mathbf{k} . \mathbf{x} - ckt)\right], \tag{17.8}$$

the μ-component of (11.22) can be written in the form

$$\hat{A}_\mu^{(+)}(x) = \sum_\lambda \varepsilon_\mu(\lambda, x)\, \hat{a}_\lambda \equiv \hat{A}^{(+)}(x) . \tag{17.9}$$

Substituting (17.9) and (17.7) into (12.48) we obtain

$$C^{(\mathcal{N})}\{y(x)\} = \int \prod_\lambda \frac{1}{\pi\langle n_\lambda\rangle} \exp\left(-\frac{|\alpha_\lambda|^2}{\langle n_\lambda\rangle}\right) .$$

$$\exp\left\{\iint\left[y(x)\sum_{\lambda'}\varepsilon^*(\lambda', x)\,\alpha_{\lambda'}^* - y^*(x)\sum_{\lambda'}\varepsilon(\lambda', x)\,\alpha_{\lambda'}\right] d^4x\right\} d^2\{\alpha_\lambda\} =$$

$$= \int \prod_\lambda \frac{1}{\pi\langle n_\lambda\rangle} \exp\left(-\frac{|\alpha_\lambda|^2}{\langle n_\lambda\rangle} + \alpha_\lambda^* y_\lambda - \alpha_\lambda y_\lambda^*\right) d^2\alpha_\lambda , \tag{17.10}$$

where

$$y_\lambda = \int y(x)\, \varepsilon^*(\lambda, x)\, d^4x . \tag{17.11}$$

(The integral over x also includes a sum over μ.) Therefore by using (16.41)

$$C^{(\mathcal{N})}\{y(x)\} = \prod_\lambda \exp\left(-\langle n_\lambda\rangle\, |y_\lambda|^2\right)$$

$$= \exp\left[-\iint y^*(x') \sum_\lambda \langle n_\lambda\rangle\, \varepsilon(\lambda, x')\, \varepsilon^*(\lambda, x)\, y(x)\, d^4x\, d^4x'\right] . \tag{17.12}$$

However the second-order correlation function is

$$\Gamma_{\mathcal{N}}^{(1,1)}(x, x') = \mathrm{Tr}\,\{\hat{\varrho}\,\hat{A}^{(+)}(x)\,\hat{A}^{(-)}(x')\} =$$

$$= \int \prod_{\lambda} \frac{1}{\pi\langle n_\lambda\rangle} \exp\left(-\frac{|\alpha_\lambda|^2}{\langle n_\lambda\rangle}\right) \sum_{\lambda}\sum_{\lambda'} \varepsilon^*(\lambda, x)\,\varepsilon(\lambda', x')\,\alpha_\lambda^*\alpha_{\lambda'}\,d^2\{\alpha_\lambda\} =$$

$$= \sum_{\lambda} \langle n_\lambda\rangle\,\varepsilon(\lambda, x')\,\varepsilon^*(\lambda, x) \tag{17.13}$$

and we finally arrive at

$$C^{(\mathcal{N})}\{y(x)\} = \exp\left[-\iint y(x)\,\Gamma_{\mathcal{N}}^{(1,1)}(x, x')\,y^*(x')\,d^4x\,d^4x'\right]. \tag{17.14}$$

The correlation function $\Gamma_{\mathcal{N}}^{(n,n)}$ is, according to (12.49),

$$\Gamma_{\mathcal{N}}^{(n,n)}(x_1, \ldots, x_{2n}) =$$

$$= (-1)^n \prod_{k=n+1}^{2n} \frac{\delta}{\delta(-y^*(x_k))} \left\{\prod_{j=1}^{n}\left[\int \Gamma_{\mathcal{N}}^{(1,1)}(x_j, x')\,y^*(x')\,d^4x'\right] C^{(\mathcal{N})}\{y(x)\}\right\}\Bigg|_{y(x)=0} =$$

$$= \sum_{\pi} \prod_{j=1}^{n} \Gamma_{\mathcal{N}}^{(1,1)}(x_j, x_{n+j}), \tag{17.15}$$

where $\sum_{\pi}$ stands for the sum of $n!$ possible permutations of the indices $n + 1, \ldots, 2n$ (or $1, \ldots, n$). It is obvious that $\Gamma^{(m,n)} = 0$ for $m \neq n$. This is just the relation (10.64a).

Making use of the complex representation (13.37) of the δ-function we can write (14.1) in the form

$$P_n(V_1, \ldots, V_n) =$$

$$= \int \Phi_{\mathcal{N}}(\{\alpha'_\lambda\}) \int \prod_{j=1}^{n} \exp(V_j z_j^* - V_j^* z_j) \exp(-V'_j z_j^* + V'^*_j z_j) \frac{d^2 z_j}{\pi^2}\,d^2\{\alpha'_\lambda\} =$$

$$= \int C^{(\mathcal{N})}(\{z_j\}) \prod_{j=1}^{n} \exp(V_j z_j^* - V_j^* z_j) \frac{d^2 z_j}{\pi^2}, \tag{17.16}$$

where the characteristic function $C^{(\mathcal{N})}(\{z_j\})$ is given by (17.14) with

$$y(x) = \sum_{j=1}^{n} z_j\,\delta(x - x_j); \tag{17.17}$$

that is

$$C^{(\mathcal{N})}(\{z_j\}) = \exp\left[-\sum_{j,k}^{n} z_j \Gamma_{\mathcal{N}}^{(1,1)}(x_j, x_k) z_k^*\right], \tag{17.18}$$

which can be written, introducing the vectors

$$\hat{Z} = \begin{pmatrix} z_1 \\ \vdots \\ z_n \end{pmatrix}, \quad \hat{V} = \begin{pmatrix} V_1 \\ \vdots \\ V_n \end{pmatrix}, \quad \hat{V}' = \begin{pmatrix} V_1' \\ \vdots \\ V_n' \end{pmatrix} \tag{17.19}$$

and the covariance matrix

$$\mathcal{R} = \langle \hat{V}' \otimes \hat{V}'^{+} \rangle, \tag{17.20}$$

in the form

$$C^{(\mathcal{N})}(\hat{Z}) = \exp(-\hat{Z}^{+}\mathcal{R}\hat{Z}). \tag{17.21}$$

Thus we have from (17.16)

$$P_n(V_1, \ldots, V_n) = \frac{1}{\pi^{2n}} \int \exp(-\hat{Z}^{+}\mathcal{R}\hat{Z}) \exp(-\hat{V}^{+}\hat{Z} + \hat{Z}^{+}\hat{V}) \prod_{j=1}^{n} d^2 z_j, \tag{17.22}$$

which finally gives us

$$P_n(V_1, \ldots, V_n) = \frac{1}{\pi^n \operatorname{Det} \mathcal{R}} \exp(-\hat{V}^{+}\mathcal{R}^{-1}\hat{V}). \tag{17.23}$$

(cf. (16.41)). This is the multivariate Gaussian distribution, so that the quantized electromagnetic chaotic field is described as a Gaussian random process.

It is obvious that very similar results are obtainable for the s-ordering. Applying the multimode analogue of (16.43) (with $s_2 = s$ and $s_1 = 1$) to (17.7) we obtain

$$\Phi(\{\alpha_\lambda\}, s) = \prod_{\lambda} \left(\langle n_\lambda \rangle + \frac{1-s}{2}\right)^{-1} \exp\left(-\frac{|\alpha_\lambda|^2}{\langle n_\lambda \rangle + \dfrac{1-s}{2}}\right), \tag{17.24}$$

which clearly shows that all the results just derived for chaotic light are also valid for the case of the s-ordering if $\langle n_\lambda \rangle$ is replaced by $\langle n_\lambda \rangle + (1-s)/2$.

Hence, both the space-time quasi-probabilities as well as the phase space quasi-distributions are multivariate Gaussian distributions with positive-definite covariance matrices for $-1 \leqq s \leqq 1$ (this includes the normal, symmetric and antinormal orderings).

The second-order correlation function

The second-order correlation function for blackbody radiation has been studied by a number of authors. Bourret [88] has used a technique employed in the theory of the isotropic turbulence of an incompressible fluid to derive expressions for the second-order electric correlation tensor. Later Sarfatt [412] rederived the main results by explicit quantum mechanical calculations.

Correlation functions suitable for the description of the temporal coherence of blackbody radiation were derived by Kano and Wolf [226] who showed that for this case the corresponding degree of coherence has no zeros in the lower half of the complex τ-plane. Mehta [295] studied the coherence time and effective bandwidth of blackbody radiation and a more complete treatment of the electric, magnetic and mixed correlation tensors of the second order for blackbody radiation (as well as of the spectral correlation tensors) was given by Mehta and Wolf in a series of papers [306, 307, 310]. In these the explicit behaviour of the second-order correlation tensors and spectral correlation tensors was investigated and the authors showed by direct calculation that the classical and quantum correlation tensors are identical for blackbody radiation. The statistical properties of blackbody radiation in an unbounded domain were investigated by Holliday [197], Holliday and Sage [198] and Keller [230] using the theory of functionals. Glauber [174, 176] derived some general properties of the correlation functions for blackbody radiation, Eberly and Kujawski [135] and Kujawski [517] studied the properties of correlation tensors in uniformly moving coordinate systems and Kujawski [240] has shown that the coherence properties of blackbody radiation can be characterized by a single scalar correlation function and discussed a resemblance between commutators of the quantized electromagnetic field and the correlation tensors. A relativistic coherence theory of blackbody radiation has been developed by Brevik and Suhonen [498, 499].

We will discuss only some simple properties of the electric correlation tensor here, particularly in connection with temporal coherence of blackbody radiation; a more detailed treatment of this topic can be found in the review article by Mandel and Wolf [285].

The quantum analogue of the electric correlation tensor $\mathscr{E}_{ij}(x_1, x_2)$ defined in (7.2) can be written in the form

$$\mathscr{E}_{ij}(x_1, x_2) = \mathrm{Tr}\,\{\hat{\varrho}\,\hat{E}_i^{(-)}(x_1)\,\hat{E}_j^{(+)}(x_2)\}\,. \tag{17.25}$$

The use of (17.13) and (11.1a) gives

$$\mathscr{E}_{ij}(x_1, x_2) = \\ = \frac{2\pi\hbar c}{L^3} \sum_{k,s} \langle n_{ks}\rangle\, k\, e_i^{(s)*}(\boldsymbol{k})\, e_j^{(s)}(\boldsymbol{k}) \exp\left[i\boldsymbol{k}\cdot(\boldsymbol{x}_2 - \boldsymbol{x}_1) - ick(t_2 - t_1)\right]. \tag{17.26}$$

As $\langle n_{ks}\rangle = \langle n_k\rangle$ (given by (11.30)) we can sum over s with the help of (11.5) and the substitution $\sum_k \to (L/2\pi)^3 \int \dots d^3\boldsymbol{k}$ giving

$$\mathscr{E}_{ij}(x_1, x_2) = \mathscr{E}_{ij}(x_2 - x_1) = \mathscr{E}_{ij}(x) = \\ = \frac{\hbar c}{(2\pi)^2} \int \frac{k}{\exp\Theta - 1} \left(\delta_{ij} - \frac{k_i k_j}{k^2}\right) \exp(ikx)\, d^3\boldsymbol{k} = \\ = \frac{\hbar c}{(2\pi)^2} \left(-\delta_{ij}\nabla^2 + \frac{\partial}{\partial x_i}\frac{\partial}{\partial x_j}\right) \int \frac{1}{k(\exp\Theta - 1)} \exp(ikx)\, d^3\boldsymbol{k}, \tag{17.27}$$

where $x \equiv (\boldsymbol{x}, \tau) = x_2 - x_1$ and the x_i are components of $\boldsymbol{x}$. Introducing spherical polar coordinates for $\boldsymbol{k}$ with the polar axis along the direction $\boldsymbol{x}$ we obtain

$$\mathscr{E}_{ij}(x) = \frac{\hbar c}{(2\pi)^2} \left(\frac{\partial}{\partial x_i}\frac{\partial}{\partial x_j} - \delta_{ij}\nabla^2\right). \\ \int_0^\infty \frac{k\, dk}{\exp\left(\frac{\hbar c k}{KT}\right) - 1} \exp(-ikc\tau) \int_0^\pi \exp(ikr\cos\vartheta)\sin\vartheta\, d\vartheta \int_0^{2\pi} d\varphi = \\ = \frac{4\hbar c}{\pi} \sum_{n=1}^\infty \left\{ \frac{\delta_{ij}}{\left[\left(n\frac{\hbar c}{KT} + ic\tau\right)^2 + r^2\right]^2} + \frac{2(x_i x_j - r^2\delta_{ij})}{\left[\left(n\frac{\hbar c}{KT} + ic\tau\right)^2 + r^2\right]^3} \right\}, \tag{17.28}$$

where $r = |\boldsymbol{x}|$. The degree of coherence is obtained in the form

$$\gamma_{ij}(x) = \frac{\mathscr{E}_{ij}(x)}{\{\mathscr{E}_{ii}(0)\,\mathscr{E}_{jj}(0)\}^{1/2}} = \frac{\mathscr{E}_{ij}(x)}{\mathscr{E}_{ii}(0)} = \\ = \frac{90}{\pi^4}\left(\frac{\hbar c}{KT}\right)^4 \sum_{n=1}^\infty \left\{ \frac{\delta_{ij}}{\left[\left(n\frac{\hbar c}{kT} + ic\tau\right)^2 + r^2\right]^2} + \frac{2(x_i x_j - r^2\delta_{ij})}{\left[\left(n\frac{\hbar c}{KT} + ic\tau\right)^2 + r^2\right]^3} \right\}, \tag{17.29}$$

where

$$\mathscr{E}_{ii}(0) = \frac{4\hbar c}{\pi}\left(\frac{KT}{\hbar c}\right)^4 \sum_{n=0}^{\infty} \frac{1}{(n+1)^4} = \frac{4\hbar c}{\pi}\left(\frac{KT}{\hbar c}\right)^4 \frac{\pi^4}{90}; \qquad (17.30)$$

here the sum is equal to the generalized Riemann ζ-function $\zeta(4,1) = \pi^4/90$ defined ([35], p. 266) by

$$\zeta(s, a) = \sum_{n=0}^{\infty} (n + a)^{-s}. \qquad (17.31)$$

As a special case we can consider temporal coherence putting $\boldsymbol{x} = \mathbf{0}$ in (17.29) and we obtain

$$\gamma_{ij}(\mathbf{0}, \tau) = \delta_{ij} \frac{90}{\pi^4} \xi \left(4,1 + i\frac{KT}{\hbar}\tau\right). \qquad (17.32)$$

Hence any two orthogonal components $E_i(\boldsymbol{x}, t_1)$ and $E_j(\boldsymbol{x}, t_2)$ are completely uncorrelated. In this connection it should be mentioned that the function (17.32) has no zeros in the lower half of the complex τ-plane [226] and therefore the phase of this function can be reconstructed uniquely from its known modulus with the help of the dispersion relations discussed in section 4.4. Thus in this case the spectrum of radiation can be determined from the visibility of the interference fringes.

It is evident that for the higher-order correlation tensors a relation of the same form as (17.15) holds, i.e. they are fully determined by the second-order correlation tensor.

Photon-counting statistics

Photon-counting statistics for Gaussian light were first considered by Glauber [172] in connection with the Hanbury Brown-Twiss experiment; in this paper he introduced, for the first time, the diagonal representation of the density matrix in terms of coherent states.

Basic formulae for the description of the photon-counting statistics of Gaussian light have already been given in Chapter 10 (e.g. equations (10.48)–(10.52)) but some more general results can also be derived.

Returning to the characteristic function in the form (14.22) we obtain for Gaussian light

$$\langle \exp ix\overline{W}\rangle_{\mathcal{N}} = \int \exp\left[-\sum_{\lambda} \frac{|\alpha_\lambda|^2}{\langle n_\lambda\rangle} + ix \sum_{\lambda,\lambda'} \alpha_\lambda^* \overline{W}_{\lambda\lambda'}\alpha_{\lambda'}\right] \prod_{\lambda} \frac{d^2\alpha_\lambda}{\pi\langle n_\lambda\rangle}, \qquad (17.33)$$

where

$$\overline{W}_{\lambda\lambda'} = \iint_0^T \iint_V \varepsilon^*(x'', \lambda)\, \mathscr{S}(\boldsymbol{x}'' - \boldsymbol{x}', t'' - t')\, \varepsilon(x', \lambda')\, d^4x'\, d^4x'' \quad (17.34)$$

and $V(x)$ corresponding to (17.9) has been used. Making use of the substitutions

$$\beta_\lambda = \frac{\alpha_\lambda}{\sqrt{\langle n_\lambda \rangle}}, \quad \sqrt{\langle n_\lambda \rangle}\, \overline{W}_{\lambda\lambda'} \sqrt{\langle n_{\lambda'} \rangle} = U_{\lambda\lambda'} \quad (17.35)$$

equation (17.33) becomes

$$\langle \exp ix\overline{W} \rangle_{\mathscr{N}} = \int \exp\Big[-\sum_\lambda |\beta_\lambda|^2 + ix \sum_{\lambda,\lambda'} \beta_\lambda^* U_{\lambda\lambda'} \beta_{\lambda'} \Big] \prod_\lambda \frac{d^2\beta_\lambda}{\pi} =$$

$$= \int \exp\Big[-\sum_\lambda |\beta_\lambda|^2 + ix \sum_\lambda \mathscr{U}_\lambda |\beta_\lambda|^2 \Big] \prod_\lambda \frac{d^2\beta_\lambda}{\pi} = \prod_\lambda \frac{1}{(1 - ix\mathscr{U}_\lambda)}, \quad (17.36)$$

where the $\mathscr{U}_\lambda$ are eigenvalues of the matrix $\hat{U} \equiv (U_{\lambda\lambda'})$. For a field in the normalization volume the quantity $W(\overline{W} = \eta W)$ given by (14.12) is suitable for the description of the field and $\mathscr{U}_\lambda = \langle n_\lambda \rangle$. If moreover all the $\langle n_\lambda \rangle$ are assumed equal so that $\mathscr{U}_\lambda = \langle n \rangle / M$, where M is the number of modes in the field, we arrive at (10.48).

The characteristic function (17.36) can be rewritten in the form

$$\langle \exp ix\overline{W} \rangle_{\mathscr{N}} = \frac{1}{\operatorname{Det}(1 - ix\hat{U})} = \exp\Big[-\sum_\lambda \log(1 - ix\mathscr{U}_\lambda) \Big] =$$

$$= \exp\Big[\sum_\lambda \sum_{n=1}^\infty \frac{(ix\mathscr{U}_\lambda)^n}{n} \Big] = \exp\Big[\sum_{n=1}^\infty \frac{(ix)^n}{n} \operatorname{Tr} \hat{U}^n \Big] = \exp\big[-\operatorname{Tr} \log(1 - ix\hat{U}) \big] \quad (17.37)$$

and comparing this result with the relation defining the cumulants $\varkappa_j^{(W)}$ (cf. (10.15)) we can write for the cumulants (for a study of cumulants see [500])

$$\varkappa_1^{(W)} = \operatorname{Tr} \hat{U} =$$

$$= \iint_0^T \iint_V \sum_\lambda \varepsilon^*(x'', \lambda)\, \mathscr{S}(\boldsymbol{x}'' - \boldsymbol{x}', t'' - t')\, \varepsilon(x', \lambda)\, \langle n_\lambda \rangle\, d^4x'\, d^4x''$$

$$= \iint_0^T \iint_V \mathscr{S}(\boldsymbol{x}'' - \boldsymbol{x}', t'' - t')\, \Gamma_{\mathscr{N}}^{(1,1)}(x'', x')\, d^4x'\, d^4x'', \quad (17.38a)$$

$$\varkappa_j^{(W)} = (j-1)!\,\mathrm{Tr}\,\hat{U}^j =$$

$$= (j-1)! \int \ldots \int \prod_{r=1}^{j} \mathscr{S}(x_r'' - x_r')\, \Gamma_{\mathscr{N}}^{(1,1)}(x_r'', x_{r+1}')\, d^4x'\, d^4x'' ,$$

$$x_{j+1}' = x_1' , \quad j \geqq 2 . \tag{17.38b}$$

Note that the first term in the series in (17.37) corresponds to the fully coherent field. The presence of the other terms violate the full coherence of the field. Equations (17.38) represent a generalization of (10.80), which can be obtained by putting $\mathscr{S}(x'' - x') = \eta\, \delta(x'' - x')\, \delta(\boldsymbol{x}' - \boldsymbol{x})$, where $\boldsymbol{x}$ is a fixed point, and assuming a stationary field.

In order to calculate the characteristic function (17.36) we must find the eigenvalues $\mathscr{U}_\lambda$ of the matrix $\hat{U}$ with elements given in (17.35), where $\overline{W}_{\lambda\lambda'}$ is given by (17.34). This problem was considered in [56, 213, 127, 410] for Gaussian Lorentzian light† and in [214, 531] for the superposition of coherent and Gaussian Lorentzian light. Considering a complete system of orthonormal functions $\{\varphi_\lambda(\boldsymbol{x}, t)\}$ over the interval $(0, T)$ and the volume V we can write

$$\hat{A}^{(+)}(x) = \sum_\lambda \hat{b}_\lambda\, \varphi_\lambda(x) , \tag{17.39a}$$

where

$$\hat{b}_\lambda = \int_0^T \int_V \hat{A}^{(+)}(x)\, \varphi_\lambda^*(x)\, d^4x . \tag{17.39b}$$

Thus for a stationary field in time and in space

$$\mathrm{Tr}\,\{\hat{\varrho}\hat{b}_\lambda^+ \hat{b}_{\lambda'}\} = \mathscr{U}_\lambda \delta_{\lambda\lambda'} =$$

$$= \iint_0^T \iint_V \mathrm{Tr}\,\{\hat{\varrho}\, \hat{A}^{(-)}(x)\, \hat{A}^{(+)}(x')\}\, \varphi_\lambda(x)\, \varphi_{\lambda'}^*(x')\, d^4x\, d^4x' . \tag{17.40}$$

Multiplying this equation by $\varphi_{\lambda'}(x'')$ and summing over λ' we finally obtain the integral equation determining $\mathscr{U}_\lambda$ and $\varphi_\lambda(x)$

$$\mathscr{U}_\lambda\, \varphi_\lambda(x'') = \int_0^T \int_V \Gamma_{\mathscr{N}}^{(1,1)*}(x'', x)\, \varphi_\lambda(x)\, d^4x , \tag{17.41}$$

since $\sum_\lambda \varphi_\lambda^*(x')\, \varphi_\lambda(x'') = \delta(x' - x'')$. The eigenvalues $\mathscr{U}_\lambda$ will depend on the spectral properties of the radiation since (17.41) involves the correlation function $\Gamma_{\mathscr{N}}^{(1,1)*}(x'', x) = \Gamma_{\mathscr{N}}^{(1,1)*}(x'' - x)$. The homogeneous integral equa-

† Jaiswal and Mehta [512] (see also [524]) have considered this problem for partially polarized light.

tion, in a more restricted form (time domain only), was solved for the Lorentzian profile of the spectral line in the previously quoted papers.

It is obvious that an exact calculation of the characteristic function is a very difficult problem although all the mathematics is relatively simple if $T \ll \tau_c \approx 1/\Delta\nu$ or $T \gg \tau_c$. In the first case the correlation function is practically constant over the time interval $(0, T)$ ($\mathscr{S}(x'' - x') = \eta\, \delta(x'' - x')\, \delta(x' - x)$ is assumed) so that $\varkappa_j^{(W)} = (j-1)!\, (\eta\langle I\rangle T)^j$ and

$$\langle \exp ix\overline{W}\rangle_{\mathcal{N}} = \exp\left[\sum_{n=1}^{\infty} \frac{(ix)^n}{n} (\eta\langle I\rangle T)^n\right] =$$

$$= \exp\left[-\log\left(1 - ix\eta\langle I\rangle T\right)\right] = \frac{1}{1 - ix\,\eta\langle I\rangle T}; \tag{17.42}$$

this corresponds to (10.45) for the Bose-Einstein distribution (10.43). In the second case we obtain (10.48) with $p(n, T)$ given by (10.50) or (10.73) where $M = T/\tau_c$. Hence formula (10.73) holds for $T \gg \tau_c$ and also for $T \ll \tau_c$ (if $M = 1$ in this case); this suggests that perhaps it represents a very good approximation for all T. Indeed it was shown by Bédard, Chang and Mandel [61] using the exact recursion relations for $p(n, T)$ derived by Bédard [56] that it represents an excellent approximation to $p(n, T)$ for all T if

$$M = \frac{T}{\xi(T)} = \frac{T^2}{2\int_0^T (T - t')\,|\gamma(t')|^2\, dt'}, \tag{17.43}$$

where (10.69) has been used. This was explicitly verified for the Lorentzian, Gaussian and rectangular forms of the spectrum. An interesting feature of Mandel's formula (10.73) is that it is independent of the form of the spectrum which enters through M only. Similar results have been obtained for the superposition of coherent and chaotic fields in [509, 510, 528, 532].

Another expression for $p(n, T)$ for a Lorentzian line shape, when

$$\langle n_\lambda\rangle\, \hbar\omega_\lambda = \frac{\text{const}}{(\omega_k - \omega_0)^2 + \Gamma^2} \tag{17.44}$$

(ω_0 is the mean frequency and Γ is halfwidth of the line), and $T \gg 1/\Gamma$, was derived by Glauber [172, 176, 177] (see also [17], p. 225). He found that

$$p(n, T) = \frac{1}{n!}\left(\frac{2\Omega T}{\pi}\right)^{1/2}\left(\frac{\Gamma\,\eta\langle I\rangle\, T}{\Omega}\right)^n K_{n-\frac{1}{2}}(\Omega T)\exp(\Gamma T), \tag{17.45}$$

where $\Omega = (\Gamma^2 + 2\eta\langle I\rangle \Gamma)^{1/2}$ and $K_{n-\frac{1}{2}}$ is a modified Hankel function of half-integral order. The asymptotic form of (17.45) is

$$p(n, T) = \frac{\langle n\rangle}{\sqrt{(2\pi\mu n^3)}} \exp\left[-\frac{1}{2\mu}\left(\sqrt{n} - \frac{\langle n\rangle}{\sqrt{n}}\right)^2\right], \qquad (17.46)$$

where $\mu = \eta\langle I\rangle/\Gamma$ and $\langle n\rangle = \eta\langle I\rangle T$. The formula (17.45) was also compared with the exact solution in [61] and was found to be in a good agreement for $T/\tau_c \gg 1$ (≈ 10). Yet another asymptotic expression for $p(n, T)$ has been suggested in [294].

As we have noted in Chapter 10 the experimental verification of the validity of (10.43) for chaotic light was given by Arecchi [41], Arecchi, Berné and Burlamacchi [44], Arecchi, Berné, Sona and Burlamacchi [46] (see also [42]), Pike [386, 387] and Freed and Haus [153]. The multimode expression (10.50) has been verified for Gaussian light with M degrees of freedom by Martienssen and Spiller [292], who also verified the validity of (10.58) for polarized ($P = 1$) and unpolarized ($P = 0$) light. A measurement of the time evolution of a stationary Gaussian field by means of joint photon-counting distributions was carried out by Arecchi, Berné and Sona [45].

In measurements of the photon-counting statistics, the so-called dead-time effect occurs; this means that for the counter there exists a time interval τ after each registration, during which no photoemission can be registered. Consequently the number of events registered during the counting interval T will be smaller than the actual number of events and the measured photon-counting distributions must be corrected. These questions have been studied by Johnson, Jones, McLean and Pike [218] and Bédard [59].

Finally we give the s-ordered form of some of the present equations. It is clear from (17.24) that the equations related to a particular s-ordering can be obtained by substituting $\langle n_\lambda\rangle \to \langle n_\lambda\rangle + (1 - s)/2$ in the corresponding equations. Thus we obtain from (16.69), considered for $s_1 = 1$ and $s_2 = s$, and (10.48)

$$\langle \exp ixW\rangle_s = \left[1 - ix\left(\frac{\langle n\rangle}{\eta M} + \frac{1-s}{2}\right)\right]^{-M}, \qquad (17.47)$$

where $\langle n\rangle$ is the mean number of counts. This leads to

$$P(W, s) = \frac{1}{\left(\dfrac{\langle n\rangle}{\eta M} + \dfrac{1-s}{2}\right)^M} \frac{W^{M-1}}{(M-1)!} \exp\left[-\frac{W}{\dfrac{\langle n\rangle}{\eta M} + \dfrac{1-s}{2}}\right] \qquad (17.48)$$

and

$$\langle W^k \rangle_s = \left(\frac{\langle n \rangle}{\eta M} + \frac{1-s}{2} \right)^k \frac{(k+M-1)!}{(M-1)!}. \tag{17.49}$$

Finally let us mention some results for Fermi particles. Corresponding to the bunching effect for photons, occurring as a consequence of the Bose-Einstein statistics with the requirement of maximum entropy, an antibunching effect for chaotic fermions can occur even though no coherent states exist for fermions. This point was discussed by Glauber [181] on the basis of wave-packets constructed for chaotic fields and by Bénard [62, 63, 494] using the theory, valid for bosons as well as fermions, of Goldberger and Watson [186]. For a discussion of the coherence properties of fermions we refer the reader to [254].

17.2 Laser radiation

The statistical properties of laser radiation (above the threshold of oscillations) are qualitatively different from those of Gaussian radiation; this fact was first recognized by Golay [182]. A complete investigation of the statistical and coherence properties of laser radiation and the laser itself requires the solution of the equations of motion for a system consisting of the radiation in interaction with the atoms of lasing active matter and reservoirs (heat baths), i.e. pumping light, cavity losses, lattice vibrations, etc. The main difference between laser radiation and thermal radiation lies in the fact that laser radiation is produced mainly by stimulated emission rather than by spontaneous emission and that very strong coupling of modes exists reflecting itself in the non-linearity of the equations of motion. This non-linearity of the laser theory is a characteristic one and only non-linear theories are able to explain successfully all properties of the laser. Moreover the non-linearity plays an important role in the stability of the laser. In the literature the statistical properties of laser radiation are treated in three ways using

a) the Langevin equations for the complex amplitudes (these are the usual equations of motion with stochastic forces describing a Markoff process added),

b) the Fokker-Planck equation for the probability distribution and

c) the master equation for the density matrix;

the last method is completely quantum-mechanical. It is advantageous to use the correspondence between functions of operators and quasi-probabilities discussed previously in the process of solving the master equation; this is done particularly in papers by Lax and his co-workers. The rather complicated analysis of such a treatment of the statistics of laser radiation lies outside the framework of this book, so we refer the reader to review articles by Haken [191, 192], Lax [246, 247], Risken [398 ,534] and Paul [339], where further references can be found, and also to papers by Lax [248, 250], Lax and Louisell [251], Lax and Yuen [252], Fleck [145–147], Scully and Lamb [416, 417] and Willis [469] (also further references cited therein). Several simplified models for devices of various sorts are treated in the book by Louisell [19] and Haken's work [192], particularly, includes a very complete analysis of the problem. Experimental verifications of the laser theory were reviewed by Armstrong and Smith [52]. An interesting approach to the problem based on an intensive use of the quantum electrodynamical formalism of Schwinger, using generalized Green functions and functional derivatives, was proposed by Korenman [238, 515]. In this section we treat only the case of the ideal laser and summarize the main results of the non-linear theory (the Risken distribution) describing the laser statistics over the whole region of operation (below, at and above the threshold of oscillations). A powerful model of the superposition of coherent and chaotic fields describing the statistical properties of the laser radiation above threshold, is applicable when a linear approximation to the theory can be used. This model will be given in the next section in greater detail and some experimental verifications will be mentioned.

Ideal laser model

As we have mentioned, Glauber [174, 176] showed that the field in a coherent state is produced by classical non-random currents. The radiation field is connected with the dipole transitions of atoms in the active medium of the laser. The polarization of these atoms oscillates with the field and they radiate energy into the field. The whole active medium has an oscillating polarization density and, since the time derivative of the polarization density gives the current distribution, the radiation field can be regarded as a product of oscillating classical (c-number) non-random currents if the laser is stabilized. Therefore the radiation field of the laser is in a coherent state $|\beta\rangle$. The density matrix describing this field is

$$\hat{\varrho} = |\beta\rangle \langle\beta| \tag{17.50}$$

and

$$\Phi_{\mathcal{N}}(\alpha) = \delta(\alpha - \beta) = \frac{\delta(|\alpha| - |\beta|)}{|\alpha|} \delta(\varphi - \psi), \tag{17.51}$$

where $\alpha = |\alpha| \exp i\varphi$ and $\beta = |\beta| \exp i\psi$; the factor $1/|\alpha|$ appears here since $d(\mathrm{Re}\,\alpha)\, d(\mathrm{Im}\,\alpha) = |\alpha|\, d|\alpha|\, d\varphi$. In the Schrödinger picture $\hat{\varrho} = |\beta \exp(-i\omega t)\rangle \langle \beta \exp(-i\omega t)|$ so that $\langle \hat{a} \rangle = \beta \exp(-i\omega t)$ is not a stationary state since it varies with time t. The distribution (17.51) expresses the fact that both the intensity $|\beta|^2$ and the phase φ of the complex amplitude are perfectly stabilized.

Unfortunately we usually have no information about the phase φ of oscillations at high frequencies and we have to assume that the phase is uniformly distributed in the interval $(0, 2\pi)$. Thus we obtain

$$\Phi_{\mathcal{N}}(\alpha) = \frac{1}{2\pi} \int_0^{2\pi} \frac{\delta(|\alpha| - |\beta|)}{|\alpha|}\, \delta(\varphi - \psi)\, d\varphi = \\ = \frac{\delta(|\alpha| - |\beta|)}{2\pi|\alpha|} = \frac{1}{\pi}\, \delta(|\alpha|^2 - |\beta|^2)\,. \tag{17.52}$$

The multimode function $\Phi_{\mathcal{N}}(\{\alpha_\lambda\})$ will have, for such a one-mode ideal laser, the form

$$\Phi_{\mathcal{N}}(\{\alpha_\lambda\}) = \frac{\delta(|\alpha_\lambda| - |\beta_\lambda|)}{2\pi|\alpha_\lambda|} \prod_{\lambda' \neq \lambda} \frac{\delta(|\alpha_{\lambda'}|)}{2\pi|\alpha_{\lambda'}|}\,. \tag{17.53}$$

Mixed states formed as a superposition of coherent states with various phases are also coherent and the field described by (17.53) fulfils the full coherence conditions (12.34) for $m = n$. The distribution (17.53) is independent of the phases $\{\varphi_\lambda\}$ and therefore it describes a stationary field. This distribution corresponds to the intensity of a perfectly stabilized laser where the phase is random (this is the case already described in classical terms by equation (10.61)); the corresponding photon-counting distribution is Poissonian and is given by (10.62). For the distribution $P(I)$ we have equation (10.61) again from (17.52) (or (17.53)) since $\langle I \rangle = (2\pi\hbar c/L^3)\, k^{-1}\,.$ $.\, |e_\mu^{(s)}(k)|^2\, |\alpha_{ks}|^2$.

The corresponding density matrix elements $\varrho(n, m)$ for the distribution (17.51) are obtained from (13.63) as

$$\varrho(n, m) = \frac{\beta^n \beta^{*m}}{\sqrt{(n!\, m!)}} \exp(-|\beta|^2) \tag{17.54}$$

and the normally ordered moments are equal to

$$\langle \hat{a}^{+k} \hat{a}^l \rangle_{\mathcal{N}} = \alpha^{*k} \alpha^l\,. \tag{17.55}$$

Putting $n = m$ in (17.54) we obtain the Poissonian probability $p(n) \equiv \varrho(n, n)$ that n photons are present in the field. The distribution $\Phi(\alpha, s)$ and the corresponding moments follow from (16.43) and (16.51) with $s_1 = 1$ and $s_2 = s$,

$$\Phi(\alpha, s) = \frac{2}{1-s} \exp\left(-\frac{2|\alpha - \beta|^2}{1-s}\right) \tag{17.56}$$

and

$$\langle \hat{a}^{+k} \hat{a}^l \rangle_s = \frac{l!}{k!}\left(\frac{1-s}{2}\right)^l \beta^{*k-l} L_l^{k-l}\left(\frac{2|\beta|^2}{s-1}\right), \quad k \geqq l. \tag{17.57}$$

The corresponding relations for the symmetric and antinormal orderings follow with $s = 0$ and $s = -1$.

To obtain the quantities $p(n)$ $[p(n, T)]$, $\langle \hat{a}^{+k} \hat{a}^l \rangle_{\mathcal{N}} = \delta_{kl} \langle \hat{a}^{+k} \hat{a}^k \rangle_{\mathcal{N}}$, $\Phi(\alpha, s)$ and $\langle \hat{a}^{+k} \hat{a}^l \rangle_s = \delta_{kl} \langle \hat{a}^{+k} \hat{a}^k \rangle_s$ corresponding to the phase averaged distribution (17.52) or (17.53) we may consider a more general case specified by

$$\Phi_{\mathcal{N}}(\{\alpha_\lambda\}) = \prod_{\lambda=1}^{M} \frac{\delta(|\alpha_\lambda|^2 - |\beta_\lambda|^2)}{\pi} \prod_{\lambda=M+1} \frac{\delta(|\alpha_\lambda|)}{2\pi|\alpha_\lambda|}. \tag{17.58}$$

The statistical properties of this model for $M = 2$ and for general M were investigated both theoretically and experimentally in [76, 78], particularly from the point of view of interference measurements of the Young type.

The distribution $P(W, s)$ in the normalization volume can be calculated as follows

$$P(W, s) = \int \Phi(\{\alpha'_\lambda\}, s)\, \delta(W - \sum_{\lambda=1}^{M} |\alpha'_\lambda|^2) \prod_{\lambda=1}^{M} \frac{d^2\alpha'_\lambda}{\pi} =$$
$$= \frac{1}{2\pi} \int_{-\infty}^{+\infty} \exp(-ixW) \langle \exp(ix \sum_{\lambda=1}^{M} |\alpha'_\lambda|^2) \rangle_s \, dx, \tag{17.59}$$

where the characteristic function is, considering M modes only in (17.58),

$$\langle \exp(ix \sum_{\lambda=1}^{M} |\alpha'_\lambda|^2) \rangle_s = \frac{2}{1-s} \int \prod_{\lambda=1}^{M} \exp\left(-\frac{2|\alpha'_\lambda - \beta_\lambda|^2}{1-s} + ix|\alpha'_\lambda|^2\right) \frac{d^2\alpha'_\lambda}{\pi} \tag{17.60}$$

(the terms $\delta(|\alpha_\lambda|)/2\pi|\alpha_\lambda|$ in (17.58) can be obtained by putting $\beta_\lambda = 0$ in some modes). Writing $|\alpha' - \beta|^2 = |\alpha'|^2 + |\beta|^2 - 2|\alpha'|\,|\beta| \cos(\varphi' - \psi)$ we obtain

$$\langle \exp(ix \sum_{\lambda=1}^{M} |\alpha'_\lambda|^2) \rangle_s = \int_0^\infty \prod_{\lambda=1}^{M} \frac{2}{1-s} \exp\left(-\frac{2|\beta_\lambda|^2}{1-s}\right).$$
$$\exp\left(-\frac{2|\alpha'_\lambda|^2}{1-s} + ix|\alpha'_\lambda|^2\right) I_0\left(4\frac{|\alpha'_\lambda|\,|\beta_\lambda|}{1-s}\right) d|\alpha'_\lambda|^2, \tag{17.61}$$

where I_0 is the modified Bessel function of zero order. Expressing I_0 in a power series and integrating over $|\alpha'_\lambda|^2$ we arrive at

$$\langle \exp ixW \rangle_s = \left(1 - ix\frac{1-s}{2}\right)^{-M} \exp\left[\frac{ix\langle n_c\rangle}{1 - ix\frac{1-s}{2}}\right], \qquad (17.62)$$

where $\langle n_c\rangle = \sum_\lambda |\beta_\lambda|^2$ is the mean number of photons in the coherent field. This expression is the generating function for the Laguerre polynomials $L_n^{M-1}(x)$ since

$$(1 - ixB)^{-M} \exp\frac{ixA}{1 - ixB} =$$

$$= \sum_{n=0}^{\infty} \frac{(ixB)^n}{\Gamma(n + M)} L_n^{M-1}\left(-\frac{A}{B}\right) \qquad (17.63a)$$

$$= (1 + B)^{-M} \exp\left(-\frac{A}{1+B}\right) .$$

$$\cdot \sum_{n=0}^{\infty} \frac{1}{\Gamma(n + M)}\left(1 + \frac{1}{B}\right)^{-n} (1 + ix)^n L_n^{M-1}\left(-\frac{A}{B(1+B)}\right). \qquad (17.63b)$$

Equation (17.62) also follows from (16.69) putting $s_2 = s$, $s_1 = 1$ and taking into account the fact that

$$P_{\mathcal{N}}(W) = \int \prod_\lambda \delta(|\alpha'_\lambda|^2 - |\beta_\lambda|^2)\, \delta(W - \sum_{\lambda'} |\alpha'_{\lambda'}|^2) \prod_{\lambda'} d|\alpha'_{\lambda'}|^2 = \delta(W - \langle n_c\rangle) . \qquad (17.64)$$

This corresponds to

$$p(n) = \sum_{\{n_\lambda\}}' \prod_\lambda \frac{|\beta_\lambda|^{2n_\lambda}}{n_\lambda!} \exp(-|\beta_\lambda|^2) = \frac{(\sum_\lambda |\beta_\lambda|^2)^n}{n!} \exp(-\sum_\lambda |\beta_\lambda|^2), \qquad (17.65)$$

where the summation $\sum'$ is restricted by the condition $\sum_\lambda n_\lambda = n$ and the multinomial theorem has been used. Substituting (17.62) into (17.59) and using the residuum theorem (or substituting (17.64) into (16.71) with $s_1 = 1$ and $s_2 = s$), we obtain

$$P(W, s) = \frac{2}{1-s}\left(\frac{W}{\langle n_c\rangle}\right)^{(M-1)/2} \exp\left[-\frac{2(W + \langle n_c\rangle)}{1-s}\right] I_{M-1}\left(4\frac{\sqrt{(W\langle n_c\rangle)}}{1-s}\right); \qquad (17.66)$$

the moments $\langle W^k \rangle_s$ are determined from (17.62) by using (17.63a)

$$\langle W^k \rangle_s = \frac{d^k}{d(ix)^k} \langle \exp ixW \rangle_s = \frac{k!}{\Gamma(k+M)} \left(\frac{1-s}{2}\right)^k L_k^{M-1}\left(\frac{2\langle n_c \rangle}{s-1}\right). \quad (17.67)$$

For the normal ordering $s = 1$ and the characteristic function (17.62) is

$$\langle \exp ixW \rangle_{\mathcal{N}} = \exp\left(ix\langle n_c \rangle\right), \quad (17.68)$$

i.e. the photon statistics are Poissonian and

$$p(n) = \frac{\langle n_c \rangle^n}{n!} \exp\left(-\langle n_c \rangle\right) \quad (17.69)$$

independently of M; the normal moments are

$$\langle W^k \rangle_{\mathcal{N}} = \langle n_c \rangle^k. \quad (17.70)$$

Equations (17.66) and (17.67) tend to (17.64) and (17.70) in the limit $s \to 1$.

The results corresponding to the one-mode distribution (17.52) can be obtained for $M = 1$.

Realistic laser model

A model more realistic than that of the ideal laser was investigated by Glauber [176] and also by Klauder and Sudarshan [17] (Sec. 9.2). This model based upon the phase diffusion of the complex amplitude leads to a Lorentzian spectrum for laser radiation.

Risken [396, 397] derived, using non-linear laser theory and the so-called rotating wave approximation of the Van der Pol oscillator, the Fokker-Planck equation for the distribution $P_{\mathcal{N}}(r)$

$$\frac{\partial P_{\mathcal{N}}}{\partial t} + \frac{1}{r}\frac{\partial}{\partial r}\left[(w - r^2)\, r^2 P_{\mathcal{N}}\right] = \frac{1}{r}\frac{\partial}{\partial r}\left(r \frac{\partial P_{\mathcal{N}}}{\partial r}\right) + \frac{1}{r^2}\frac{\partial^2 P_{\mathcal{N}}}{\partial \varphi^2} \quad (17.71)$$

(r and φ are the modulus and the phase of the complex field amplitude). Equation (17.71) is written for the normalized quantities; w is the pumping parameter ($w < 0$ below threshold, $w = 0$ at threshold and $w > 0$ above the threshold of oscillations). The steady-state solution is

$$P_{\mathcal{N}}(r) = \frac{N}{2\pi} \exp\left(-\frac{r^4}{4} + w\frac{r^2}{2}\right) = \frac{N}{2\pi} \exp\left(\frac{w^2}{4}\right) \exp\left[-\tfrac{1}{4}(r^2 - w^2)^2\right], \quad (17.72)$$

where

$$\frac{1}{N} = \int_0^\infty \exp\left(-\frac{r^4}{4} + w\frac{r^2}{2}\right) r\, dr\,. \tag{17.73}$$

Substituting $r^2 = I/\sqrt{(\pi)}\, I_0$, where I is the intensity of the field and I_0 is the average intensity at threshold, we can rewrite (17.72) in the form

$$P_{\mathcal{N}}(I) = \frac{2}{\pi I_0} \frac{\exp(-w^2)}{1 + \operatorname{erf} w} \exp\left(-\frac{I^2}{\pi I_0^2} + 2\frac{wI}{\sqrt{(\pi)}\, I_0}\right) =$$

$$= \frac{2}{\pi I_0} \frac{1}{1 + \operatorname{erf} w} \exp\left[-\frac{1}{\pi I_0^2}(I - \sqrt{(\pi)}\, w I_0)^2\right], \quad I \geqq 0\,,$$

$$P_{\mathcal{N}}(I) = 0, \quad I < 0\,, \tag{17.74}$$

where $\operatorname{erf} w = (2/\sqrt{\pi}) \int_0^w \exp(-x^2)\, dx$. The same result was also obtained by Hemstead and Lax [195], Fleck [145, 146], Scully and Lamb [415], Weidlich, Risken and Haken [464] and Lax [247].

For the mean intensity we obtain

$$\langle I \rangle_{\mathcal{N}} = I_0 \left(\sqrt{(\pi)}\, w + \frac{\exp(-w^2)}{1 + \operatorname{erf} w}\right). \tag{17.75}$$

The photon-counting distribution has been calculated by Smith and Armstrong [425] (see also [52]) in the form

$$p(n, T) = \frac{1}{\sqrt{\pi}} \frac{D^n}{n!} \frac{\exp(-wD + \frac{1}{4}D^2)}{1 + \operatorname{erf} w}$$

$$\cdot \sum_{m=0}^{n} \binom{n}{m} c^m \left[\Gamma\left(\frac{n-m+1}{2}\right) + (-1)^{n-m}\, \Gamma\left(\frac{n-m+1}{2}, c^2\right)\right],$$

$$c \geqq 0\,,$$

$$= \frac{1}{\sqrt{\pi}} \frac{D^n}{n!} \frac{\exp(-wD + \frac{1}{4}D^2)}{1 + \operatorname{erf} w}$$

$$\cdot \sum_{m=0}^{n} \binom{n}{m} (-1)^m |c|^m \left[\Gamma\left(\frac{n-m+1}{2}\right) - \Gamma\left(\frac{n-m+1}{2}, |c|^2\right)\right],$$

$$c < 0\,, \tag{17.76}$$

where $D = \sqrt{(\pi)}\, \eta I_0 T$, $c = w - D/2$, η is the detector efficiency, T is the counting interval, $\Gamma(a, x) = \int_x^\infty \exp(-t)\, t^{a-1}\, dt$ is the incomplete gamma function and $\Gamma(a, 0) \equiv \Gamma(a)$ is the gamma function. Another form of

$p(n, T)$ was obtained by Bédard [57]

$$p(n, T) = \frac{N\mu^n}{\sqrt{(2\pi)}} \exp\left[\tfrac{1}{4}(\mu^2 - 2J - 2w^2)\right] D_{-n-1}(\mu - \sqrt{(2)}\, w), \qquad (17.77)$$

where $\mu = \sqrt{(\pi/2)}\, \eta I_0 T$, $J = \sqrt{(\pi)}\, \eta I_0 T w$, $N = 2/(1 + \operatorname{erf} w)$ and $D_n(z)$ is the parabolic cylinder function ([14], p. 1064). The factorial moments are (see [57, 51])

$$\langle \overline{W}^k \rangle_{\mathcal{N}} = \frac{N\mu^k}{\sqrt{(2\pi)}}\, k! \exp\left(-\tfrac{1}{2}w^2\right) D_{-k-1}(-\sqrt{(2)}\, w). \qquad (17.78)$$

There are three typical regions of operation for the laser. Below threshold $w < 0$ and if $|w|$ is sufficiently large, then

$$P_{\mathcal{N}}(I) \approx \text{const.} \exp\left(-\frac{2|w|\, I}{\sqrt{(\pi)}\, I_0}\right), \qquad (17.79)$$

which is an exponential distribution; thus the laser radiation below threshold is Gaussian. Near and at threshold the correct distribution $P_{\mathcal{N}}(I)$ is given by (17.74). Well above threshold, where the linearized theory is appropriate and $w > 0$, we may write $\sqrt{I} = \sqrt[4]{(\pi)}\, \sqrt{(wI_0)}\, (1 + \varepsilon)$ where ε is small, since the function (17.74) is sharply peaked and so we have from (17.74)

$$\begin{aligned} P_{\mathcal{N}}(I) &= \frac{2}{\pi I_0} \frac{1}{1 + \operatorname{erf} w} \exp\left[-w^2((1 + \varepsilon)^2 - 1)^2\right] \approx \\ &\approx \text{const.} \exp\left[-4w^2 \frac{[\sqrt{(I)} - \sqrt[4]{(\pi)}\, \sqrt{(wI_0)}]^2}{\sqrt{(\pi)}\, wI_0}\right] = \\ &= \text{const.} \exp\left[-\frac{4w}{\sqrt{(\pi)}\, I_0}\, [\sqrt{(I)} - \sqrt{(\sqrt{(\pi)}\, wI_0)}]^2\right]. \end{aligned} \qquad (17.80)$$

The amplitude distribution $P_{\mathcal{N}}(\sqrt{I})$ is a Gaussian function centered at $\sqrt{I} = \sqrt{(\sqrt{(\pi)}\, wI_0)}$ and it tends to the distribution $\delta(I - \langle I \rangle)$ in the limit $w \to \infty$, valid for an ideal laser. The distribution (17.74) can be regarded as the "smooth" δ-distribution $\delta(I - \sqrt{(\pi)}\, wI_0)$ in the form of a Gaussian distribution; such a model of laser statistics was also proposed in [55]. Some further laser models were discussed in [424] and [452].

Two- and multimode models for the laser were discussed in [276, 450, 76, 78, 313]. A quantum-mechanical description of laser beams and their interaction with matter was treated in [339] and in references cited therein as well as in [342, 394, 344, 343, 99, 100, 101]. The transient solution of the Fokker-Planck equation (17.71) has been investigated in [399].

Now we mention some experimental studies of laser statistic, and those verifying the present model of the laser photon statistics can also be divided into three groups – a) below threshold, b) near and at threshold and c) above threshold.

a) Below threshold

As we have seen the photon statistics in this region are Bose-Einstein. The primary work on laser light below threshold was done by Freed and Haus [153] and Smith and Armstrong [426] who verified the validity of the Bose-Einstein distribution (10.43) for $T < \tau_c$. Freed and Haus [153] also measured sub-threshold counting statistics for $T > \tau_c$ which leads to the verification of (17.46) and (10.50) with M given by (17.43). A review of these experimental results can be found in [52]. Similar results were obtained by Arecchi [41] (see also [44, 46]) who started with a stabilized laser field from a laser far above threshold and then scattered this beam from a moving scattering plate.

b) Near and at threshold

Near the threshold the non-linear theory must be used for the description of the photon statistics and equations (17.74), (17.76), (17.77) and (17.78)

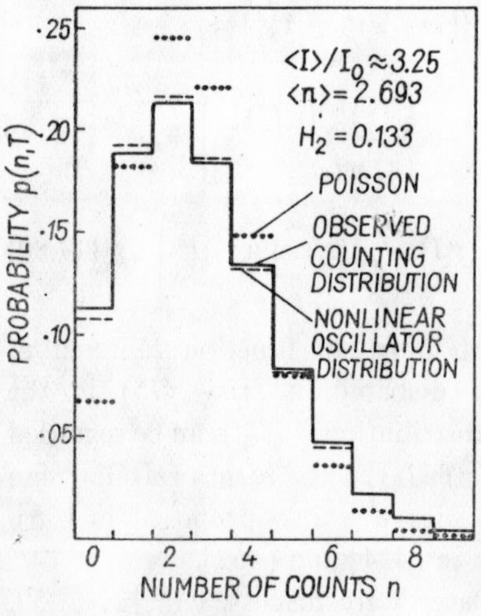

Fig. 17.1. Counting distribution observed just above threshold (solid line). The Poisson distribution (dotted lines) and the nonlinear oscillator distribution giving the best fit (dashed line) are also shown. For $n = 7$, 8 and 9 the non-linear oscillator and observed distributions are coincident. [After A. W. Smith and J. A. Armstrong, *Phys. Rev. Lett.* **16**, 1169 (1966); reprinted with permission.]

are appropriate. Equation (17.76) was shown to be valid by Smith and Armstrong [425] (see Fig. 17.1) who also measured the reduced factorial moment $H_2 = \langle \hat{n}(\hat{n}-1)\rangle/\langle \hat{n}\rangle^2 - 1 = \langle W^2\rangle_{\mathcal{N}}/\langle W\rangle_{\mathcal{N}}^2 - 1 = \varkappa_2^{(W)}/\varkappa_1^{(W)2}$,

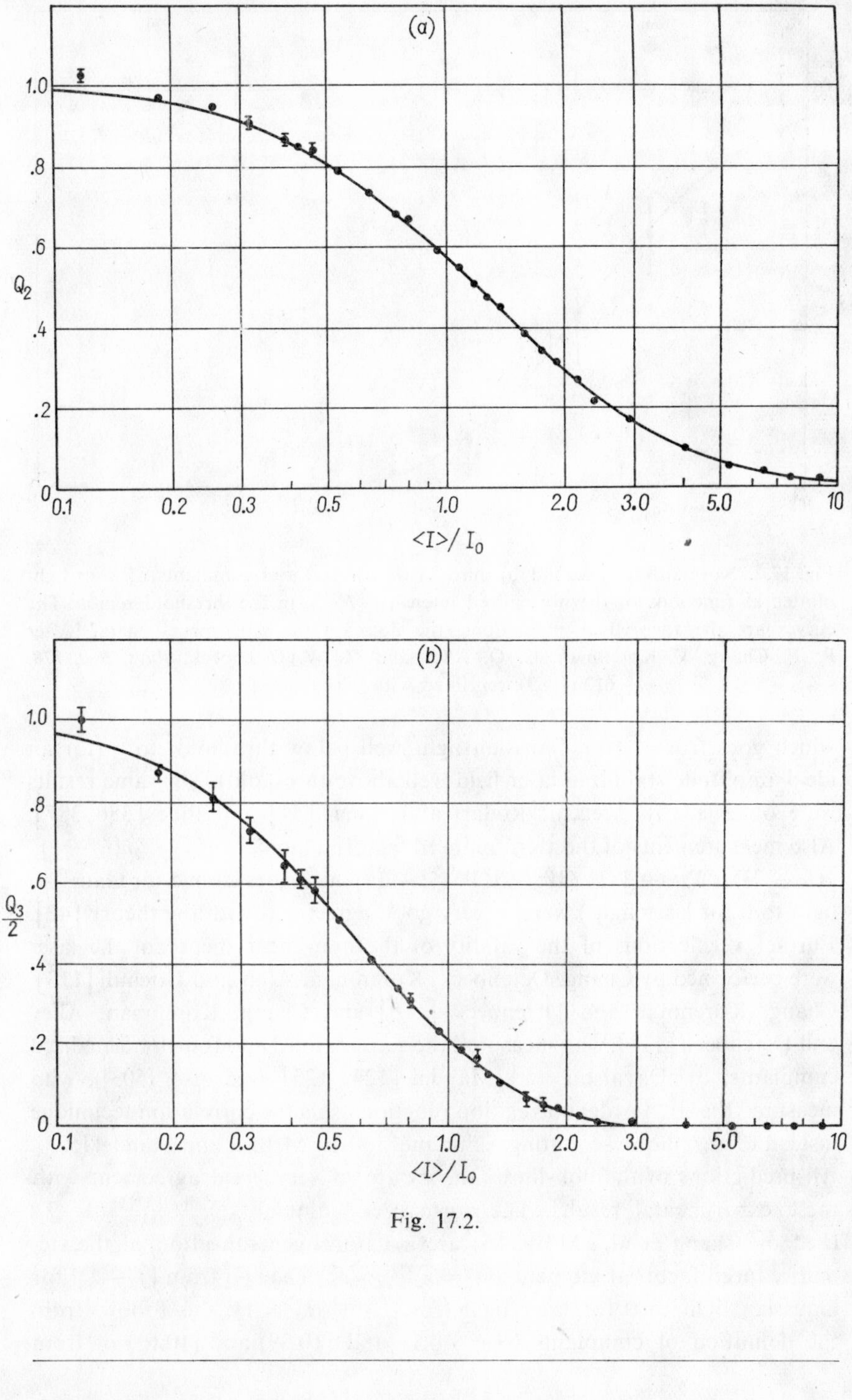
(a)
1.0
.8
.6
Q_2
.4
.2
0
0.1
0.2
0.3
0.5
1.0
2.0
3.0
5.0
10
$\langle I\rangle/I_0$
(b)
1.0
.8
.6
$\frac{Q_3}{2}$
.4
.2
0
0.1
0.2
0.3
0.5
1.0
2.0
3.0
5.0
10
$\langle I\rangle/I_0$

Fig. 17.2.

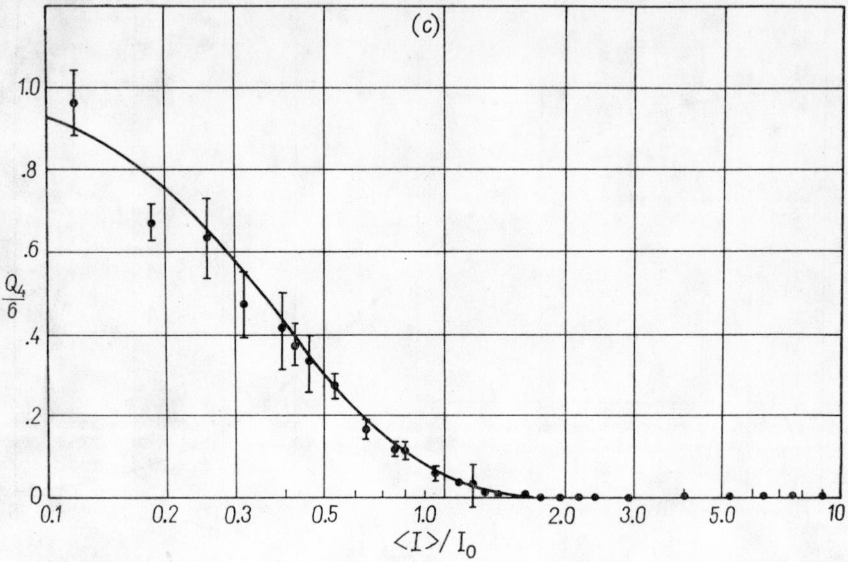

Fig. 17.2. Normalized a) second b) third c) fourth factorial cumulants of laser light plotted as functions of the normalized intensity $\langle I \rangle / I_0$ in the threshold region. The curves are the theoretical predictions, the dots are the experimental data. [After R. F. Chang, V. Korenman, C. O. Alley and R. W. Detenbeck, *Phys. Rev.* ***178***, 612 (1969); reprinted with permission.]

which goes from 1 (for Gaussian light well below threshold) to 0 (for an ideal amplitude stabilized laser field well above threshold). The same results were obtained by Arecchi, Rodari and Sona [51] and Pike [386, 387]. Also measurements of the third reduced factorial moment $H_3 = \langle \hat{n}(\hat{n} - 1) . . (\hat{n} - 2) \rangle / \langle \hat{n} \rangle^3 - 1 = \langle W^3 \rangle_{\mathcal{N}} / \langle W \rangle^3_{\mathcal{N}} - 1$ (which goes from 5 for Gaussian light to 0 for laser light) were in very good agreement with the theory [42]. Further verifications of the validity of the non-linear theory of the laser were performed by Chang, Detenbeck, Korenman, Alley and Hochuli [113], Chang, Korenman and Detenbeck [115] and Chang, Korenman, Alley and Detenbeck [114], who measured the second, third and fourth normalized cumulants, by Davidson and Mandel [124, 125] (see also [503]), who measured the sixth-order correlation function using the correlation technique instead of the photon-counting technique, and by Meltzer and Mandel [525]. All predictions of the non-linear theory are in very good agreement with these experimental results. The normalized cumulants $\varkappa_j^{(W)} / \varkappa_1^{(W)j}$ $(\equiv Q_j)$ used by Chang et al. [113–115] are a natural generalization of the second reduced factorial moment $H_2 = \varkappa_2^{(W)} / \varkappa_1^{(W)2}$. They go from $(j - 1)!$ for Gaussian light to 0 for laser light (for $j > 1$, $H_1 = 1$); this follows from the definition of cumulants (Sec. 10.1) and (10.39) and (10.61) or from the definition of cumulants and characteristic functions (10.45) and (17.68). The measurements of Chang et al. [113–115] verify the theory in the threshold region from 1/10 to 10 times the threshold intensity. Their results are shown in Fig. 17.2.

Some measurements of the dynamics of laser radiation at threshold have also been carried out [49].

c) Above threshold

In the region above threshold ($I/I_0 > 5$) the model of the superposition of fully coherent and Gaussian radiation described by the distribution (17.80) (which will be studied in greater detail in the next section) is appropriate for the description of laser statistics. Important work has been

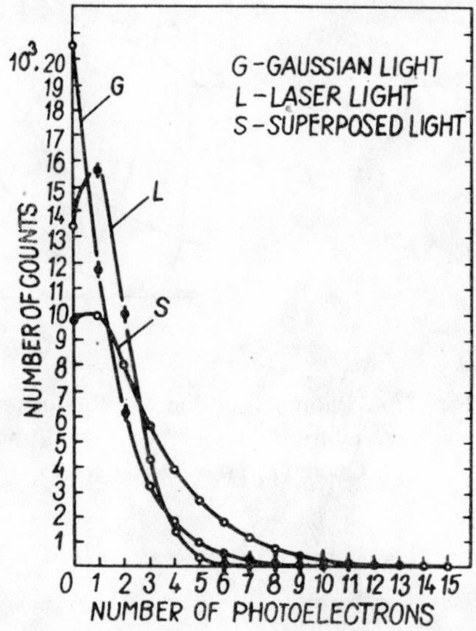

Fig. 17.3. Photon-counting distribution for Gaussian (*G*), laser (*L*) and superimposed fields (*S*). [After F. T. Arecchi, A. Berné and P. Burlamacchi, *Phys. Rev. Lett.* ***16***, 32 (1966); reprinted with permission.]

done on this region by Arecchi, Berné and Burlamacchi [44] (see also [46]) and also by Smith and Armstrong [425], Freed and Haus [154], Martienssen and Spiller [293] and Magill and Soni [260]. Photon-counting distributions observed by Arecchi, Berné and Burlamacchi [44] are shown in Fig. 17.3. These results verify the validity of the superposition model of coherent and Gaussian light for the laser light above threshold where the linear approximation to the non-linear laser theory is appropriate ($I/I_0 > 5$). The corresponding equations which have been verified by experiment are derived in the next section. The curve *L* in Fig. 17.3 confirms that well above threshold $p(n, T)$ is Poissonian so that in this region the laser field is in a phase averaged coherent state. The ideal laser model previously given is then appropriate.

The photon-counting distribution of modulated laser beam was obtained and experimentally determined by Pearl and Troup [345] (for further

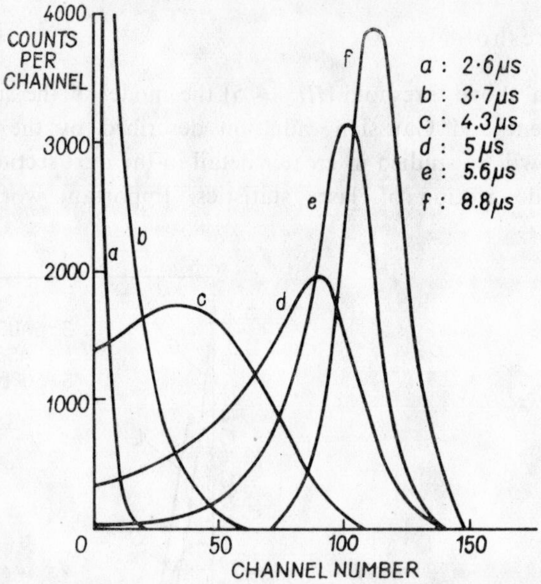

Fig. 17.4. Photon-counting distributions with different time delays for a single-mode gas laser in transient operating conditions. [After F. T. Arecchi, V. Degiorgio and B. Querzola, *Phys. Rev. Lett.* **19**, 1168 (1967); reprinted with permission.]

studies of the photon statistics of modulated beams see [495, 504, 532] and references cited therein).

The transient solution of the laser Fokker-Planck equation (17.71) was investigated by Risken and Vollmer [399] and the transient statistics was measured by Arecchi, Degiorgio and Querzola [47] (see also [42]). The results are shown in Fig. 17.4. The statistics vary from the Bose-Einstein distributions a, b, corresponding to operation below threshold, to the stabilized Poisson distribution f well above threshold.

17.3 Superposition of coherent and chaotic fields as a model of laser light statistics

Because of their importance in describing laser-light statistics above the threshold of operation, the statistics of a superposition of coherent and chaotic fields have been studied theoretically by a number of authors; this work is also relevant to other branches of physics as a model of the superposition of signal and noise. The first papers dealing quantum-mechanically with this subject were published by Lachs [241], Troup [451] and Glauber [177]; in these the photon-counting distribution and its factorial moments for the superposition of one-mode coherent and narrow-band chaotic fields with the same mean frequencies were derived. Another approach to this problem, based on calculation of the correlation functions, was developed by Morawitz [324, 325]. These results were generalized to multimode fields by Peřina [358, 361, 360] and by Peřina and Mišta [370] (see also [518, 507]). An extension with respect to the s-ordering was given by Peřina and Horák [365, 367]. In a recent paper by Jakeman and Pike [214] it was pointed out that additional spectral information can be obtained using heterodyne detection the chaotic field being superimposed, before detection, on a known coherent component. In this way the central frequency of the chaotic field can be determined. However, the general heterodyne detection problem for chaotic (thermal) light includes the case when both the central frequency of the chaotic field and the frequency of the coherent field are arbitrary and differ from one another. Some one-mode results obtained by these authors were completed and generalized to multimode fields (with regard to the s-ordering) by Peřina and Horák [366]†.

The function $\Phi_{\mathcal{N}}(\{\alpha_\lambda\})$ for the superposition of coherent and chaotic fields can be derived as the convolution of the Gaussian function (17.7)

† Recently an experimental study [514] devoted to the statistics of heterodyne detection of Gaussian (Lorentzian) light and papers [501, 513, 529] and to the study of cumulants for the superposition of coherent and chaotic fields has appeared. A detailed analysis of formulae derived in this section for the superposition of multimode coherent and chaotic fields has been performed and their use as approximate formulae (for an arbitrary spectrum of chaotic light, particularly for a Lorentzian spectrum and arbitrary counting time intervals) has been investigated in [509, 510]. The similaɹ results for two-photon absorpiton have been given in [528] and for partially polarized light in [532]. A review of the statistical and coherence properties of the superposition of coherent and chaotic fields has been given in [530] including multiphoton absorption, modulation, propagation through turbulent atmosphere and Gaussian media and time development of the photon-counting statistics. The last work represents an extension of the present section.

and the δ-function distribution $\delta(\{\alpha_\lambda\} - \{\beta_\lambda\}) = \prod_\lambda^M \delta(\alpha_\lambda - \beta_\lambda)$ according to the quantum superposition law (13.147), i.e.

$$\Phi_{\mathcal{N}}(\{\alpha_\lambda\}) = \iint \prod_\lambda^M \delta(\alpha_\lambda - \alpha'_\lambda - \alpha''_\lambda)\, \delta(\alpha'_\lambda - \beta_\lambda) \frac{\exp\left[-\frac{|\alpha''_\lambda|^2}{\langle n_{ch\lambda}\rangle}\right]}{\pi\langle n_{ch\lambda}\rangle}\, d^2\alpha'_\lambda\, d^2\alpha''_\lambda =$$

$$= \prod_\lambda^M \frac{1}{\pi\langle n_{ch\lambda}\rangle} \exp\left[-\frac{|\alpha_\lambda - \beta_\lambda|^2}{\langle n_{ch\lambda}\rangle}\right], \tag{17.81}$$

this is the multimode form of (13.95) (we write $\langle n_{ch\lambda}\rangle$ instead of $\langle n_\lambda\rangle$ for the mean number of photons in the mode λ of the chaotic field in this section to distinguish this from the mean number $\langle n_{c\lambda}\rangle$ of photons in the coherent field). The multimode form of (16.43) for $s_2 = s$ and $s_1 = 1$ gives

$$\Phi(\{\alpha_\lambda\}, s) = \prod_\lambda^M \frac{1}{\langle n_{ch\lambda}\rangle + \frac{1-s}{2}} \exp\left[-\frac{|\alpha_\lambda - \beta_\lambda|^2}{\langle n_{ch\lambda}\rangle + \frac{1-s}{2}}\right]. \tag{17.82}$$

This equation reduces to (17.24) if $\{\beta_\lambda\} = 0$. For $s = -1$ we can obtain the function $\Phi_{\mathcal{A}}(\{\alpha_\lambda\})$ for the superposition of coherent and chaotic fields (cf. (13.98)).

One-mode field

First we assume a one-mode field and the mode index λ will be omitted for simplicity. Substituting $\Phi_{\mathcal{N}}(\alpha)$ into (13.63) we find, performing the integration, $\varrho(n, m)$ in the form (13.129), which led to the correct $\Phi_{\mathcal{N}}(\alpha)$ in the form (13.95). Another way of calculating $\varrho(n, m)$ is to introduce the function $R(z^*, z, \varepsilon)$ as

$$R(z^*, z, \varepsilon) = \int \frac{\exp\left[-\frac{|\alpha - \beta|^2}{\langle n_{ch}\rangle}\right]}{\pi\langle n_{ch}\rangle} \exp\left[-\varepsilon|\alpha|^2 + z^*\alpha + z\alpha^*\right] d^2\alpha =$$

$$= \frac{1}{\varepsilon\langle n_{ch}\rangle + 1} \exp\left[-\frac{\varepsilon|\beta|^2}{\varepsilon\langle n_{ch}\rangle + 1}\right] \exp\left[\frac{\langle n_{ch}\rangle}{\varepsilon\langle n_{ch}\rangle + 1}|z|^2 + \frac{\beta^* z + \beta z^*}{\varepsilon\langle n_{ch}\rangle + 1}\right], \tag{17.83}$$

where (16.41) has been used and $\varepsilon \geqq 0$ is a real number. Using relations of the form (16.52) and (16.54) with $(s_2 - s_1)/2$ replaced by $\langle n_{ch}\rangle/(\varepsilon\langle n_{ch}\rangle$

$+ 1)$, $\hat{a}^+$ by $\beta^*/(\varepsilon\langle n_{ch}\rangle + 1)$ and $\hat{a}$ by $-\beta/(\varepsilon\langle n_{ch}\rangle + 1)$ we obtain from (16.54b)

$$R(z^*, z, \varepsilon) =$$

$$= \frac{1}{\varepsilon\langle n_{ch}\rangle + 1} \exp\left[-\frac{\varepsilon|\beta|^2}{\varepsilon\langle n_{ch}\rangle + 1}\right] \sum_{j=0}^{\infty} \sum_{l=0}^{\infty} \frac{z^j}{j!\,j!} \left(\frac{\beta^*}{\varepsilon\langle n_{ch}\rangle + 1}\right)^{j-l} \cdot$$

$$\cdot \left(\frac{\langle n_{ch}\rangle}{\varepsilon\langle n_{ch}\rangle + 1}\right)^l z^{*l} L_l^{j-l}\left(-\frac{|\beta|^2}{\langle n_{ch}\rangle(\varepsilon\langle n_{ch}\rangle + 1)}\right). \qquad (17.84)$$

For $\varrho(n, m)$, [322], we have

$$\varrho(n, m) = \frac{1}{\sqrt{(n!\,m!)}} \frac{\partial^n}{\partial z^{*n}} \frac{\partial^m}{\partial z^m} R(z^*, z, 1)\Big|_{z=z^*=0} =$$

$$= \frac{1}{m!} \sqrt{\left(\frac{n!}{m!}\right)} \frac{1}{\langle n_{ch}\rangle + 1} \exp\left[-\frac{|\beta|^2}{\langle n_{ch}\rangle + 1}\right] \frac{\langle n_{ch}\rangle^n}{(\langle n_{ch}\rangle + 1)^m} \beta^{*m-n} .$$

$$. L_n^{m-n}\left(-\frac{|\beta|^2}{\langle n_{ch}\rangle(\langle n_{ch}\rangle + 1)}\right), \quad m \geqq n , \qquad (17.85)$$

which is just (13.129). If $n > m$, the condition $\varrho(n, m) = \varrho^*(m, n)$ can be used. The normal moments $\langle \hat{a}^{+k}\hat{a}^l\rangle_{\mathcal{N}}$ can be calculated from (17.84) as follows [322, 370],

$$\langle \hat{a}^{+k}\hat{a}^l\rangle_{\mathcal{N}} = \frac{\partial^k}{\partial z^k} \frac{\partial^l}{\partial z^{*l}} R(z^*, z, 0)\Big|_{z=z^*=0} =$$

$$= \frac{l!}{k!} \langle n_{ch}\rangle^l \beta^{*k-l} L_l^{k-l}\left(-\frac{|\beta|^2}{\langle n_{ch}\rangle}\right), \quad k \geqq l ; \qquad (17.86)$$

if $k < l$, then $\langle \hat{a}^{+k}\hat{a}^l\rangle_{\mathcal{N}} = \langle \hat{a}^{+l}\hat{a}^k\rangle_{\mathcal{N}}^*$. This is in agreement with (17.57) with $(1 - s)/2$ replaced by $\langle n_{ch}\rangle$.

All these equations can easily be extended to the s-ordering by the substitution $\langle n_{ch}\rangle \to \langle n_{ch}\rangle + (1 - s)/2$, as follows from (17.82).

Equations (17.85) and (17.86) also yield $\varrho(n,m)$ and $\langle \hat{a}^{+k}\hat{a}^l\rangle_{\mathcal{N}}$ for the time dependent function $\phi_{\mathcal{N}}$ of the parametric amplifier given by (13.155) if β is replaced by $\bar{\alpha}(\alpha_0, \beta_0, t)$ and $\langle n_{ch}\rangle$ by $(\sinh \varkappa t)^2$.

The special cases of coherent and chaotic fields can be obtained in the limits $\langle n_{ch}\rangle \to 0$ and $\beta \to 0$ respectively.

Equation (17.85) for $m = n$ gives the probability that n photons are in the field. The moments (17.86) for $k = l$ are measurable by photodetectors.

Multimode field

Employing calculations (17.60) – (17.62) for the distribution function (17.82) we obtain

$$\langle \exp ixW \rangle_s =$$

$$= \prod_{\lambda} \left[1 - ix\left(\langle n_{ch\lambda} \rangle + \frac{1-s}{2} \right) \right]^{-1} \exp \left[\frac{ix \langle n_{c\lambda} \rangle}{1 - ix\left(\langle n_{ch\lambda} \rangle + \frac{1-s}{2} \right)} \right] \tag{17.87a}$$

$$= \sum_{n=0}^{\infty} (ix)^n \sum_{\substack{\sum n_\lambda = n \\ \lambda}} \prod_{\lambda} \frac{\left(\langle n_{ch\lambda} \rangle + \frac{1-s}{2} \right)^{n_\lambda}}{n_\lambda!} L_{n_\lambda}^0 \left(- \frac{\langle n_{c\lambda} \rangle}{\langle n_{ch\lambda} \rangle + \frac{1-s}{2}} \right) \tag{17.87b}$$

$$= \sum_{n=0}^{\infty} (1 + ix)^n \sum_{\substack{\sum n_\lambda = n \\ \lambda}} \prod_{\lambda} \left(1 + \langle n_{ch\lambda} \rangle + \frac{1-s}{2} \right)^{-1} .$$

$$. \exp \left[- \frac{\langle n_{c\lambda} \rangle}{1 + \langle n_{ch\lambda} \rangle + \frac{1-s}{2}} \right] \frac{1}{n_\lambda!} \left(1 + \frac{1}{\langle n_{ch\lambda} \rangle + \frac{1-s}{2}} \right)^{-n_\lambda} .$$

$$. L_{n_\lambda}^0 \left(- \frac{\langle n_{c\lambda} \rangle}{\left(\langle n_{ch\lambda} \rangle + \frac{1-s}{2} \right)\left(\langle n_{ch\lambda} \rangle + \frac{1-s}{2} + 1 \right)} \right), \tag{17.87c}$$

where we have also used (17.63a, b) and $\langle n_{ch\lambda} \rangle$ and $\langle n_{c\lambda} \rangle$ are the mean numbers of photons in the mode λ of the chaotic and coherent fields respectively.

Thus we arrive at [370]

$$\langle W^k \rangle_s = \frac{d^k}{d(ix)^k} \langle \exp ixW \rangle_s \bigg|_{ix=0} =$$

$$= k! \sum_{\substack{\sum n_\lambda = k \\ \lambda}} \prod_{\lambda} \frac{1}{n_\lambda!} \left(\langle n_{ch\lambda} \rangle + \frac{1-s}{2} \right)^{n_\lambda} L_{n_\lambda}^0 \left(- \frac{\langle n_{c\lambda} \rangle}{\langle n_{ch\lambda} \rangle + \frac{1-s}{2}} \right) \tag{17.88}$$

and

$$p(n) = \frac{1}{n!} \frac{d^n}{d(ix)^n} \langle \exp ixW \rangle_{\mathcal{N}} \bigg|_{ix=-1} =$$

$$= \sum_{\substack{\sum n_\lambda = n \\ \lambda}} \prod_\lambda (1 + \langle n_{ch\lambda} \rangle)^{-1} \exp \left[- \frac{\langle n_{c\lambda} \rangle}{\langle n_{ch\lambda} \rangle + 1} \right] \frac{1}{n_\lambda!} \cdot$$

$$\cdot \left(1 + \frac{1}{\langle n_{ch\lambda} \rangle}\right)^{-n_\lambda} L^0_{n_\lambda} \left(- \frac{\langle n_{c\lambda} \rangle}{\langle n_{ch\lambda} \rangle (1 + \langle n_{ch\lambda} \rangle)} \right). \qquad (17.89)$$

The distribution $P(W, s)$ may in principle be obtained from

$$P(W, s) = W^{M-1} \exp(-W) \sum_{n=0}^{\infty} \frac{n!\, L_n^{M-1}(W)}{\Gamma(n + M)} \sum_{j=0}^{n} \frac{(-1)^j \langle W^j \rangle_s}{j!\,(n-j)!\,\Gamma(j + M)}, \qquad (17.90)$$

which is analogous to (10.31), under the assumption that

$$\int_0^\infty P^2(W, s)\, W^{1-M} \exp(W)\, dW = \sum_{n=0}^{\infty} \frac{[\Gamma(n + M)]^3}{n!} c_n^2 < \infty, \qquad (17.91)$$

where $c_n = (n!/\Gamma(n + M)) \sum_{j=0}^{n} (-1)^j \langle W^j \rangle_s / [j!\,(n-j)!\,\Gamma(j + M)]$.

These formulae can be made simpler if all mean occupation numbers per mode in the chaotic field are equal, i.e. it is $\langle n_{ch\lambda} \rangle = \langle n_{ch} \rangle / M$, where $\langle n_{ch} \rangle$ is the mean total number of photons in the chaotic field and M is the number of modes. Equations (17.87a), (17.88) and (17.89) give in this case [365, 367]

$$\langle \exp ixW \rangle_s = \left[1 - ix \left(\frac{\langle n_{ch} \rangle}{M} + \frac{1-s}{2} \right) \right]^{-M} \cdot$$

$$\cdot \exp \left[\frac{ix \langle n_c \rangle}{1 - ix \left(\frac{\langle n_{ch} \rangle}{M} + \frac{1-s}{2} \right)} \right], \qquad (17.92)$$

$$\langle W^k \rangle_s = \frac{k!}{\Gamma(k + M)} \left(\frac{\langle n_{ch} \rangle}{M} + \frac{1-s}{2} \right)^k L_k^{M-1} \left(- \frac{\langle n_c \rangle}{\frac{\langle n_{ch} \rangle}{M} + \frac{1-s}{2}} \right) \qquad (17.93)$$

and [358, 361, 360]

$$p(n) = \frac{1}{\Gamma(n + M)} \left(1 + \frac{M}{\langle n_{ch} \rangle}\right)^{-n} \left(1 + \frac{\langle n_{ch} \rangle}{M}\right)^{-M} \exp\left[-\frac{\langle n_c \rangle M}{M + \langle n_{ch} \rangle}\right] \cdot$$

$$\cdot L_n^{M-1}\left(-\frac{\langle n_c \rangle M^2}{\langle n_{ch} \rangle (M + \langle n_{ch} \rangle)}\right), \tag{17.94}$$

where $\langle n_c \rangle = \sum_\lambda |\beta_\lambda|^2$ and the following identity for the Laguerre polynomials ([14], Sec. 8.97)

$$\sum_{\sum_\lambda^M n_\lambda = j} \prod_\lambda^M \frac{1}{\Gamma(n_\lambda + \mu_\lambda + 1)} L_{n_\lambda}^{\mu_\lambda}(x_\lambda) = \frac{1}{\Gamma(j + \sum_\lambda^M \mu_\lambda + M)} L_j^{\mathscr{P}}(\sum_\lambda^M x_\lambda) \tag{17.95}$$

with

$$\mathscr{P} = \sum_\lambda^M \mu_\lambda + M - 1,$$

has been used.

The photon-counting distribution can be obtained using the substitutions $\langle n_{ch} \rangle \to \eta \langle n_{ch} \rangle$ and $\langle n_c \rangle \to \eta \langle n_c \rangle$, where η is the photoefficiency.

The distribution (17.94) is given in Fig. 17.5 for $M = 1$ and $M = 5$ and various $\langle n_{ch} \rangle$ and $\langle n_c \rangle$. The curves shown in Fig. 17.5a correspond to Fig. 17.3 based on measurements of Arecchi et al. [44]. Fig. 17.5b shows the tendency of $p(n)$ to go to the Poisson distribution when M increases. Indeed, from (17.92), $\lim_{M \to \infty} \langle \exp ix\hat{n} \rangle_{\mathscr{N}} = \exp[ix(\langle n_{ch} \rangle + \langle n_c \rangle)]$ which leads to the Poisson distribution $p(n) = \langle n \rangle^n \exp(-\langle n \rangle)/n!$, where $\langle n \rangle = \langle n_{ch} \rangle + \langle n_c \rangle$.

The Fourier transform of (17.92) gives the distribution $P(W, s)$ [365, 367]

$$P(W, s) = \left(\frac{\langle n_{ch} \rangle}{M} + \frac{1 - s}{2}\right)^{-1} \left(\frac{W}{\langle n_c \rangle}\right)^{(M-1)/2} \cdot$$

$$\cdot \exp\left[-\frac{W + \langle n_c \rangle}{\frac{\langle n_{ch} \rangle}{M} + \frac{1 - s}{2}}\right] I_{M-1}\left(2 \frac{\sqrt{(\langle n_c \rangle W)}}{\frac{\langle n_{ch} \rangle}{M} + \frac{1 - s}{2}}\right) \tag{17.96}$$

in analogy to (17.66).

All families of distributions $P(W, s)$ tend to $P(W, 1) = P_{\mathscr{N}}(W)$ in the strong field limit and the special cases $\langle n_{ch} \rangle \to 0$ and $\langle n_c \rangle \to 0$ lead to the corresponding equations for coherent and chaotic fields respectively.

For the heterodyne detection of chaotic light, previously mentioned, formulae describing the superposition of coherent and chaotic fields with different mean frequencies are needed. These can be derived from equations (17.87), (17.88) and (17.89) and we give them for the normal ordering only.

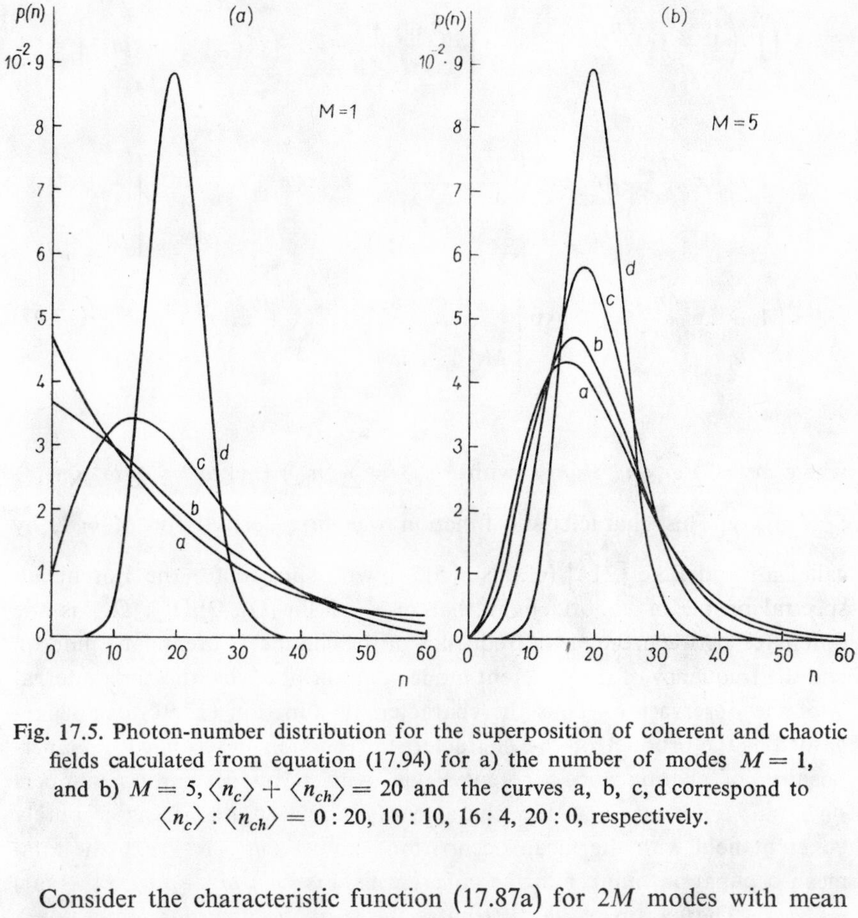

Fig. 17.5. Photon-number distribution for the superposition of coherent and chaotic fields calculated from equation (17.94) for a) the number of modes $M = 1$, and b) $M = 5$, $\langle n_c \rangle + \langle n_{ch} \rangle = 20$ and the curves a, b, c, d correspond to $\langle n_c \rangle : \langle n_{ch} \rangle = 0:20,\ 10:10,\ 16:4,\ 20:0$, respectively.

Consider the characteristic function (17.87a) for $2M$ modes with mean occupation numbers $\langle n_{ch\lambda} \rangle = \langle n_{ch} \rangle / M$ per mode for $\lambda = 1, 2, \ldots, M$ and $\langle n_{ch\lambda} \rangle = 0$ for $\lambda = M + 1, \ldots, 2M$; $\langle n_{ch} \rangle$ is the mean occupation photon number in the whole chaotic field. Let $\omega_{\lambda\mu}$ characterize the frequency shift between the λ-mode of the chaotic field and the μ-mode of the coherent field and $\langle n_{c\lambda\mu} \rangle$ be the mean number in the coherent mode μ which can be superimposed on the mode λ of the chaotic field. We introduce the shift parameters $\omega_\lambda^2 = \sum_\mu \langle n_{c\lambda\mu} \rangle \, \omega_{\lambda\mu}^2 / \langle n_{c\lambda} \rangle$, where $\langle n_{c\lambda} \rangle = \sum_\mu \langle n_{c\lambda\mu} \rangle$ and make the replacements $\langle n_{c\lambda}\rangle \to \langle n_{c\lambda}\rangle\,\omega_\lambda^2$ and $\langle n_{c\lambda+M}\rangle \to \langle n_{c\lambda}\rangle\,(1-\omega_\lambda^2)$ ($\lambda = 1, \ldots, M$) in (17.87a) considered for $s = 1$. We then arrive at the following characteristic function for the superposition of M-mode coherent and chaotic fields with frequency shifts described by ω_λ

$$\langle \exp ixW\rangle_{\mathcal{N}} =$$

$$= \prod_{\lambda=1}^{M}\left(1 - ix\frac{\langle n_{ch}\rangle}{M}\right)^{-1} \exp\left[\frac{ix\,\omega_\lambda^2\langle n_{c\lambda}\rangle}{1 - ix\dfrac{\langle n_{ch}\rangle}{M}} + ix\langle n_{c\lambda}\rangle\,(1-\omega_\lambda^2)\right] =$$

$$= \left(1 - ix\frac{\langle n_{ch}\rangle}{M}\right)^{-M} \exp\left[\frac{ix\,\omega^2\langle n_c\rangle}{1 - ix\dfrac{\langle n_{ch}\rangle}{M}} + ix\langle n_c\rangle\,(1-\omega^2)\right] =$$

$$= \left(1 - ix\frac{\langle n_{ch}\rangle}{M}\right)^{-M} \exp\left[\frac{(ix)^2\,\omega^2\langle n_c\rangle\,\langle n_{ch}\rangle}{M\left(1 - ix\dfrac{\langle n_{ch}\rangle}{M}\right)} + ix\langle n_c\rangle\right], \qquad (17.97)$$

where $\omega^2 = \sum_\lambda^M \omega_\lambda^2\langle n_{c\lambda}\rangle/\langle n_c\rangle$ with $\langle n_c\rangle = \sum_\lambda \langle n_{c\lambda}\rangle$ (i.e. $\omega^2 = \sum_{\lambda,\mu}\omega_{\lambda\mu}^2\langle n_{c\lambda\mu}\rangle$ $: \sum_{\lambda,\mu}\langle n_{c\lambda\mu}\rangle$). This characteristic function was first derived for $M = 1$, by Jakeman and Píke [214] (cf. also [513]), who showed for the Lorentzian spectral profile of chaotic light that $\omega_{\lambda\mu} = 2\sin(\Omega_{\lambda\mu}/2)/\Omega_{\lambda\mu}$; $\Omega_{\lambda\mu}$ is the difference between the mean frequency of the chaotic (Lorentzian) mode λ and the frequency of the coherent mode μ, multiplied by the time interval T of the observation. Thus the characteristic function (17.97) can be regarded as the product of the characteristic function describing the superposition of chaotic and coherent fields, with mean occupation numbers $\langle n_{ch}\rangle$ and $\langle n_c\rangle\,\omega^2$, and the characteristic function describing a purely coherent field with the mean occupation number $\langle n_c\rangle\,(1-\omega^2)$ (the total mean occupation number in the coherent field is $\langle n_c\rangle\,\omega^2 + \langle n_c\rangle\,(1-\omega^2) = \langle n_c\rangle$). The parameter ω distributes the mean occupation number $\langle n_c\rangle$ between the coherent part of the superposition and the purely coherent field. For $\omega = 1$, $\langle n_c\rangle$ belongs fully to the superposition and we obtain the characteristic function for the superposition with the same frequencies; for $\omega = 0$, $\langle n_c\rangle$ belongs fully to the purely coherent field and we obtain the product of the characteristic functions for the purely chaotic and purely coherent fields.

The distribution $P_{\mathcal{N}}(W)$ can be calculated from (17.97) by means of the Fourier transform and

$$P_{\mathcal{N}}(W) =$$

$$= \frac{M}{\langle n_{ch}\rangle}\left[\frac{W - \langle n_c\rangle(1-\omega^2)}{\langle n_c\rangle\,\omega^2}\right]^{(M-1)/2} \exp\left[-\frac{W + \langle n_c\rangle(2\omega^2 - 1)}{\langle n_{ch}\rangle} M\right].$$

$$I_{M-1}\left(2|\omega|\, M\,\frac{[\langle n_c\rangle(W - \langle n_c\rangle(1-\omega^2))]^{1/2}}{\langle n_{ch}\rangle}\right),$$

$$W \geqq \langle n_c\rangle(1-\omega^2),$$

$$P_{\mathcal{N}}(W) = 0\,, \qquad W < \langle n_c\rangle(1-\omega^2)\,. \tag{17.98}$$

The moments $\langle W^k\rangle_{\mathcal{N}}$ and the distribution $p(n)$ can be obtained using (17.98), the substitution $W = W' + \langle n_c\rangle(1-\omega^2)$, (17.93) for $s = 1$ and (17.94) [366] or simply by using the above substitutions in (17.88) for $s = 1$ and (17.89) considered for $2M$ modes. In this way we obtain

$$\langle W^k\rangle_{\mathcal{N}} = k! \sum_{\sum_{\lambda}^{2M} n_\lambda = k} \left(\frac{\langle n_{ch}\rangle}{M}\right)^{\sum_{\lambda}^{M} n_\lambda} \prod_{\lambda}^{M} \frac{1}{n_\lambda!} L^0_{n_\lambda}\left(-\frac{\langle n_{c\lambda}\rangle M}{\langle n_{ch}\rangle}\right).$$

$$\cdot \prod_{\lambda' = M+1}^{2M} \frac{\langle n_{c\lambda'}\rangle^{n_{\lambda'}}}{n_{\lambda'}!}\,. \tag{17.99}$$

Writing

$$\sum_{\sum_{\lambda}^{2M} n_\lambda = k} = \sum_{j=0}^{k} \sum_{\sum_{\lambda}^{M} n_\lambda = j} \sum_{\sum_{\lambda = M+1}^{2M} n_\lambda = k - j}$$

and using the identity (17.95) and the polynomial theorem we arrive at

$$\langle W^k\rangle_{\mathcal{N}} = \left\langle \frac{\hat{n}!}{(\hat{n} - k)!}\right\rangle =$$

$$= \sum_{j=0}^{k} \frac{k!}{(k-j)!\,\Gamma(j+M)} [\langle n_c\rangle(1-\omega^2)]^{k-j}\,.$$

$$\left(\frac{\langle n_{ch}\rangle}{M}\right)^{j} L_j^{M-1}\left(-\frac{\langle n_c\rangle\,\omega^2 M}{\langle n_{ch}\rangle}\right). \tag{17.100}$$

Similarly one obtains

$$p(n) = \left(1 + \frac{\langle n_{ch} \rangle}{M}\right)^{-M} \exp\left[-\frac{\langle n_c \rangle [M + \langle n_{ch} \rangle (1 - \omega^2)]}{M + \langle n_{ch} \rangle}\right] .$$

$$\sum_{j=0}^{n} \frac{1}{(n-j)!\,\Gamma(j+M)} [\langle n_c \rangle (1 - \omega^2)]^{n-j} .$$

$$\left(1 + \frac{M}{\langle n_{ch} \rangle}\right)^{-j} . L_j^{M-1}\left(-\frac{\langle n_c \rangle \omega^2 M^2}{\langle n_{ch} \rangle (\langle n_{ch} \rangle + M)}\right). \tag{17.101}$$

For $M \to \infty$ the distribution $p(n)$ again tends to the Poisson distribution with $\langle n \rangle = \langle n_{ch} \rangle + \langle n_c \rangle$ since from (17.97) $\lim_{M\to\infty} \langle \exp ixW \rangle_{\mathcal{N}} = \exp [ix(\langle n_{ch} \rangle + \langle n_c \rangle)]$.

One can easily write down the corresponding equations when the $\langle n_{ch\lambda} \rangle$ are different from one another [366, 532]. However, they are of a rather more complicated structure.

For the second moment $\langle W^2 \rangle_{\mathcal{N}}$ we obtain

$$\langle W^2 \rangle_{\mathcal{N}} = \langle n_{ch} \rangle^2 \frac{M+1}{M} + 2\langle n_{ch} \rangle \langle n_c \rangle \left(1 + \frac{\omega^2}{M}\right) + \langle n_c \rangle^2 , \tag{17.102a}$$

that is

$$\langle (\Delta W)^2 \rangle_{\mathcal{N}} = \frac{\langle n_{ch} \rangle^2 + 2\omega^2 \langle n_{ch} \rangle \langle n_c \rangle}{M} \tag{17.102b}$$

and

$$\langle \hat{n}^2 \rangle = \langle W \rangle_{\mathcal{N}} + \langle W^2 \rangle_{\mathcal{N}} = \langle n_{ch} \rangle + \langle n_c \rangle + \langle n_{ch} \rangle^2 \left(1 + \frac{1}{M}\right) +$$
$$+ 2\langle n_{ch} \rangle \langle n_c \rangle \left(1 + \frac{\omega^2}{M}\right) + \langle n_c \rangle^2 , \tag{17.103a}$$

giving

$$\langle (\Delta \hat{n})^2 \rangle = \langle W \rangle_{\mathcal{N}} + \langle (\Delta W)^2 \rangle_{\mathcal{N}} =$$
$$= \langle n_{ch} \rangle + \langle n_c \rangle + \frac{\langle n_{ch} \rangle^2}{M} + \frac{2\omega^2 \langle n_{ch} \rangle \langle n_c \rangle}{M} . \tag{17.103b}$$

The first and the second terms in (17.103b) correspond to the Poisson distribution (with $\langle n \rangle = \langle n_{ch} \rangle + \langle n_c \rangle$), the third term represents the photon bunching of the chaotic field and the fourth term represents the photon bunching due to interference effects. This term is zero when $\omega = 0$ while the last two terms are zero when $M \to \infty$. The latter case gives the Poissonian variance $\langle(\Delta\hat{n})^2\rangle = \langle n_{ch}\rangle + \langle n_c\rangle$ $(\langle(\Delta W)^2\rangle_{\mathcal{N}} = 0)$.

As a special case we obtain for $\omega = 1$ equations (17.92), (17.93), (17.94) and (17.96) (for $s = 1$) and for $\langle n_{ch}\rangle \to 0$ and $\langle n_c\rangle \to 0$ we arrive at the previously given formulae for coherent and chaotic fields.

The present formulae for the superposition of coherent and chaotic fields describe generally the superposition of an M-mode chaotic field with an N-mode coherent field ($M \geqq N$). The coherent N-mode field can be generated by a laser operating on N modes well above threshold or by N one-mode lasers operating in this region with their fields superimposed. The statistical properties of light from an M-mode laser operating in the region where the linearized laser theory is appropriate (as well as some scattering experiments using an M-mode laser) can also be described by the present formulae. In principle the formulae with different mean frequencies are suitable for the description of heterodyne detection of a chaotic M-mode field, when the chaotic field is superimposed on a coherent N-mode field ($M \geqq N$) with frequency shifts characterized by ω_λ. However, the characteristic function (17.97) is of the same form regardless of the number N ($\leqq M$) of coherent modes. Hence, the most important case in practice, the detection of an M-mode chaotic field superimposed on a one-mode coherent field is described by the same characteristic function and the corresponding formulae by putting $\langle n_{c\lambda}\rangle = 0$ for all modes except the considered mode.

The most general method of treating the superposition of coherent and chaotic fields uses the correlation function technique. Denoting the s-ordered correlation function as $\Gamma_s^{(n,n)}$ we can write

$$\Gamma_s^{(n,n)}(x_1, \ldots, x_n, x_n, \ldots, x_1) = \\ = \int \prod_\lambda^M \left[\pi\left(\langle n_{ch\lambda}\rangle + \frac{1-s}{2}\right)\right]^{-1} \exp\left[-\frac{|\alpha_\lambda - \beta_\lambda|^2}{\langle n_{ch\lambda}\rangle + \frac{1-s}{2}}\right] . \\ V^*(x_1) \ldots V^*(x_n)\, V(x_n) \ldots V(x_1)\, d^2\{\alpha_\lambda\} . \qquad (17.104)$$

The substitution $\alpha_\lambda - \beta_\lambda = \gamma_\lambda$ leads to

$$\Gamma_s^{(n,n)}(x_1, \ldots, x_n, x_n, \ldots, x_1) = \int \prod_\lambda^M \left[\pi\left(\langle n_{ch\lambda}\rangle + \frac{1-s}{2}\right)\right]^{-1} . \\ . \exp\left[-\frac{|\gamma_\lambda|^2}{\langle n_{ch\lambda}\rangle + \frac{1-s}{2}}\right] [V^*(x_1) + B^*(x_1)] \ldots [V^*(x_n) + B^*(x_n)] . \\ . [V(x_n) + B(x_n)] \ldots [V(x_1) + B(x_1)]\, d^2\{\gamma_\lambda\} , \qquad (17.105)$$

where $V(x) \equiv V(x, \{\gamma_\lambda\})$ is a Gaussian variable and $B(x) \equiv B(x, \{\beta_\lambda\})$ is a coherent field. Applying the formula (17.15) modified to the s-ordering ($\langle n_\lambda \rangle$ is replaced by $\langle n_{ch\lambda} \rangle + (1 - s)/2$) we can in principle calculate all $\Gamma_s^{(n,n)}$. A graphical method for this has been developed in [370]. Denoting the second-order correlation function of the chaotic field as ${}^{ch}\Gamma_s^{(1,1)}$ we obtain successively

$$\Gamma_s^{(1,1)}(x_1, x_1) = {}^{ch}\Gamma_s^{(1,1)}(x_1, x_1) + |B(x_1)|^2 , \tag{10.106a}$$

$$\begin{aligned} \Gamma_s^{(2,2)}(x_1, x_2, x_2, x_1) = \\ &= ({}^{ch}\Gamma_s^{(1,1)}(x_1, x_1) + |B(x_1)|^2) ({}^{ch}\Gamma_s^{(1,1)}(x_2, x_2) + |B(x_2)|^2) + \\ &+ |{}^{ch}\Gamma_s^{(1,1)}(x_1, x_2)|^2 + 2 \operatorname{Re} \{{}^{ch}\Gamma_s^{(1,1)}(x_1, x_2) B^*(x_2) B(x_1)\} , \end{aligned} \tag{17.106b}$$

etc. The variance of the counts of a photodetector is, [324, 286],

$$\begin{aligned} \langle(\Delta \hat{n})^2\rangle = \langle n \rangle + \frac{\langle n_{ch} \rangle^2}{T^2} \iint_0^T |{}^{ch}\gamma_{\mathcal{N}}^{(1,1)}(t_1 - t_2)|^2 \, dt_1 \, dt_2 + \\ + 2 \frac{\langle n_{ch} \rangle \langle n_c \rangle}{T^2} \operatorname{Re} \left\{ \iint_0^T {}^{ch}\gamma_{\mathcal{N}}^{(1,1)}(t_1 - t_2) \, {}^{c}\gamma_{\mathcal{N}}^{(1,1)*}(t_1 - t_2) \, dt_1 \, dt_2 \right\} , \end{aligned} \tag{17.107}$$

where $\langle n \rangle = \langle n_{ch} \rangle + \langle n_c \rangle$, $\langle n_{ch} \rangle = \eta \langle I_{ch} \rangle T$, $\langle n_c \rangle = \eta I_c T$ and ${}^{ch}\gamma_{\mathcal{N}}^{(1,1)}$ and ${}^{c}\gamma_{\mathcal{N}}^{(1,1)}$ are the degrees of coherence for chaotic and coherent fields respectively. In (17.107) the first term is Poissonian, the second describes the Hanbury Brown - Twiss effect of chaotic light and the third is the interference term.

The comparison of (17.103b) with (17.107) provides the expression for M (for $\langle n_c \rangle = 0$ we obtain (17.43)). The substitution of this M in the formulae (17.98), (17.100) and (17.101) makes it possible to use them as an approximation for arbitrary spectra and counting time intervals [509, 510] (for partially polarized fields see [532], where also exact formulae, obtained on the basis of (17.87) — (17.89) with $\langle n_{ch\lambda} \rangle$ corresponding to the spectrum of chaotic field through the integral equation (17.41), have been obtained; for two-photon absorption see [528]). The comparison of the exact and approximate third factorial moments for the Lorentzian spectrum of chaotic light and one-mode coherent light shows very good accuracy of the approximate formulae.

Chapter 18

INTERFERENCE OF INDEPENDENT LIGHT BEAMS

An important question related to the coherence of optical fields is the interference of independent light beams. Although such interference effects have been observed in the region of radiowaves it was relatively more difficult to observe them in the optical region. However, it is clear that in principle no limitations exist apart from the difficulty even in the optical region. An outline of an interference experiment with independent sources is given in Fig. 18.1. Two sources S_1 and S_2 produce slightly convergent light

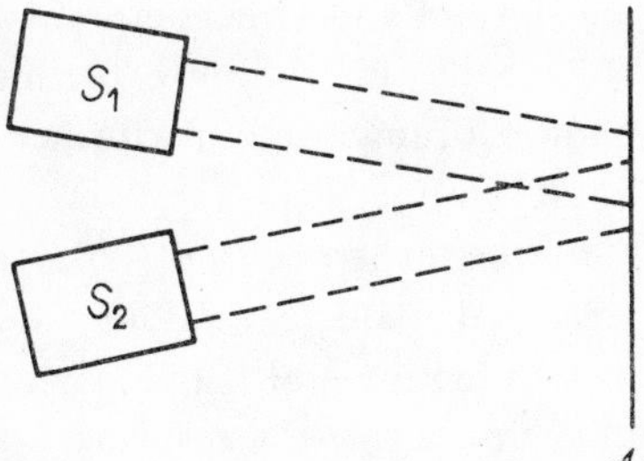

Fig. 18.1. A scheme of the interference experiment with independent sources.

beams which are projected on to the screen $\mathscr{A}$. In part of the screen common to both beams one observes the resulting field by means of a photodetector system. In general, interference fringes will be present (since both the fields are linear and must interfere) but they will vary with time. If the resolving time of the detector is long the fringes will be washed out and no interference effect is observed; if the resolving time is sufficiently short fringes are visible. In this connection the question arises of whether this result contradicts the well-known remark of Dirac [9] (p. 9) that "each photon interferes only with itself. Interference between different photons never occurs".

The first demonstration of beats resulting from the superposition of incoherent light beams was given by Forrester, Gudmundsen and Johnson [150], who used the two spectral components of a Zeemann doublet. It is interesting to note that this experiment was performed with non-degenerate light and the mean number of photons received on a coherence area, during the time in which a steady beat was observed, was much less than one. With the development of the laser, producing light beams with degeneracy parameter $\delta \gg 1$, such beat experiments became easier to perform (Javan, Ballik and Bond [217] and Lipsett and Mandel [256, 257]). Magyar and Mandel [261, 262] have shown that interference fringes may also be observed by superposing two independent laser beams.

The significance of a large value of the degeneracy parameter δ for these experiments can easily be seen. The number of photons defining the interference pattern in the receiving plane in a time less than the coherence time is limited by δ. When $\delta \ll 1$ it is difficult to think of interference fringes.

All these experiments can be explained classically (cf. [285]) and the characteristic feature of the description of these effects is that, since the effects are transient, an averaging process must be avoided. A quantum-mechanical description, based on the coherent state formalism, was given by Paul, Brunner and Richter [341], Paul [338], Mandel [274], Richter, Brunner and Paul [395], Paul [339, 340] and also by Korenman [237] and Jordan and Ghielmetti [219] among others.

Recently interesting experiments on the interference of independent photon beams were performed by Pfleegor and Mandel [378–380] and by Radloff [390].

In the experiment of Pfleegor and Mandel interference fringes were measured under conditions where the light intensity was so low that the mean time interval between photons was large compared with their transit time through the measuring apparatus, i.e. there was a high probability that one photon was absorbed before the next one was emitted by one or other of the (laser) sources. Since the intensities were very low, the photon correlation technique was required to observe the interference fringes. The interference pattern was received on a stack of thin glass plates, each of which had a thickness corresponding to about a half fringe width. The plates were cut and arranged so that any light falling on the 1st, 3rd, 5th, etc. plate was fed to one photomultiplier, while light falling on the 2nd, 4th, 6th, etc. plate was fed to the other. When the half fringe spacing coincides with the plate thickness, and for example the fringe maxima fall on the odd-numbered plates, one phototube will register nearly all the photons and the other almost none. The position of the fringe maxima is unpredictable and random, but if the number of photons registered by one phototube increases, the number registered by the other must decrease, provided the fringe spacing is right for the plates. Thus there must be a negative correlation between the numbers of counts from the two phototubes and such a negative correlation was indeed observed. This result shows (since the experimental conditions were arranged so that one photon was absorbed before the next one was emitted) that the above mentioned Dirac statement is applicable to these experiments. In general experiments with independent light beams are in agreement with the Dirac statement since any "localization" of a photon in space-time automatically rules out the possibility of knowing its momentum, as a consequence of the uncertainty principle. Thus one cannot say to which beam a given photon belongs – each photon is to be considered as being partly in both beams and interfering only with itself.

Radloff [390] performed an analogous experiment to the Pfleegor-Mandel spatial interference experiment – a temporal interference experiment for observing beats between two independent laser beams. A discussion can be found in [437].

In principle, interference experiments of this sort are also possible (although more difficult) with chaotic sources. An interference experiment with two independent thermal sources was performed by Haig and Sillitto [190]. They achieved partial success in observing spatial modulation in the mutual coherence pattern by using the intensity correlation technique.

Let the field generated by the source S_1 be described by the function $\phi_{\mathcal{N}}^{(1)}(\{\alpha_{\lambda 1}\})$ and the field generated by the source S_2 be described by the function $\Phi_{\mathcal{N}}^{(2)}(\{\alpha_{\lambda 2}\})$. The superposition of both these fields is then described by the convolution (13.147). The expectation value of the intensity at a point $x \equiv (\boldsymbol{x}, t)$ on the screen $\mathscr{A}$ is given by (13.148), that is

$$I(x) = \langle \hat{A}^{(-)}(x)\, \hat{A}^{(+)}(x) \rangle = I^{(1)}(x) + I^{(2)}(x) +$$

$$+ 2\,\mathrm{Re}\left\{ \int \Phi_{\mathcal{N}}^{(1)}(\{\alpha_{\lambda 1}\})\, V^*(x, \{\alpha_{\lambda 1}\})\, d^2\{\alpha_{\lambda 1}\} \int \Phi_{\mathcal{N}}^{(2)}(\{\alpha_{\lambda 2}\})\, V(x, \{\alpha_{\lambda 2}\})\, d^2\{\alpha_{\lambda 2}\} \right\}, \tag{18.1}$$

where $I^{(j)}(x) \equiv \langle \hat{A}^{(-)}(x, \{\alpha_{\lambda j}\})\, \hat{A}^{(+)}(x, \{\alpha_{\lambda j}\}) \rangle$, $j = 1, 2$ are averaged intensities produced by each source if the other is absent. We restrict ourselves here and in the following to linearly polarized light for simplicity. The third term in (18.1) represents an interference term and if this term does not vanish one can observe the interference effect. In general this term will not vanish if the fields produced by both the sources contain non-zero coherent components. As an example we can give the case when the fields of both the sources are in coherent states, that is

$$\Phi_{\mathcal{N}}^{(j)}(\{\alpha_{\lambda j}\}) = \prod_{\lambda} \delta(\alpha_{\lambda j} - \beta_{\lambda j}), \quad j = 1, 2. \tag{18.2}$$

In this case the interference term is equal to

$$2 \operatorname{Re} \{V^*(x, \{\beta_{\lambda 1}\}) V(x, \{\beta_{\lambda 2}\})\}, \tag{18.3}$$

and so it is non-vanishing. However this assumes a knowledge of the phases of $\{\beta_\lambda\}$ which is not the case in practice, since the phase information about optical vibrations is almost always lost. Then the distributions (18.2) must be integrated over the phases of $\{\alpha_{\lambda j}\}$ giving distributions of the form (17.58) independent of the phases describing the stationary fields. The interference term in (18.1) is zero now and no interference effect can be observed in the intensity pattern. This conclusion is in agreement with our earlier results that such an interference effect can occur only when there is at least partial coherence between the beams.

The interference pattern can be detected by two quadratic detectors (photodetectors). In this case we have to examine the correlation of intensities at two space-time points $\langle I(x_1) I(x_2)\rangle$, as in describing the Hanbury Brown-Twiss effect. Assuming that the beams are described by $\Phi_{\mathcal{N}}^{(1)}$ and $\Phi_{\mathcal{N}}^{(2)}$ and that the phases of $\{\alpha_\lambda\}$ are uniformly distributed in the interval $(0, 2\pi)$ we obtain

$$\begin{aligned}
&\langle I(x_1) I(x_2)\rangle = \\
&= \langle [V^*(x_1, \{\alpha_{\lambda 1}\}) + V^*(x_1 \{\alpha_{\lambda 2}\})][V(x_1, \{\alpha_{\lambda 1}\}) + V(x_1, \{\alpha_{\lambda 2}\})] \,. \\
&\quad [V^*(x_2, \{\alpha_{\lambda 1}\}) + V^*(x_2, \{\alpha_{\lambda 2}\})] [V(x_2, \{\alpha_{\lambda 1}\}) + V(x_2, \{\alpha_{\lambda 2}\})]\rangle = \\
&= \langle |V(x_1, \{\alpha_{\lambda 1}\})|^2 |V(x_2, \{\alpha_{\lambda 1}\})|^2\rangle + \langle |V(x_1, \{\alpha_{\lambda 1}\})|^2 |V(x_2, \{\alpha_{\lambda 2}\})|^2\rangle + \\
&+ \langle |V(x_1, \{\alpha_{\lambda 2}\})|^2 |V(x_2, \{\alpha_{\lambda 1}\})|^2\rangle + \langle |V(x_1, \{\alpha_{\lambda 2}\})|^2 |V(x_2, \{\alpha_{\lambda 2}\})|^2\rangle + \\
&+ \langle V^*(x_1, \{\alpha_{\lambda 1}\}) V(x_1, \{\alpha_{\lambda 2}\}) V^*(x_2, \{\alpha_{\lambda 2}\}) V(x_2, \{\alpha_{\lambda 1}\})\rangle + \\
&+ \langle V^*(x_1, \{\alpha_{\lambda 2}\}) V(x_1, \{\alpha_{\lambda 1}\}) V^*(x_2, \{\alpha_{\lambda 1}\}) V(x_2, \{\alpha_{\lambda 2}\})\rangle,
\end{aligned} \tag{18.4}$$

where terms such as

$$\langle |V(x_1, \{\alpha_{\lambda 1}\})|^2 V^*(x_2, \{\alpha_{\lambda 1}\}) V(x_2, \{\alpha_{\lambda 2}\})\rangle,$$
$$\langle V^*(x_1, \{\alpha_{\lambda 1}\}) V(x_1, \{\alpha_{\lambda 2}\}) V^*(x_2, \{\alpha_{\lambda 1}\}) V(x_2, \{\alpha_{\lambda 2}\})\rangle,$$

etc. are equal to zero. The fifth and sixth terms in this expression generally lead to an almost periodic variation of the correlation $\langle I(x_1) I(x_2)\rangle$ with

$|\boldsymbol{x}_2 - \boldsymbol{x}_1|$ and $|t_2 - t_1|$. Assuming quasi-monochromatic beams with a spread Δk much less than the mean wave number k_0 and if $|\boldsymbol{x}_1 - \boldsymbol{x}_2| \ll \ll 1/\Delta k$ and $|t_1 - t_2| \ll 1/c\,\Delta k$, we obtain

$$\langle I(x_1)\, I(x_2)\rangle = \langle [|V(x_1, \{\alpha_{\lambda 1}\})|^2 + |V(x_1, \{\alpha_{\lambda 2}\})|^2]^2\rangle$$
$$+ 2\langle |V(x_1, \{\alpha_{\lambda 1}\})|^2\rangle \langle |V(x_1, \{\alpha_{\lambda 2}\})|^2\rangle \,.$$
$$\cos\left[(\boldsymbol{k}_0^{(1)} - \boldsymbol{k}_0^{(2)})(\boldsymbol{x}_2 - \boldsymbol{x}_1) - c(k_0^{(1)} - k_0^{(2)})(t_2 - t_1)\right], \qquad (18.5)$$

where $\boldsymbol{k}_0^{(j)}$ and $k_0^{(j)}$ refer to the jth beam ($j = 1, 2$). Thus, over a limited space-time region, the intensity correlation shows a cosinusoidal dependence on space and time, which can be interpreted both in terms of interference fringes and light beats. The generating function has been calculated in a close form for this case in [17] (section 10-2).

CONCLUSION AND OUTLOOK

In this book we have studied the coherence properties of optical fields as well as electromagnetic fields by classical and quantum methods and the relation between the classical and quantum methods of describing coherence has been investigated by using the formalism of the coherent states and the Glauber-Sudarshan representation of the density matrix. Particular attention has been paid to the detection of optical fields and many theoretical results have been shown to be in very good agreement with experimental results obtained by a number of authors. The main purpose of this book was the investigation of the coherence properties of free electromagnetic fields and, although we have mentioned some results of coherence theory for processes involving the interaction of the electromagnetic field with matter (particularly in connection with the laser), no systematic approach to this problem has been developed here. However we have given references to important papers treating this problem.

In conclusion it is worthwhile to point out a few other topics which are undergoing development at the present time and some possible applications of the methods of the theory of coherence in other branches of physics. In connection with practical applications the problem of the propagation of light in non-free space, in random and turbulent media, in non-linear media, etc. is of increasing importance. From papers devoted to this problem we quote a series by Beran and his co-workers [65, 66, 68, 488, 489, 196] as well as by Arecchi et al [43]. Questions of light scattering by turbulent media and statistical media have been discussed by Bertolotti, Crosignani, Di Porto and Sette [77, 79, 80], Di Porto, Crosignani and Bertolotti [133], Arecchi, Giglio and Tartari [50], Arecchi [42] and Pike [386, 387], Mandel [522], Diament and Teich [505], Korenman [516] and Bertolotti, Crosignani and Di Porto [496], among others. Also the possibility of deducing information on the scattering properties of sources, including phase information, from correlations of light scattered by the sources has been investigated in the series of papers by Goldberger, Lewis and Watson [183, 184], Goldberger and Watson [185, 186] and Fetter [143]. Quantum statistics in non-linear optics has been studied by Shen [421]. At present increasing interest exists in the statistics of multiphoton processes (cf. reference, on p. 217) and of superradiant states [497, 506].

The technique of coherent states has been extended to other branches of physics. Carruthers and Dy [108] have discussed an application of this technique to solid state physics, in particular to the theory of irreversible processes in anharmonic crystals. An application to the theory of superfluidity has been suggested by López [258], Langer [244, 519] and Carruthers and Nieto [109] and to the study of ferromagnetism by Rezende and Zagury [393].

Some approaches to the study of the statistical and correlation properties of systems of fermions have also been developed as we have mentioned. Finally we should mention a possibility of extending the coherent state formalism to the pseudo-oscillator systems for which the commutator is $[\hat{a}, \hat{a}^{+}] = -1$ [210, 490]. The possibility of taking into account the structure of sources (such as their multipolarity) has been discussed by Zardecki [486].

All these examples show that the methods and results of the theory of coherence of the electromagnetic field will also be fruitful in other branches of physics. This is particularly true for the coherent state formalism which permits a formulation of quantum electrodynamics in a form applicable to systems with a large number of particles since such a formulation has a classical limit and retains phase information of the field.

REFERENCES

Books

1. AKHIEZER, N. I. *Classical Moment Problem and some Related Questions*, Chap. I, Oliver and Boyd, Edinburgh, 1970.
2. AKHIEZER, A. I. and BERESTETSKY V. B. *Quantum Electrodynamics*, Interscience, New York, 1965.
3. BERAN, M. J. and PARRENT, G. B. *Theory of Partial Coherence*, Prentice-Hall, Englewood Cliffs, New Jersey, 1964.
4. BEREZANSKY, J. M. *Expansions in Eigenfunctions of Selfconjugate Operators*, American Math. Soc., New York, 1968.
5. BLOEMBERGEN, N. *Non-Linear Optics*, W. A. Benjamin, New York, 1965.
6. BOGOLYUBOV, N. N., MEDVEDEV, B. V. and POLIVANOV, M. K. *Questions of the Theory of Dispersion Relations*, Gos. izd fiz. mat. lit., Moscow, 1958.
7. BOGOLYUBOV, N. N. and SHIRKOV, D. V. *Introduction to the Theory of Quantized Fields*, Interscience, New York, 1959.
8. BORN, M. and WOLF, E. *Principles of Optics*, Pergamon Press, Oxford, 1965.
9. DIRAC, P. A. M. *Principles of Quantum Mechanics* 4th edn., Clarendon Press, Oxford, 1958.
10. FEYNMAN, R. P. and HIBBS, A. R. *Quantum Mechanics and Path Integrals*, McGraw-Hill, New York, 1965.
11. FRANÇON, M. and SLANSKY, S. *Cohérence en optique*, Ed. Centre Nat. Rech. Sci., Paris, 1965.
12. GELFAND, I. M. and SHILOV, G. E. *Generalized Functions*, Vol. I, Academic Press, New York, 1964.
13. GELFAND, I. M. and VILENKIN N. YA. *Generalized Functions*, Vol. 4, *Applications of Harmonic Analysis*, Academic Press, New York, 1964.
14. GRADSHTEYN, I. S. and RYZHIK, I. M. *Table of Integrals, Series and Products*, Academic Press, New York, 1965.
15. JEFFREYS, H. and JEFFREYS, B. S. *Methods of Mathematical Physics*, 2nd edn., p. 73, Cambridge Univ. Press, New York, 1950.
16. KHURGYN, J. I. and YAKOVLEV, V. P. *Methods of Entire Functions in Radiophysics, Theory of Communications and Optics*, Gos. izd. fiz. mat. lit., Moscow, 1962.
17. KLAUDER, J. R. and SUDARSHAN, E. C. G. *Fundamentals of Quantum Optics*, W. A. Benjamin, New York, 1968.
18. LANDAU, L. D. and LIFSHITZ,, E. M. *Statistical Physics*, 2nd edn., Pergamon, Oxford, 1959.
19. LOUISELL, W. H. *Radiation and Noise in Quantum Electronics*, McGraw - Hill, New York, 1964.

20. Maréchal, A. and Françon, M. *Diffraction. Structure des images*, Ed. Rev. d'Opt. Théor. et Instr., Paris, 1960.
21. Messiah, A. *Quantum Mechanics*, Vol. I, II, John Wiley and Sons, New York, 1961, 1962.
22. Morse, P. M. and Feshbach, H. *Methods of Theoretical Physics*, Vol. I, McGraw-Hill, New York, 1953.
23. Muskhelishvili, N. I. *Singular Integral Equations*, Noordhoff, Groningen, 1953.
24. O'Neill, E. L. *Introduction to Statistical Optics*, Addison-Wesley, Reading, Mass,. 1963.
25. Paley, R. E. and Wiener, N. *Fourier Transforms in Complex Domain*, New York, 1934.
26. Petrov, A. Z. *Einstein Spaces*, Pergamon, Oxford, 1969.
27. Schweber, S. S. *An Introduction to Relativistic Quantum Field Theory*, Row, Peterson and Co. Evanston, Ill., Elmsford, New York, 1961.
28. Shannon, C. E. *Mathematical Theory of Communications*, Univ. of Illinois Press, 1949.
29. Stroke, G. W. *An Introduction to Coherent Optics and Holography*, Academic Press, New York, 1966.
30. Thirring, W. E. *Principles of Quantum Electrodynamics*, Academic Press, New York, 1962.
31. Titchmarsh, E. C. *Introduction to the Theory of Fourier Integrals*, 2nd edn, Clarendon Press, Oxford, 1948.
32. Troup, G. J. *Optical Coherence Theory, Recent Developments*, Methuen, London, 1967.
33. Verdet, E. *Leçons d'Optique Physique I*, p. 106, L'Imprimerie Impériale, Paris, 1869.
34. Volterra, V. *Theory of Functionals and of Integral and Integro-Differential Equations*, Dover Publications, New York, 1959.
35. Whittaker, E. and Watson, G. N. *A Course of Modern Analysis*, 4th edn., p. 266, Cambridge Univ. Press, Cambridge, 1940.

Research Papers and Reviews

36. Agarwal, G. S. *Phys. Rev.* ***178***, 2025 (1969).
37. Agarwal, G. S. and Wolf, E. *Phys. Lett.* ***26A***, 485 (1968).
38. Agarwal, G. S. and Wolf, E. *Phys. Rev. Lett.* ***21***, 180 (1968).
39. Agarwal, G. S. and Wolf, E. *Phys. Rev. Lett.* ***21***, 656 (E) (1968).
40. Agarwal, G. S. and Wolf, E. *Phys. Rev. D2*, 2161 and 2187 and 2206 (1970).
41. Arecchi, F. T. *Phys. Rev. Lett.* ***15***, 912 (1965).
42. Arecchi, F. T. *Photocount Distributions and Field Statistics*, Lectures at the Inter. School of Phys. E. Fermi, Varena, 1967.
43. Arecchi, F. T. et al.*Rivista Nuovo Cim.* (I) ***1***, 181 (1969).
44. Arecchi, F. T., Berné, A. and Burlamacchi, P. *Phys. Rev. Lett.* ***16***, 32 (1966).
45. Arecchi, F. T., Berné, A. and Sona, A. *Phys. Rev. Lett.* ***17***, 260 (1966).
46. Arecchi, F. T., Berné, A., Sona, A. and Burlamacchi, P. *IEEE J. Quant. Electr., QE-2*, 341 (1966).

47. Arecchi, F. T., Degiorgio, V. and Querzola, B. *Phys. Rev. Lett.* ***19***, 1168 (1967).
48. Arecchi, F. T., Gatti, E. and Sona, A. *Phys. Lett.* ***20***, 27 (1966).
49. Arecchi, F. T., Giglio, M. and Sona, A. *Phys. Lett.* ***25A***, 341 (1967).
50. Arecchi, F. T., Giglio, M. and Tartari, U. *Phys. Rev.* ***163***, 186 (1967).
51. Arecchi, F. T., Rodari, G. S. and Sona, A. *Phys. Lett.* ***25A***, 59 (1967).
52. Armstrong, J. A. and Smith, A. W. *Progress in Optics*, Vol. VI, p. 213, Ed. E. Wolf, North-Holland, Amsterdam, 1967.
53. Barakat, R. *J. opt. Soc. Am.* ***53***, 317 (1963).
54. Barakat, R. *J. opt. Soc. Am.* ***56***, 739 (1966).
55. Bédard, G. *Phys. Lett.* ***21***, 32 (1966).
56. Bédard, G. *Phys. Rev.* ***151***, 1038 (1966).
57. Bédard, G. *Phys. Lett.* ***24A***, 613 (1967).
58. Bédard, G. *J. Opt. Soc. Am.* ***57***, 1201 (1967).
59. Bédard, G. *Proc. phys. Soc.* ***90***, 131 (1967).
60. Bédard, G. *Phys. Rev.* ***161***, 1304 (1967).
61. Bédard, G., Chang, J. C. and Mandel, L. *Phys. Rev.* ***160***, 1496 (1967).
62. Bénard, C. M. *Quantum Optics* (Proc. Tenth Session Scottish Univ. Summer School in Phys) p. 535, Eds S. M. Kay and A. Maitland, Academic Press, London and New York, 1970.
63. Bénard, C. M. *C. R. Acad. Sci., Paris* ***268***, 1504 (1969).
64. Bendjaballah, C. *C. R. Acad. Sci., Paris* ***268***, 1719 (1969).
65. Beran, M. J. *J. opt. Soc. Am.* ***56***, 1475 (1966).
66. Beran, M. J. *IEEE Trans. Ant. Propag.* ***AP-15***, 66 (1967).
67. Beran, M. J. and Corson, P. *J. math. Phys.* ***6***, 271 (1965).
68. Beran, M. J. and DeVelis, J. B. *J. opt. Soc.* Am. ***57***, 186 (1967).
69. Beran, M. J., DeVelis, J. and Parrent, G. *Phys. Rev.* ***154***, 1224 (1967).
70. Beran, M. J. and Parrent, G. B. *J. opt. Soc. Am.* ***52***, 98 (1962).
71. Beran, M. J. and Parrent, G. B. *Nuovo Cim.* (10) ***27***, 1049 (1963).
72. Berek, M. J. *Z. Phys.* ***36***, 675 (1926).
73. Berek, M. J. *Z. Phys.* ***36***, 824 (1926).
74. Berek, M. J. *Z. Phys.* ***37***, 287 (1926).
75. Berek, M. J. *Z. Phys.* ***40***, 420 (1926).
76. Bertolotti, M., Crosignani, B., Di Porto, P. and Sette, D. *Phys. Rev.* ***150***, 1054 (1966).
77. Bertolotti, M., Crosignani, B., Di Porto, P. and Sette, D. *Phys. Rev.* ***157***, 146 (1967).
78. Bertolotti, M., Crosignani, B., Di Porto, P. and Sette, D. *Z. Phys.* ***205***, 129 (1967).
79. Bertolotti, M., Crosignani, B., Di Porto, P. and Sette, D. *J. Phys. A* ***2***, 126 (1969).
80. Bertolotti, M., Crosignani, B., Di Porto, P. and Sette, D. *J. Phys. A* ***2***, 473 (1969).
81. Bertrand, P. P. and Mishkin, E. A. *Phys. Lett.* ***25A***, 204 (1967).
82. Bialynicka-Birula, Z. *Phys. Rev.* ***173***, 1207 (1968).
83. Blanc-Lapierre, A. and Dumontet, P. *Rev. Opt.* ***34***, 1 (1955).
84. Boileau, E. and Picinbono, B. *J. opt. Soc. Am.* ***58***, 1238 (1968).
85. Bonifacio, R., Narducci, L. M. and Montaldi, E. Abstracts of the Second Roch. Conf. on Coherence and Quant. Opt., 1966, p. 46.

86. Bonifacio, R., Narducci, L. M. and Montaldi, E. *Phys. Rev. Lett.* ***16***, 1125 (1966).
87. Bothe, W. *Z. Phys.* ***41***, 345 (1927).
88. Bourret, R. C. *Nuovo Cim.* ***18***, 347 (1960).
89. Brannen, E., Ferguson, H. I. S. and Wehlau, W. *Can. J. Phys.* ***36***, 871 (1958).
90. Brown, R. Hanbury *Sky Telesc.* ***28***, 64 (1964).
91. Brown, R. Hanbury and Twiss, R. Q. *Phil. Mag.* ***45***, 663 (1954).
92. Brown, R. Hanbury and Twiss, R. Q. *Nature, Lond.* ***177***, 27 (1956).
93. Brown, R. Hanbury and Twiss, R. Q. *Nature, Lond.* ***178***, 1046 (1956).
94. Brown, R. Hanbury and Twiss, R. Q. *Nature, Lond.* ***178***, 1447 (1956).
95. Brown, R. Hanbury and Twiss, R. Q. *Proc. R. Soc.* (A) ***242***, 300 (1957).
96. Brown, R. Hanbury and Twiss, R. Q. *Proc. R. Soc.* (A) ***243***, 291 (1957).
97. Brown, R. Hanbury and Twiss, R. Q. *Proc. R. Soc.* (A) ***248***, 199 (1958).
98. Brown, R. Hanbury and Twiss, R. Q. *Proc. R. Soc.* (A) ***248***, 222 (1958).
99. Brunner, W. *Ann. Physik* ***20***, 53 (1967).
100. Brunner, W. *Ann. Physik* ***22***, 67 (1968).
101. Brunner, W. and Paul, H. *Ann. Physik* ***23***, 152 (1969).
102. Brunner, W., Paul, H. and Richter, G. *Ann. Physik* ***16***, 343 (1965).
103. Cahill, K. E. *Phys. Rev.* ***138***, B 1566 (1965).
104. Cahill, K. E. *Phys. Rev.* ***180***, 1244 (1969).
105. Cahill, K. E. and Glauber, R. J. *Phys. Rev.* ***177***, 1857 (1969).
106. Cahill, K. E. and Glauber, R. J. *Phys. Rev.* ***177***, 1882 (1969).
107. Campagnoli, G. and Zambotti, G. *Nuovo Cim.* ***57A***, 468 (1968).
108. Carruthers, P. and Dy, K. S. *Phys. Rev.* ***147***, 214 (1966).
109. Carruthers, P. and Nieto, M. M. *Rev. mod. Phys.* ***40***, 411 (1968).
110. Carusotto, S. *Nuovo Cim.* ***70B***, 73 (1970).
111. Carusotto, S., Fornaca, G. and Polacco, E. *Phys. Rev.* ***157***, 1207 (1967).
112. Carusotto, S., Fornaca, G. and Polacco, E. *Phys. Rev.* ***165***, 1391 (1968).
113. Chang, R. F., Detenbeck, R. W., Korenman, V., Alley, C. O. and Hochuli, U. *Phys. Lett.* ***25A***, 272 (1967).
114. Chang, R. F., Korenman, V., Alley, C. O. and Detenbeck, R. W. *Phys. Rev.* ***178***, 612 (1969).
115. Chang, R. F., Korenman, V. and Detenbeck, R. W. *Phys. Lett.* ***26A***, 417 (1968).
116. Cohen - Tannoudji, C. and Kastler, A. *Progress in Optics,* Vol. V, p. 1, Ed. E. Wolf, North-Holland, Amsterdam, 1966.
117. Conway, J. M. *Phys. Rev.* ***156***, 1365 (1967).
118. Crosignani, B. and Di Porto, P. *Phys. Lett.* ***24A***, 69 (1967).
119. Crosignani, B., Di Porto, P. and Engelmann, F. *Z. Naturf.* ***23a***, 743 (1968).
120. Crosignani, B., Di Porto, P. and Solimeno, S. *Phys. Lett.* ***27A***, 568 (1968).
121. Crosignani, B., Di Porto, P. and Solimeno, S. *Phys. Lett.* ***28A***, 271 (1968).
122. Crosignani, B., Di Porto, P. and Solimeno, S. *Quantum Harmonic Oscillator with Time-Dependent Frequency*, preprint, 1968 (see also Solimeno, S., Di Porto, P. and Crosignani, B. *J. math. Phys.* ***10***, 1922 (1969)).
123. Crosignani, B., Di Porto, P. and Solimeno, S. *Phys. Rev.* ***186***, 1342 (1969).
124. Davidson, F. and Mandel, L. *Phys. Lett.* ***25A***, 700 (1967).
125. Davidson, F. and Mandel, L. *Phys. Lett.* ***27A***, 579 (1968).
126. Dialetis, D. *J. math. Phys.* ***8***, 1641 (1967).
127. Dialetis, D. *J. Phys. A* ***2***, 229 (1969).

128. DIALETIS, D. *J. opt. Soc. Am.* **59**, 74 (1969).
129. DIALETIS, D. and WOLF, E. *Nuovo Cim.* (10) **47**, 113 (1967).
130. DIAMENT, P. and TEICH, M. C. *J. opt. Soc. Am.* **59**, 661 (1969).
131. DICKE, R. H. *Phys. Rev.* **93**, 99 (1954).
132. DICKE, R. H. *Quantum Electronics* (Proc. of the Third Inter. Congress), p. 35, Eds N. Bloembergen and P. Grivet, Dunod et Cie., Paris, 1964.
133. DI PORTO, P., CROSIGNANI, B. and BERTOLOTTI, M. *J. appl. Phys.* **40**, 5083 (1969).
134. EBERLY, J. H. and KUJAWSKI, A. *Phys. Lett.* **24A**, 426 (1967).
135. EBERLY, J. H. and KUJAWSKI, A. *Phys. Rev.* **155**, 10 (1967).
136. EDWARDS, S. F. and PARRENT, G. B. *Opt. Acta* **6**, 367 (1959).
137. EINSTEIN, A. *Phys. Z.* **10**, 185 (1909).
138. EINSTEIN, A. *Phys. Z.* **10**, 817 (1909).
139. EINSTEIN, A. *La théorie du rayonnement et les quanta* (Instituts Solvay, Brussels, Conseil de Physique 1er 1911) p. 407, Eds. P. Langevin and L. de Broglie, Gauthier-Villars, Paris, 1912.
140. FANO, U. *Phys. Rev.* **93**, 121 (1954).
141. FANO, U. *Rev. mod. Phys.* **29**, 74 (1957).
142. FANO, U. *Am. J. Phys.* **29**, 539 (1961).
143. FETTER, A. L. *Phys. Rev.* **139**, A 1616 (1965).
144. FEYNMAN, R. P. *Rev. mod. Phys.* **20**, 367 (1948).
145. FLECK, J. A. *Phys. Rev.* **149**, 309 (1966).
146. FLECK, J. A. *Phys. Rev.* **149**, 322 (1966).
147. FLECK, J. A. *Phys. Rev.* **152**, 278 (1966).
148. FORRESTER, A. T. *J. opt. Soc. Am.* **51**, 253 (1961).
149. FORRESTER, A. T. *Advances in Quantum Electronics*, p. 233, Ed. J. R. Singer, Columbia Univ. Press, New York, 1961.
150. FORRESTER, A. T., GUDMUNDSEN, R. A. and JOHNSON, P. O. *Phys. Rev.* **99**, 1691 (1955).
151. FRANÇON, M. and MALLICK, S. *Progress in Optics*, Vol. VI, p. 71, Ed. E. Wolf, North-Holland, Amsterdam, 1967.
152. FRAY, S., JOHNSON, F. A., JONES, R., MCLEAN, T. P. and PIKE, E. R. *Phys. Rev.* **153**, 357 (1967).
153. FREED, C. and HAUS, H. A. *Phys. Rev. Lett.* **15**, 943 (1965).
154. FREED, C. and HAUS, H. A. *IEEE J. Quant. Electr.*, *QE-2*, 190 (1966).
155. FÜRTH, R. *Z. Phys.* **48**, 323 (1928).
156. FÜRTH, R. *Z. Phys.* **50**, 310 (1928).
157. GABOR, D. *J. Instn Elect. Engrs* **93**, 429 (1946).
158. GABOR, D. *Nature* **161**, 777 (1948).
159. GABOR, D. *Proc. Symp. on Astr. Optics and Rel. Subj.*, p. 17, Ed. Z. Kopal, North-Holland, Amsterdam, 1956.
160. GABOR, D. *Progress in Optics*, Vol. I, p. 109, Ed. E. Wolf, North-Holland, Amsterdam, 1961
161. GABOR, D. *Opt. Acta* **13**, 299 (1966).
162. GAMO, H. *Advances in Quantum Electronics*, p. 252, Ed. J. R. Singer, Columbia Univ. Press, New York, 1961.
163. GAMO, H. *Electromagnetic Theory and Antennas* Part 2, p. 801, Ed. E. C. Jordan, Macmillan, New York, 1963.
164. GAMO, H. *J. appl. Phys.* **34**, 875 (1963).

165. Gamo, H. *Progress in Optics*, Vol. III, p. 187, Ed. E. Wolf, North-Holland, Amsterdam, 1964.
166. Gelfand, I. M. and Minlos, P. A. *Dokl. Acad. Sci.* (*USSR*) ***97***, 209 (1954).
167. Gelfand, I. M. and Yaglom, A. M. *Usp. mat. nauk* ***11***, 77 (1956).
168. Germey, K. *Ann. Physik* ***10***, 141 (1963).
169. Ghielmetti, F. *Phys. Lett.* ***12***, 210 (1964).
170. Givens, M. P. *J. opt. Soc. Am.* ***51***, 1032 (1961).
171. Givens, M. P. *J. opt. Soc. Am.* ***52***, 225 (1962).
172. Glauber, R. J. *Phys. Rev. Lett.* ***10***, 84 (1963).
173. Glauber, R. J. *Phys. Rev.* ***130***, 2529 (1963).
174. Glauber, R. J. *Phys. Rev.* ***131***, 2766 (1963).
175. Glauber, R. J. *Quantum Electronics* (Proc. of the Third Inter. Congress), p. 111, Eds. N. Bloembergen and P. Grivet, Dunod et Cie., Paris, 1964.
176. Glauber, R. J. *Quantum Optics and Electronics*, p. 144, Eds. C. De Witt, A. Blandin and C. Cohen-Tannoudji, Gordon and Breach, New York, 1965).
177. Glauber, R. J. *Physics of Quantum Electronics*, p. 778, Eds. P. L. Kelley, B. Lax and P. E. Tannenwald, McGraw-Hill, New York, 1966.
178. Glauber, R. J. *Phys. Lett.* ***21***, 650 (1966).
179. Glauber, R. J. *Proc. Symp. Modern. Optics*, p. 1, Polytechnic Press, New York, 1967.
180. Glauber, R. J. *Fundamental Problems in Statistical Mechanics II*, p. 140, Ed. E. G. D. Cohen, North-Holland, Amsterdam, 1968.
181. Glauber, R. J. *Quantum Optics* (Proc. Tenth Session Scottish Univ. Summer School in Phys.) p. 53, Eds. S. M. Kay and A. Maitland, Academic Press, London and New York, 1970.
182. Golay, M. J. E. *Proc. IRE* ***49***, 958 (1961).
183. Goldberger, M. L., Lewis, H. W. and Watson, K. M. *Phys. Rev.* ***132***, 2764 (1963).
184. Goldberger, M. L., Lewis, H. W. and Watson, K. M. *Phys. Rev.* ***142***, 25 (1966).
185. Goldberger, M. L. and Watson, K. M. *Phys. Rev.* ***134***, B 919 (1964).
186. Goldberger, M. L. and Watson, K. M. *Phys. Rev.* ***137***, B 1396 (1965).
187. Graham, R. *Z. Phys.* ***210***, 319 (1968).
188. Graham, R. and Haken, H. *Z. Phys.* ***210***, 276 (1968).
189. Güttinger, W. *Fortschr. Phys.* ***14***, 483 (1966).
190. Haig, N. D. and Sillitto, R. M. *Phys. Lett.* ***28A***, 463 (1968).
191. Haken, H. *Dynamical Processes in Solid State Optics*, Part. I, p. 168, Eds. R. Kubo and H. Kamimura, W. A. Benjamin, New York, 1967.
192. Haken, H. *Quantum Optics* (Proc. Tenth Session Scottish Univ. Summer School in Phys.) p. 201, Eds. S. M. Kay and A. Maitland, Academic Press, London and New York, 1970.
193. Haken, H., Risken, H. and Weidlich, W. *Z. Phys.* ***206***, 355 (1967).
194. Harwit, M. *Phys. Rev.* ***120***, 1551 (1960).
195. Hempstead, R. D. and Lax, M. *Phys. Rev.* ***161***, 350 (1967).
196. Ho, T. L. and Beran, M. J. *J. opt. Soc. Am.* ***58***, 1335 (1968).
197. Holliday, D. *Phys. Lett.* ***8***, 250 (1964).
198. Holliday, D. and Sage, M. L. *Annls. Phys.* ***29***, 125 (1964).
199. Holliday, D. and Sage, M. L. *Phys. Rev.* ***138***, B 485 (1965).
200. Hopkins, H. H. *Proc. R. Soc.* (A) ***208***, 263 (1951).

201. HOPKINS, H. H. *Proc. R. Soc.* (A) ***217***, 408 (1953).
202. HOPKINS, H. H. *J. opt. Soc. Am.* ***47***, 508 (1957).
203. HOPKINS, H. H. *Proc. R. Soc. B* ***70***, 1002 (1957).
204. HORÁK, R. *Opt. Acta* ***16***, 111 (1969).
205. HORÁK, R. *Czech. J. Phys. B* ***19***, 827 (1969).
206. HORÁK, R. *Czech. J. Phys. B* ***21***, 7 (1971).
207. INGARDEN, R. S. *Fortschr. Phys.* ***13***, 755 (1965).
208. INGARDEN, R. S. *Higher Order Temperatures and Coherence of Light,* (preprint, 1967).
209. JACQUINOT, P. *Rep. Prog. Phys.* ***23***, 267 (1960).
210. JAISWAL, A. K. and MEHTA, C. L. *Phys. Lett.* ***29A***, 245 (1969).
211. JAKEMAN, E., OLIVER, C. J. and PIKE, E. R. *J. Phys. A* ***1***, 406 (1968).
212. JAKEMAN, E., OLIVER, C. J. and PIKE, E. R. *J. Phys. A* ***1***, 497 (1968).
213. JAKEMAN, E. and PIKE, E. R. *J. Phys. A* ***1***, 128 (1968).
214. JAKEMAN, E. and PIKE, E. R. *J. Phys. A* ***2***, 115 (1969).
215. JANOSSY, L. *Nuovo Cim.* ***6***, 125 (1957).
216. JANOSSY, L. *Nuovo Cim.* ***12***, 369 (1959).
217. JAVAN, A., BALLIK, E. A. and BOND, W. L. *J. opt. Soc. Am.* ***52***, 96 (1962).
218. JOHNSON, F. A., JONES, R., McLEAN, T. P. and PIKE, E. R. *Phys. Rev. Lett.* ***61***, 589 (1966).
219. JORDAN, F. T. and GHIELMETTI, F. *Phys. Rev. Lett.* ***12***, 607 (1964).
220. KAHN, F. D. *Opt. Acta* ***5***, 93 (1958).
221. KANO, Y. *Nuovo Cim.* (10), ***23***, 328 (1962).
222. KANO, Y. *Annls. Phys.* ***30***, 127 (1964).
223. KANO, Y. *J. phys. Soc. Japan* ***19***, 1555 (1964).
224. KANO, Y. *J. math. Phys.* ***6***, 1913 (1965).
225. KANO, Y. *Nuovo Cim.* (10) ***43***, 1 (1966).
226. KANO, Y. and WOLF, E. *Proc. phys. Soc.* ***80***, 1273 (1962).
227. KARCZEWSKI, B. *Phys. Lett.* ***5***, 191 (1963).
228. KARCZEWSKI, B. *Nuovo Cim.* (10) ***30***, 906 (1963).
229. KASTLER, A. *Quantum Electronics* (Proc. of the Third Inter. Congress), p. 3, Eds. N. Bloembergen and P. Grivet, Dunod et Cie., Paris, 1964.
230. KELLER, E. F. *Phys. Rev.* ***139***, B 202 (1965).
231. KELLEY, P. L. and KLEINER, W. H. *Phys. Rev.* ***136***, A 316 (1964).
232. KEPRT, J. *Optik* ***27***, 213 (1968).
233. KHALFIN, L. A. *Dokl. Acad. Sci.* (*USSR*) ***132***, 1051 (1960).
234. KLAUDER, J. R. *Annls. Phys.* ***11***, 123 (1960).
235. KLAUDER, J. R. *Phys. Rev. Lett.* ***16***, 534 (1966).
236. KLAUDER, J. R., McKENNA, J. and CURRIE, D. G. *J. math. Phys.* ***6***, 734 (1965).
237. KORENMAN, V. *Phys. Rev. Lett.* ***14***, 293 (1965).
238. KORENMAN, V. *Annls. Phys.* ***39***, 72 (1966).
239. KUJAWSKI, A. *Nuovo Cim.* (10) ***44***, 326 (1966).
240. KUJAWSKI, A. *Acta phys. Pol.* ***34***, 957 (1968).
241. LACHS, G. *Phys. Rev.* ***138***, B 1012 (1965).
242. LAMBROPOULOS, P. *Phys. Rev.* ***168***, 1418 (1968).
243. LAMBROPOULOS, P., KIKUCHI, C. and OSBORN, R. K. *Phys. Rev.* ***144***, 1081 (1966).
244. LANGER, J. S. *Phys. Rev.* ***167***, 183 (1968).
245. LAUE, M. *Ann. Physik* ***23***, 1 (1907).
246. LAX, M. *Brandeis Summer Institute Lectures*, Gordon and Breach, New York 1966.

247. Lax, M. *Dynamical Processes in Solid State Optics,* Part. I, p. 195, Eds. R. Kubo and H. Kamimura, W. A. Benjamin, New York, 1967.
248. Lax, M. *Phys. Rev.* ***157***, 213 (1967).
249. Lax, M. *Phys. Rev.* ***160***, 290 (1967).
250. Lax, M. *Phys. Rev.* ***172***, 350 (1968).
251. Lax, M. and Louisell, W. H. *IEEE J. Quant. Electr.* ***QE-3***, 47 (1967).
252. Lax, M. and Yuen, H. *Phys. Rev.* ***172***, 362 (1968).
253. Ledinegg, E. *Z. Phys.* ***191***, 177 (1966).
254. Ledinegg, E. *Z. Phys.* ***205***, 25 (1967).
255. Linfoot, E. H. *Proc. Symp. Astr. Optics and Rel. Subj.*, p. 38, Ed. Z. Kopal, North-Holland, Amsterdam, 1956.
256. Lipsett, M. S. and Mandel, L. *Nature* ***199***, 553 (1963).
257. Lipsett, M. S. and Mandel, L. *Quantum Electronics* (Proc. of the Third Inter. Congress), p. 1271, Eds. N. Bloembergen and P. Grivet, Dunod et Cie., Paris, 1964.
258. López, A. *Phys. Lett.* ***25A***, 83 (1967).
259. Louisell, W. H. *Quantum Optics* (Proc. Tenth Session Scottish Univ. Summer School in Phys.) p. 177, Eds. S. M. Kay and A. Maitland, Academic Press, London and New York, 1970.
260. Magill, P. J. and Soni, R. P. *Phys. Rev. Lett.* ***16***, 911 (1966).
261. Magyar, G. and Mandel, L. *Nature* ***198***, 255 (1963).
262. Magyar, G. and Mandel, L. *Quantum Electronics* (Proc. of the Third Inter. Congress), p. 1247, Eds. N. Bloembergen and P. Grivet, Dunod et Cie., Paris, 1964.
263. Mandel, L. *Proc. phys. Soc.* ***72***, 1037 (1958).
264. Mandel, L. *Proc. phys. Soc.* ***74***, 233 (1959).
265. Mandel, L. *J. opt. Soc. Am.* ***51***, 797 (1961).
266. Mandel, L. *J. opt. Soc. Am.* ***51***, 1342 (1961).
267. Mandel, L. *J. opt. Soc. Am.* ***52***, 1335 (1962).
268. Mandel, L. *Electromagnetic Theory and Antennas*, Part 2, p. 811, Ed. E. C. Jordan, Macmillan, New York, 1963.
269. Mandel, L. *Progress in Optics* Vol. II, p. 181, Ed. E. Wolf, North-Holland, Amsterdam, 1963.
270. Mandel, L. *Proc. phys. Soc.* ***81***, 1104 (1963).
271. Mandel, L. *Phys. Lett.* ***7***, 117 (1963).
272. Mandel, L. *Quantum Electronics* (Proc. of the Third Inter. Congress), p. 101. Eds. N. Bloembergen and P. Givet, Dunod et Cie., Paris, 1964.
273. Mandel, L. *Phys. Lett.* ***10***, 166 (1964).
274. Mandel, L. *Phys. Rev.* ***134***, A 10 (1964).
275. Mandel, L. *Phys. Rev.* ***136***, B 1221 (1964).
276. Mandel, L. *Phys. Rev.* ***138***, B 753 (1965).
277. Mandel, L. *Phys. Rev.* ***144***, 1071 (1966).
278. Mandel, L. *Phys. Rev.* ***152***, 438 (1966).
279. Mandel, L. *Proc. Symp. Modern Optics*, p. 143, Polytechnic Press, New York, 1967.
280. Mandel, L. *J. opt. Soc. Am.* ***57***, 613 (1967).
281. Mandel, L., Sudarshan, E. C. G. and Wolf, E. *Proc. phys. Soc.* ***84***, 435 (1964).
282. Mandel, L. and Wolf, E. *J. opt. Soc. Am.* ***51***, 815 (1961).
283. Mandel, L. and Wolf, E. *Phys. Rev.* ***124***, 1696 (1961).
284. Mandel, L. and Wolf, E. *Proc. Phys. Soc.* ***80***, 894 (1962).

285. MANDEL, L. and WOLF, E. *Rev. mod. Phys.* ***37***, 231 (1965).
286. MANDEL, L. and WOLF, E. *Phys. Rev.* ***149***, 1033 (1966).
287. MANDELSTAM, L. I. *Sochineniya (Collected papers)*, Vol. I, p. 229, 1948.
288. MANDELSTAM, L. I. *Sochineniya (Collected papers)*, Vol. II, p. 388, 1947.
289. MARATHAY, A. S. *J. opt. Soc. Am.* ***56***, 619 (1966).
290. MARÉCHAL, A. and CROCE, P. *C. R. Acad. Sci.* (Paris) ***237***, 607 (1953).
291. MARTIENSSEN, W. and SPILLER, E. *Am. J. Phys.* ***32***, 919 (1964).
292. MARTIENSSEN, W. and SPILLER, E. *Phys. Rev. Lett.* ***16***, 531 (1966).
293. MARTIENSSEN, W. and SPILLER, E. *Phys. Rev.* ***145***, 285 (1966).
294. MCLEAN, T. P. and PIKE, E. R. *Phys. Lett.* ***15***, 318 (1965).
295. MEHTA, C. L. *Nuovo Cim.* ***28***, 401 (1963).
296. MEHTA, C. L. *J. math. Phys.* ***5***, 677 (1964).
297. MEHTA, C. L. *Nuovo Cim.* (10) ***36***, 202 (1965).
298. MEHTA, C. L. *Nuovo Cim.* (10) ***45***, 280 (1966).
299. MEHTA, C. L. *Phys. Rev. Lett.* ***18***, 752 (1967).
300. MEHTA, C. L. *J. opt. Soc. Am.* ***58***, 1233 (1968).
301. MEHTA, C. L. *J. Phys. A* ***1***, 385 (1968).
302. MEHTA, C. L., CHAND, P., SUDARSHAN, E. C. G. and VEDAM, R. *Phys. Rev.* ***157***, 1198 (1967).
303. MEHTA, C. L. and MANDEL, L. *Electromagnetic Wave Theory*, Part 2, p. 1069. Proc. Symp., Pergamon Press, Oxford, 1967.
304. MEHTA, C. L. and SUDARSHAN, E. C. G. *Phys. Rev.* ***138***, B 274 (1965).
305. MEHTA, C. L. and SUDARSHAN, E. C. G. *Phys. Lett.* ***22***, 574 (1966).
306. MEHTA, C. L. and WOLF, E. *Phys. Rev.* ***134***, A 1143 (1964).
307. MEHTA, C. L. and WOLF, E. *Phys. Rev.* ***134***, A 1149 (1964).
308. MEHTA, C. L. and WOLF, E. *Phys. Rev.* ***157***, 1183 (1967).
309. MEHTA, C. L. and WOLF, E. *Phys. Rev.* ***157***, 1188 (1967).
310. MEHTA, C. L. and WOLF, E. *Phys. Rev.* ***161***, 1328 (1967).
311. MEHTA, C. L., WOLF, E. and BALACHANDRAN, A. P. *J. math. Phys.* ***7***, 133 (1966).
312. MICHELSON, A. A. *Phil. Mag.* (5) ***30***, 1 (1890).
313. MILLER, M. M. *Phys. Lett.* ***27A***, 185 (1968).
314. MILLER, M. M. and MISHKIN, E. A. *Phys. Rev.* ***152***, 1110 (1966).
315. MILLER, M. M. and MISHKIN, E. A. *Phys. Lett.* ***24A***, 188 (1967).
316. MILLER, M. M. and MISHKIN, E. A. *Phys. Rev.* ***164***, 1610 (1967).
317. MIŠTA, L. *Phys. Lett.* ***25A***, 646 (1967).
318. MIŠTA, L. *Czech. J. Phys. B* ***19***, 443 (1969).
319. MÖLLER, B. *Opt. Acta* ***15***, 223 (1968).
320. MOLLOW, B. R. *Phys. Rev.* ***162***, 1256 (1967).
321. MOLLOW, B. R. *Phys. Rev.* ***175***, 1555 (1968).
322. MOLLOW, B. R. and GLAUBER, R. J. *Phys. Rev.* ***160***, 1076 (1967).
323. MOLLOW, B. R. and GLAUBER, R. J. *Phys. Rev.* ***160***, 1097 (1967).
324. MORAWITZ, H. *Phys. Rev.* ***139***, A 1072 (1965).
325. MORAWITZ, H. *Z. Phys.* ***195***, 20 (1966).
326. MORGAN, B. L. and MANDEL, L. *Phys. Rev. Lett.* ***16***, 1012 (1966).
327. MOYAL, J. E. *Proc. Cambridge Phil. Soc.* ***45***, 99 (1949).
328. NUSSENZVEIG, H. M. *J. math. Phys.* ***8***, 561 (1967).
329. O'NEILL, E. L. and ASAKURA, T. *J. phys. Soc. Japan* ***16***, 301 (1961).
330. O'NEILL, E. L. and WALTHER, A. *Opt. Acta* ***10***, 33 (1963).
331. PANCHARATNAM, S. *Proc. Ind. Acad. Sci.* ***57***, 218 (1963).

332. Pancharatnam, S. *Proc. Ind. Acad. Sci.* **57**, 231 (1963).
333. Paoli, T. L. *Phys. Rev.* **163**, 1348 (1967).
334. Parrent, G. B. *J. opt. Soc. Am.* **49**, 787 (1959).
335. Parrent, G. B. *Opt. Acta* **6**, 285 (1959).
336. Parrent, G. B. *J. opt. Soc. Am.* **51**, 143 (1961).
337. Parrent, G. B. and Roman, P. *Nuovo Cim.* **15**, 370 (1960).
338. Paul, H. *Ann. Physik* **14**, 147 (1964).
339. Paul, H. *Fortschr. Phys.* **14**, 141 (1966).
340. Paul, H. *Ann. Physik* **19**, 210 (1967).
341. Paul, H., Brunner, W. and Richter, G. *Ann. Physik* **12**, 325 (1963).
342. Paul, H., Brunner, W. and Richter, G. *Ann. Physik* **17**, 262 (1966).
343. Paul, H. and Frahm, J. *Ann. Physik* **19**, 354 (1967).
344. Paul, H., Frahm, J. and Rauh, D. *Ann. Physik.* **19**, 344 (1967).
345. Pearl, P. and Troup, G. J. *Phys. Lett.* **27 A**, 560 (1968).
346. Peřina, J. *Opt. Acta* **10**, 333 (1963).
347. Peřina, J. *Opt. Acta* **10**, 337 (1963).
348. Peřina, J. *Proc. Symp. Interkamera,* p. 139, Ed. J. Morávek, STN, Prague, 1963.
349. Peřina, J. *Acta Universitatis Palackianae* **15**, 87 (1964).
350. Peřina, J. *Phys. Lett.* **12**, 194 (1964).
351. Peřina, J. *Phys. Lett.* **14**, 34 (1965).
352. Peřina, J. *Phys. Lett.* **15**, 129 (1965).
353. Peřina, J. *Phys. Lett.* **19**, 195 (1965).
354. Peřina, J. *Acta Universitatis Palackianae* **18**, 49 (1965).
355. Peřina, J. *Thesis,* Palacký Univ., 1965 (unpublished).
356. Peřina, J. *Czech. J. Phys.* **B 16**, 907 (1966).
357. Peřina, J. *Czech. J. Phys.* **B 17**, 1086 (1967).
358. Peřina, J. *Phys. Lett.* **24A**, 333 (1967).
359. Peřina, J. *Phys. Lett.* **24A**, 698 (1967).
360. Peřina, J. *Czech. J. Phys.* **B 18**, 197 (1968).
361. Peřina, J. *Acta Universitatis Palackianae* **27**, 227 (1968).
362. Peřina, J. *Czech. J. Phys.* **B 19**, 151 (1969).
363. Peřina, J. *Opt. Acta* **16**, 289 (1969).
364. Peřina, J. *Quantum Optics* (Proc. Tenth Session Scottish Univ. Summer School in Phys.) p. 513, Eds. S. M. Kay and A. Maitland, Academic Press, London and New York, 1970.
365. Peřina, J. and Horák, R. *Optics Comm.* **1**, 91 (1969).
366. Peřina, J. and Horák, R. *J. Phys.* **A 2**, 702 (1969).
367. Peřina, J. and Horák, R. *Czech. J. Phys.* **B 20**, 149 (1970).
368. Peřina, J. and Kvapil, J. *Optik* **28**, 575 (1968/69).
369. Peřina, J. and Mišta, L. *Phys. Lett.* **27A**, 217 (1968).
370. Peřina, J. and Mišta, L. *Czech. J. Phys.* **B 18**, 697 (1968).
371. Peřina, J. and Mišta, L. *Ann. Physik* **22**, 372 (1969).
372. Peřina, J. and Peřinová, V. *Opt. Acta* **12**. 333 (1965).
373. Peřina, J. and Peřinová, V. *Opt. Acta* **16**, 309 (1969).
374. Peřina, J. and Stroke, G. W. *Holographic Method of "Deconvolution" and Analytic Continuation,* (preprint, 1968) (See also Peřina, J., *Czech. J. Phys.* **B 21**, 731 (1971).
375. Peřina, J. and Tillich, J. *Acta Universitatis Palackianae* **21**, 153 (1966).
376. Peřinová, V. *Čas. pěst. mat.* **94**, 253 (1969).
377. Peřinová, V. *Čas. pěst. mat.* **94**, 297 (1969).

378. PFLEEGOR, R. L. and MANDEL, L. *Phys. Lett.* ***24 A***, 766 (1967).
379. PFLEEGOR, R. L. and MANDEL, L. *Phys. Rev.* ***159***, 1084 (1967).
380. PFLEEGOR, R. L. and MANDEL, L. *J. opt. Soc. Am.* ***58***, 946 (1968).
381. PHILLIPS, D. T., KLEIMAN, H. and DAVIS, S. P. *Phys. Rev.* ***153***, 113 (1967).
382. PICARD, R. H. and WILLIS, C. R. *Phys. Rev.* ***139***, A 10 (1965).
383. PICINBONO, B. *Proc. Symp. Modern Optics*, p. 167, Polytechnic Press, New York, 1967.
384. PICINBONO, B. *Phys. Lett.* ***29A***, 614 (1969).
385. PICINBONO, B. and BOILEAU, E. *J. opt. Soc. Am.* ***58***, 784 (1968).
386. PIKE, E. R. *Rivista Nuovo Cim.* (special issue) ***1***, 277 (1969).
387. PIKE, E. R. *Quantum Optics* (Proc. Tenth Session Scottish Univ. Summer School in Phys.) p. 127, Eds. S. M. Kay and A. Maitland, Academic Press, London and New York, 1970.
388. PIOVOSO, M. J. and BOLGIANO, L. P. *Proc. IEEE* ***55***, 1519 (1967).
389. PURCELL, E. M. *Nature* ***178***, 1449 (1956).
390. RADLOFF, W. *Phys. Lett.* ***27A***, 366 (1968); *Ann. Physik* ***26***, 178 (1971).
391. REBKA, G. A. and POUND, R. V. *Nature* ***180***, 1035 (1957).
392. REED, I. S. *IRE Trans. on Inform. Theory* ***IT-8***, 194 (1962).
393. REZENDE, S. M. and ZAGURY, N. *Phys. Lett.* ***29A***, 47 and 616 (1969).
394. RICHTER, G. *Ann. Physik* ***18***, 331 (1966).
395. RICHTER, G., BRUNNER, W. and PAUL, H. *Ann. Physik* ***14***, 239 (1964).
396. RISKEN, H. *Z. Phys.* ***186***, 85 (1965).
397. RISKEN, H. *Z. Phys.* ***191***, 302 (1966).
398. RISKEN, H. *Fortschr. Phys.* ***16***, 261 (1968).
399. RISKEN, H. and VOLLMER, H. D. *Z. Phys.* ***204***, 240 (1967).
400. ROBL, H. R. *Phys. Lett.* ***24A***, 288 (1967).
401. ROBL, H. R. *Phys. Rev.* ***165***, 1426 (1968).
402. ROCCA, F. *J. Phys.* ***28***, 113 (1967).
403. ROMAN, P. *Nuovo Cim.* (10) ***13***, 974 (1959).
404. ROMAN, P. *Nuovo Cim.* (10) ***20***, 759 (1961).
405. ROMAN, P. *Nuovo Cim.* (10), ***22***, 1005 (1961).
406. ROMAN, P. and MARATHAY, A. S. *Nuovo Cim.* ***30***, 1452 (1963).
407. ROMAN, P. and WOLF, E. *Nuovo Cim.* (10) ***17***, 462 (1960).
408. ROMAN, P. and WOLF, E. *Nuovo Cim.* (10) ***17***, 477 (1960).
409. ROSENFELD, L. *Niels Bohr and Development of Physics*, Ed. W. Pauli, (1958).
410. ROUSSEAU, M. *C. R. Acad. Sci., Paris* ***268***, 1477 (1969).
411. ROUSSEAU, M. *J. Phys.* ***30***, 675 (1969).
412. SARFATT, J. *Nuovo Cim.* ***27***, 1119 (1963).
413. SCHMEIDLER, W. *Ann. d. Finnischen Akad. Wiss., I. Math.-Phys.* ***I***, 220 (1956).
414. SCHRÖDINGER, E. *Naturwissenschaften* ***14***, 644 (1927).
415. SCULLY, M. O. and LAMB, W. E. *Phys. Rev. Lett.* ***16***, 853 (1966).
416. SCULLY, M. O. and LAMB, W. E. *Phys. Rev.* ***159***, 208 (1967).
417. SCULLY, M. O. and LAMB, W. E. *Phys. Rev.* ***166***, 246 (1968).
418. SENITZKY, I. R. *Phys. Rev.* ***111***, 3 (1958).
419. SENITZKY, I. R. *Phys. Rev.* ***127***, 1638 (1962).
420. SERIES, G. W. *Quantum Optics* (Proc. Tenth Session Scottish Univ. Summer School in Phys.) p. 395, Eds. S. M. Kay and A. Maitland, Academic Press, London and New York, 1970.
421. SHEN, Y. R. *Phys. Rev.* ***155***, 921 (1967).

422. SHIGA, F. and INAMURA, S. *Phys. Lett.* ***25A***, 706 (1967).
423. SILLITTO, R. M. *Proc. R. Soc. Edinburgh A* ***66***, 93 (1963).
424. SILLITTO, R. M. *Phys. Lett.* ***27 A***, 624 (1968).
425. SMITH, A. W. and ARMSTRONG, J. A. *Phys. Rev. Lett.* ***16***, 1169 (1966).
426. SMITH, A. W. and ARMSTRONG, J. A. *Phys. Lett.* ***19***, 650 (1966).
427. STEEL, W. H. *J. opt. Soc. Am.* ***47***, 405 (1957).
428. STREIFER, W. *Ann. Phys.* ***27***, 72 (1964).
429. STREIFER, W. *J. opt. Soc. Am.* ***56***, 1481 (1966).
430. STROKE, G. W. *Photograf. Korrespondenz* ***104***, 82 (1968).
431. STROKE, G. W. *Opt. Acta* ***16***, 401 (1969).
432. STROKE, G. W., INDEBETOUW, G. and PUECH, C. *Phys. Lett.* ***26A***, 443 (1968).
433. STROKE, G. W. and ZECH, R. G. *Phys. Lett.* ***25A***, 89 (1967).
434. STRONG, J. and VANASSE, G. A. *J. opt. Soc. Am.* ***49***, 844 (1959).
435. SUDARSHAN, E. C. G. *Proc. Symp. Optical Masers*, p. 45, John Wiley, New York, 1963.
436. SUDARSHAN, E. C. G. *Phys. Rev. Lett.* ***10***, 277 (1963).
437. TEICH, M. C. *Appl. Phys. Lett.* ***14***, 201 (1969).
438. TEICH, M. C. and DIAMENT, P. *J. appl. Phys.* ***40***, 625 (1969).
439. TEICH, M. C. and WOLGA, G. J. *Phys. Rev. Lett.* ***16***, 625 (1966).
440. TER HAAR, D. *Rep. Prog. Phys.* ***24***, 304 (1961).
441. TITULAER, U. M. and GLAUBER, R. J. *Phys. Rev.* ***140***, B 676 (1965).
442. TITULAER, U. M. and GLAUBER, R. J. *Phys. Rev.* ***145***, 1041 (1966).
443. TOLL, J. *Phys. Rev.* ***104***, 1760 (1956).
444. TORALDO DI FRANCIA, G. *J. opt. Soc. Am.* ***45***, 497 (1955).
445. TORALDO DI FRANCIA, G. *Opt. Acta* ***13***, 323 (1966).
446. TORALDO DI FRANCIA, G. *J. opt. Soc. Am.* ***59***, 799 (1969).
447. TORALDO DI FRANCIA, G. *Proc. Conf. Europ. Phys. Soc.*, 1969 (Florence), ***Rivista*** *Nuovo Cim.* (special issue) ***1***, (1969).
448. TORALDO DI FRANCIA, G. *Quantum Optics* (Proc. Tenth Session Scottish Univ. Summer School in Phys.) p. 323, Eds. S. M. Kay and A. Maitland, Academic Press, London and New York, 1970.
449. TROUP, G. J. *Proc. phys. Soc.* ***86***, 39 (1965).
450. TROUP, G. J. *Phys. Lett.* ***17***, 264 (1965).
451. TROUP, G. J. *Nuovo Cim.* (10) ***42***, 79 (1966).
452. TROUP, G. J. *Phys. Lett.* ***28A***, 251 (1968).
453. TUCKER, J. and WALLS, D. F. *Phys. Rev.* ***178***, 2036 (1969).
454. TWISS, R. Q. *Opt. Acta* ***16***, 423 (1969).
455. TWISS, R. Q. and LITTLE, A. G. *Australian J. Phys.* ***12***, 77 (1959).
456. TWISS, R. Q., LITTLE, A. G. and BROWN, R. HANBURY *Nature* ***180***, 324 (1957).
457. VAN CITTERT, P. H. *Physica* ***1***, 201 (1934).
458. VAN CITTERT, P. H. *Physica* ***6***, 1129 (1939).
459. VAN KAMPEN, N. G. *Phys. Rev.* ***89***, 1072 (1953).
460. WALTHER, A. *Opt. Acta* ***10***, 41 (1963).
461. WANG, M. C. and UHLENBECK, G. E. *Rev. mod. Phys.* ***17***, 323 (1945).
462. WEBBER, J. C. *Phys. Lett.* ***27A***, 5 (1968).
463. WEBBER, J. C. *Canad. J. Phys.* ***47***, 363 (1969).
464. WEIDLICH, W., RISKEN, H. and HAKEN, H. *Z. Phys.* ***201***, 396 (1967).
465. WIENER, N. *J. Math. Phys.* ***7***, 109 (1928).
466. WIENER, N. *J. Franklin Inst.* ***207***, 525 (1929).

467. WIENER, N. *Acta Math.* ***55***, 117 (1930).
468. WIGNER, E. *Phys. Rev.* ***40***, 749 (1932).
469. WILLIS, C. R. *Phys. Rev.* ***147***, 406 (1966).
470. WOLF, E. *Nuovo Cim.* ***12***, 884 (1954).
471. WOLF, E. *Proc. R. Soc.* (*A*) ***225***, 96 (1954).
472. WOLF, E. *Proc. R. Soc.* (*A*) ***230***, 246 (1955).
473. WOLF, E. *Proc. Symp. Astr. Optics and Rel. Subj.*, p. 177, Ed. Z. Kopal, North-Holland, Amsterdam, 1956.
474. WOLF, E. *Phil. Mag.* (8), ***2***, 351 (1957).
475. WOLF, E. *Proc. phys. Soc.* ***71***, 257 (1958).
476. WOLF, E. *Nuovo Cim.* ***13***, 1165 (1959).
477. WOLF, E. *Proc. phys. Soc.* ***76***, 424 (1960).
478. WOLF, E. *Proc. phys. Soc.* ***80***, 1269 (1962).
479. WOLF, E. *Proc. Symp. Optical Masers*, p. 29, John Wiley, New York, 1963.
480. WOLF, E. *Phys. Lett.* ***3***, 166 (1963).
481. WOLF, E. *Quantum Electronics* (Proc. of the Third Inter. Congress), p. 13, Eds, N. Bloembergen and P. Grivet, Dunod et Cie., Paris, 1964.
482. WOLF, E. *Jap. J. appl. Phys.* ***4***, Suppl. I, 1 (1965).
483. WOLF, E. *Opt. Acta* ***13***, 281 (1966).
484. WOLF, E. and AGARWAL, G. S. *Polarisation, Matière et Rayonnement*, p. 541, Société Française de Physique, Presses Universitaires de France, 1969.
485. WOLF, E. and MEHTA, C. L. *Phys. Rev. Lett.* ***13***, 705 (1964).
486. ZARDECKI, A. *Acta phys. Pol.* ***35***, 271 (1969).
487. ZERNIKE, F. *Physica* ***5***, 785 (1938).
488. ZUCKERMAN, J. L. and DEVELIS, J. B. *Proc. Symp. Modern Optics*, p. 193, Polytechnic Press, New York, 1967.
489. ZUCKERMAN, J. L. and DEVELIS, J. B. *J. opt. Soc. Am.* ***58***, 175 (1968).
490. AGARWAL, G. S. *Nuovo Cim.* ***65B*** (10), 266 (1970).
491. AGARWAL, G. S. *Phys. Rev.* ***A1***, 1445 (1970).
492. BARASHEV, P. P. *Phys. Lett.* ***32A***, 291 (1970).
493. BARASHEV, P. P. *J. Exp. Theor. Phys.* ***59***, 1318 (1970).
494. BÉNARD, CH. *Phys. Rev.* ***A2***, 2140 (1970).
495. BENDJABALLAH, M. C. and PERROT, F. *C. R. Acad. Sci., Paris* ***271***, 1085 (1970).
496. BERTOLOTTI, M., CROSIGNANI, B. and DI PORTO, P. *J. Phys. A* (Gen. Phys.) ***3***, L 37 (1970).
497. BIALYNICKA-BIRULA, Z. *Phys. Rev.* ***D1***, 400 (1970).
498. BREVIK, I. and SUHONEN, E. *Physica Norvegica* ***3***, 135 (1968).
499. BREVIK, I. and SUHONEN, E. *Nuovo Cim.* ***65B*** (10), 187 (1970).
500. CANTRELL, C. D. *Phys. Rev.* ***A1***, 672 (1970).
501. CANTRELL, C. D. *Phys. Rev.* ***A3***, 728 (1971).
502. CLARK, W. G., ESTES, L. E. and NARDUCCI, L. M. *Phys. Lett.* ***33A***, 517 (1970).
503. DAVIDSON, F. *Phys. Rev.* ***185***, 446 (1969).
504. DIAMENT, P. and TEICH, M. C. *J. opt. Soc. Am.* ***60***, 682 (1970).
505. DIAMENT, P. and TEICH, M. C. *J. opt. Soc. Am.* ***60***, 1489 (1970).
506. EBERLY, J. H. and REHLER, N. E. *Phys. Rev.* ***A2***, 1607 (1970).
507. FILLMORE, G. L. *Phys. Rev.* ***182***, 1384 (1969).
508. HELSTROM, C. W. *Quasi-Classical Analysis of Coupled Oscillators*, preprint, Westinghouse Research Laboratories, 1966.
509. HORÁK, R., MIŠTA, L. and PEŘINA, J. *J. Phys. A* ***4***, 231 (1971).

510. Horák, R., Mišta, L. and Peřina, J. *Czech. J. Phys.* ***B21***, 614 (1971).
511. Jaiswal, A. K. and Agarwal, G. S. *J. opt. Soc. Am.* ***59***, 1446 (1969).
512. Jaiswal, A. K. and Mehta, C. L. *Phys. Rev.* ***186***, 1355 (1969).
513. Jaiswal, A. K. and Mehta, C. L. *Phys. Rev.* ***A2***, 168 and 2570 (1970).
514. Jakeman, E., Oliver, C. J. and Pike, E. R. *Phys. Lett.* ***34A***, 101 (1971).
515. Korenman, V. *Phys. Rev.* ***154***, 1233 (1967).
516. Korenman, V. *Phys. Rev.* ***A2***, 449 (1970).
517. Kujawski, A. *Bull. Acad. Pol. Sci.* ***17***, 467 (1969).
518. Lachs, G. *J. appl. Phys.* ***38***, 3439 (1967).
519. Langer, J. S. *Phys. Rev.* ***184***, 219 (1969).
520. Lehmberg, R. H. *Phys. Rev.* ***167***, 1152 (1968).
521. Lyons, J. and Troup, G. J. *Phys. Lett.* ***32A***, 352 (1970).
522. Mandel, L. *Phys. Rev.* ***181***, 75 (1969).
523. Mandel, L. and Meltzer, D. *Phys. Rev.* ***188***, 198 (1969).
524. Mehta, C. L. *Progress in Optics*, Vol. VIII, p. 375, Ed. E. Wolf, North-Holland, Amsterdam, 1970 .
525. Meltzer, D. and Mandel, L. *Phys. Rev. Lett.* ***25***, 1151 (1970).
526. Millet, J. and Usselio-La-Verna, W. *Opt. Comm.* ***2***, 120 (1970).
527. Millet, J. and Varnier, B. *Opt. Comm.* ***1***, 211 (1969).
528. Mišta, L. and Peřina, J. *Opt. Comm.* ***2***, 441 (1971).
529. Mišta, L., Peřina, J. and Peřinová, V. *Phys. Lett.* ***35A***, 197 (1971).
530. Peřina, J. *Statistical and Coherence Properties of the Superposition of Coherent and Chaotic Fields*, Palacký University, Olomouc, 1971 (unpublished).
531. Peřina, J. and Peřinová, V. *Phys. Lett.* ***35A***, 283 (1971).
532. Peřina, J., Peřinová, V. and Mišta, L. *Opt. Comm.* ***3***, 89 (1971).
533. Picinbono, B. and Rousseau, M. *Phys. Rev.* ***A1***, 635 (1970).
534. Risken, H. *Progress in Optics*, Vol. VIII, p. 241, Ed. E. Wolf, North-Holland, Amsterdam, 1970.
535. Teich, M. C., Abrams, R. L. and Gandrud, W. B. *Opt. Comm.* ***2***, 206 (1970).
536. Tunkin, V. G. and Tchirkin, A. S. *J. Exp. Theor. Phys.* ***58***, 191 (1970).

Index